Plant Responses to Stress and Environmental Stimulus

Plant Responses to Stress and Environmental Stimulus

Editor

Santiago Signorelli

MDPI • Basel • Beijing • Wuhan • Barcelona • Belgrade • Manchester • Tokyo • Cluj • Tianjin

Editor
Santiago Signorelli
Plant Biology
Universidad de la República
Montevideo
Uruguay

Editorial Office
MDPI
St. Alban-Anlage 66
4052 Basel, Switzerland

This is a reprint of articles from the Special Issue published online in the open access journal *Agronomy* (ISSN 2073-4395) (available at: www.mdpi.com/journal/agronomy/special_issues/environmental_stimulus).

For citation purposes, cite each article independently as indicated on the article page online and as indicated below:

LastName, A.A.; LastName, B.B.; LastName, C.C. Article Title. *Journal Name* **Year**, *Volume Number*, Page Range.

ISBN 978-3-0365-5780-9 (Hbk)
ISBN 978-3-0365-5779-3 (PDF)

Contents

About the Editor

Santiago Signorelli

Santiago Signorelli was born in 1987 in Uruguay. In 2005, he started his degree in Biochemistry at Universidad de la República (UdelaR, Uruguay). In 2007, he began research in academia and took a teaching and research position at a laboratory of computational chemistry in the School of Sciences. In 2009, he graduated from UdelaR doing his honours project in Molecular Biology, and the same year he took a new position as teaching and research assistant of Biochemistry in the School of Agronomy (UdelaR). In the same institution, Signorelli took postgraduate studies related to drought stress response in plants. In 2015, Signorelli took a postdoctoral position at Universidad the Western Australia (UWA, Australia), where he studied the role of oxygen and light in grapevine bud dormancy. In 2017, he got a fellowship to performed another postdoc about gene editing in banana Cavendish at KU Leuven (Belgium). In 2019, returned to Australia to work as a researcher in a series of projects related to abiotic stress and plant development. From 2021, Signorelli is an Associate Professor of Biochemistry from the School of Agronomy (UdelaR) where he is consolidating his research group "Food and Plant Biology". Signorelli's group aims to understand how plants respond to environmental stimulus and use that knowledge to generate improved crops and optimise agricultural practices.

Preface to "Plant Responses to Stress and Environmental Stimulus"

This Book is a collection of articles studying the response of plants to diverse environmental stimuli, as well as methodological papers describing new protocols for plant phenotyping. The book is addressed to anyone in the scientific community interested in plant biology, especially in the field of abiotic stress.

Santiago Signorelli
Editor

agronomy

Editorial

Plant Responses to Stress and Environmental Stimulus

Santiago Signorelli [1,2]

[1] Food and Plant Biology Group, Departamento de Biología Vegetal, Facultad de Agronomía, Universidad de la República, Av. Garzón 780, Sayago CP, Montevideo 12900, Uruguay; ssignorelli@fagro.edu.uy; Tel.: +598-2354-0229

[2] The Centre of Excellence in Plant Energy Biology, School of Molecular Sciences, The University of Western Australia, 35 Stirling Highway, Crawley, WA 6009, Australia

Citation: Signorelli, S. Plant Responses to Stress and Environmental Stimulus. *Agronomy* **2022**, *12*, 2250. https://doi.org/10.3390/agronomy12102250

Academic Editor: Alfonso Albacete

Received: 7 September 2022
Accepted: 14 September 2022
Published: 20 September 2022

Plants respond to diverse environmental stimuli such as light, nutrients, temperature, and oxygen, which shape their growth and fate. When these stimuli are suboptimal for adequate plant growth, they cause stress. This Special Issue aimed to collect research articles providing evidence about plant responses to stresses and environmental stimuli, as well as some methodological papers describing new models or methodologies for plant phenotyping.

One type of environmental stress, often overlooked, is low temperatures at night. Rajametov et al. [1] studied the effect of low night temperatures on diverse growth and productivity parameters of thirty-nine *Capsicum annuum* L. accessions, including chili and bell fruit varieties, as a means to assist in the identification of low-temperature-tolerant cultivars to assist breeding programs. Using 10 °C as the night-low temperature and 15 °C as a control, the authors found that low temperature reduced plant height; the number of flowers; fruit weight, length, and diameter; and the number of seeds per fruit in all accessions. However, a few parameters, such as stem diameter and length of main axis, were differentially affected in the different accessions. By performing correlations, principal component, and hierarchical cluster analysis, the authors observed that the group of best-performing accessions was mainly discriminated by its positive influence on most reproductive traits, such as number of fruits, fruit length, fruit diameter and fruit fresh weight. The authors suggested these parameters to be critical to identify night-low-temperature-tolerant genotypes in breeding programs. Finally, the authors identified bell and chili pepper accessions with contrasting performances in low temperatures [1].

Aiming to understand the mechanisms involved in saline stress tolerance in rice (*Oryza sativa* L.), Jahan et al. [2] analyzed the transcriptomic response to saline stress in rice seedlings of elite mega-hybrid rice (LYP9) and its parents (PA64s and 93-11). The authors found that the mega-hybrid LYP9 outperformed the parental lines in terms of salt stress tolerance (100 mM NaCl). The transcriptomic response to salt of these genotypes varied over time, initially (7 days) mainly relating to photosynthesis, but later (14 days) relating to hydrogen peroxide metabolic processes and cell wall organization. In addition, the authors found that the transcription factors belonging to the bHLH family were the most abundant among the differentially expressed genes, suggesting that this family of transcription factors may play a prominent role in salt stress tolerance.

In another study on rice, Ahmad et al. [3] evaluated the performance of 2030 japonica rice accessions, based on six agronomic traits, under lowland and upland conditions, to identify drought-tolerant genotypes. With these traits, they determined a drought-resistant grade (DRG) score that was used to classify the accessions, 10% of them being classified as drought-tolerant. Based on the drought-resistant grade, 42 elite genotypes, including upland and lowland genotypes, were selected. These genotypes may be an essential source of material for rice-breeding programs, or could even be used as such by producers, in areas susceptible to drought conditions.

Under drought conditions and other abiotic stress conditions, reactive oxygen species are over-produced. Ascorbate and glutathione play a relevant role as soluble antioxidants to prevent oxidative damage caused by excessive ROS [4]. In this Special Issue, Hoang et al. [5] characterized arabidopsis ascorbate- (*vtc2-4* and *vtc5-2*) and glutathione-(*cad2-1*) deficient mutants under abiotic stress. These mutants had a reduced sensitivity to ABA during germination and a lower germination rate under osmotic and salt stress. Moreover, the glutathione-deficient mutants showed lower tolerance to osmotic, salt, oxidative, and cadmium stress according to leaf area and root growth parameters. However, the ascorbate-deficient mutants showed, in some traits such as primary root length, number of lateral roots, and leaf area, a better performance than the wild type in some stress conditions. Under more severe stress, the ascorbate deficiency resulted in poorer performance in all stress conditions tested. This study evidences that ascorbate and glutathione are not antioxidants with redundant functions.

Given the importance of root architecture responses under drought stress, Urbanavičiūtė et al. [6] tested the root system diversity in six durum wheat genotypes with contrasting performance under drought. Even though a small number of genotypes were used, the authors found a large variability among them in terms of development, distribution, and architecture of the root system when subjected to drought. Interestingly, even the drought-tolerant genotypes showed contrasting strategies in response to drought. The authors concluded that high-throughput scanners are a valuable tool to identify interesting root traits in response to drought and speed up the selection of genotypes for plant breeding programs.

Stomatal conductance is one of the most important physiological responses to drought. Thus, stomatal conductance can be very informative of the capability of different genotypes to respond to drought. However, monitoring stomatal conductance during drought imposition can be a tedious and laborious activity. Simondi et al. [7] developed a mathematical model to predict stomatal conductance and water consumption kinetics with low sampling requirements. In particular, the authors developed this model using soybean plants in controlled conditions, but the approach they used can be reproduced in other conditions and plant species to determine the parameters B and k that feed the model in the newer conditions, in order to simplify the monitoring of stomatal conductance and water consumption.

Also in drought conditions, Berriel et al. [8] studied crop performance indexes, determined through isotopic analysis, in four legumes used as summer cover crops, *Crotalaria juncea, Crotalaria spectabilis, Crotalaria ochroleuca*, and *Cajanus cajan*. Based on the analysis of the parameters used as crop performance indexes, *C. cajan* was the most promising legume to be used as a cover crop among the four cover crops tested. Furthermore, the authors proposed the ratio between fixed nitrogen and transpired water, and the ratio between fixed nitrogen and ^{13}C isotopic discrimination, as good indicators of drought tolerance in fixing legumes.

Kulczycki et al. [9] also studied the effect of drought and its combination with sulfur fertilization on growth and yield parameters in maize and wheat. In these plants, in the absence of sulfur fertilization, moderate drought (45% FWC, relative to 60% FWC in control) had a small impact on grain yield, whereas a more severe drought condition (30%) significantly affected grain yield. Interestingly, the authors showed that in these crops, sulfur fertilization only had a positive impact in the absence of severe drought. These results show that using fertilizers in the context of drought may not be recommended, as it may not have beneficial impacts on growth-related parameters. Therefore, the investment in applying fertilizers will not pay off in terms of productivity. From a methodological point of view, this study also evidences that works intended to show the effect of mineral fertilization on growth-related parameters need to ensure the absence of drought episodes that can modify the response of plants to the nutrients attenuating the positive effects of fertilization.

However, in a different context, a similar message is obtained from the work reported by Signorelli et al. [10], who evaluated the effect of soil water content on bud burst rate using grapevine single node cuttings. The authors showed that soil water content (% of field capacity) significantly affects bud burst rate, concluding that soil water content has to be controlled to assess bud burst in perennial plants. Alternatively, when it is impossible to monitor water content during the assay, the authors recommend using a high soil water content to avoid the effect of water availability on bud burst rate. The manuscript has a methodological scope and illustrates a protocol to determine field capacity and monitor it. This protocol is helpful for works researching in the field of drought stress.

Although water restrictions are generally detrimental to plant growth, excessive irrigation can also be detrimental, especially for plants domesticated in arid regions. This is the case with olive trees, which were domesticated in the Eastern Mediterranean but now grow in more humid environments such as the one found in Rio de la Plata (South America), and where irrigation practices are very common. Conde et al. [11] compared the effect of full irrigation (supplying the water equivalent to the average of 100% evapotranspiration), partial irrigation (equivalent to 50% evapotranspiration), and water deprivation on fruit yield, oil content, and other productivity-related parameters, in two cultivars, Arbequina and Frantoio. The authors found that irrigation resulted in a significant increase in fruit weight and pulp/pit ratio but did not reduce the oil content in either cultivar. On the other hand, water restriction induced the content of polyphenols in the fruits in both cultivars, revealing a possible management practice by limiting water availability to improve olive oil quality without affecting oil productivity.

Another work dealing with agronomical management practices in tree corps was presented by Severino et al. [12], who studied pre- and post-harvest management of sunburn in apples (Granny Smith) under neotropical climate conditions. In multiple seasons, the authors evaluated the effect of sunburn protectors, white nets (20% translucent), black nets (35 and 50% translucent), and control without a netting system or sunburn protector applications. The authors found that the black net-50% treatment had the most positive effect on preventing sunburns, whereas the use of sunburn protectors did not affect any parameter tested in the neotropical climate of Uruguay. Importantly, the netting did not affect growth and leaf carbon assimilation, meaning that it can be recommended as a management practice for apples susceptible to sunburn such as Granny Smith.

Other environmental stimuli may be due to the presence of microbes in the surrounding environment of plants. These microbes can be beneficial or detrimental to plant growth, and understanding these possible interactions is relevant to proposing environmentally friendly strategies to promote plant growth and quality. In this sense, some studies have evaluated arbuscular mycorrhiza fungi (AMF) and seaweed extracts (SWE) as biofertilizers, however, little is known about the interaction of these biofertilizers. Rasouli et al. [13] studied the effect of these biofertilizers, separately and in combination, on plant growth and antioxidant capacity of lettuce plants. The authors found that the use of *Glomus mosseae* (20 g pot^{-1}) as AMF and *Ascophyllum nodosum* (0.5, 1.5, and 3.0 g pot^{-1}) as SWE had positive effects on plant growth. Moreover, the combination of these biofertilizers had the most positive results on plant growth and total antioxidant capacity, a property that can be indicative of crop quality.

Plants themselves produce compounds that can affect other plant growth. This is the case with walnut, which produces high amounts of juglone, a phenolic compound that can be toxic or growth-stunting to other plants. Medic et al. [14] evaluated the effect of pure juglone and walnut leaf extracts on the plant growth of four different crop species, *Beta vulgaris* L., *Brassica rapa* L. var. japonica, *Lactuca sativa* L., and *Valerianella locusta* Laterr. The authors found that beetroot and lettuce were less susceptible to juglone and other allelochemicals than turnip and mache, which presented lower yield and reduced quality. This research is useful to understand what crops can be grown in soils where walnuts were grown, particularly in the early years post-planting the new crop.

Together, these manuscripts help us to understand the multiple variables affecting crop productivity and quality, and the magnitude of the consequences of the different variables (such as water, salt, sunlight, temperature, microbes, and even other plant compounds). In some of these cases, the studies have proposed better agronomical management practices to avoid the effect of possible stressors, and in others, it is likely that the information generated will assist other studies to translate the basic knowledge into applicable knowledge for the agronomic field.

Acknowledgments: S.S. is a grateful active member of the Uruguayan System of Researchers (SNI, Uruguay), and the Basic Research Development Program (PEDECIBA, Uruguay). S.S. is supported by the CPR scheme of the International Centre of Genetic Engineering and Biotechnology (ICGEB), project number ICGEB_URY21_04_EC_2021, the CSIC I+D program of CSIC (Uruguay), project number CSIC_I+D_2020_21.

Conflicts of Interest: The author declares no conflict of interest.

References

1. Rajametov, S.N.; Lee, K.; Jeong, H.B.; Cho, M.C.; Nam, C.W.; Yang, E.Y. The effect of night low temperature on agronomical traits of thirty-nine pepper accessions (*Capsicum annuum* L.). *Agronomy* **2021**, *11*, 1986. [CrossRef]
2. Jahan, N.; Lv, Y.; Song, M.; Zhang, Y.; Shang, L.; Lu, Y.; Ye, G.; Qian, Q.; Gao, Z.; Guo, L. Transcriptomic analysis of short-term salt-stress response in mega hybrid rice seedlings. *Agronomy* **2021**, *11*, 1328. [CrossRef]
3. Ahmad, M.S.; Wu, B.; Wang, H.; Kang, D. Identification of drought tolerance on the main agronomic traits for rice (*Oryza sativa* L. ssp. japonica) germplasm in China. *Agronomy* **2021**, *11*, 1740. [CrossRef]
4. Signorelli, S.; Corpas, F.J.; Rodríguez-Ruiz, M.; Valderrama, R.; Barroso, J.B.; Borsani, O.; Monza, J. Drought stress triggers the accumulation of NO and SNOs in cortical cells of Lotus japonicus L. roots and the nitration of proteins with relevant metabolic function. *Environ. Exp. Bot.* **2019**, *161*, 228–241. [CrossRef]
5. Hoang, M.T.T.; Doan, M.T.A.; Nguyen, T.; Tra, D.P.; Chu, T.N.; Dang, T.P.T.; Quach, P.N.D. Phenotypic characterization of arabidopsis ascorbate and glutathione deficient mutants under abiotic stresses. *Agronomy* **2021**, *11*, 764. [CrossRef]
6. Urbanavičiūtė, I.; Bonfiglioli, L.; Pagnotta, M.A. Diversity in Root Architecture of Durum Wheat at Stem Elongation under Drought Stress. *Agronomy* **2022**, *12*, 1329. [CrossRef]
7. Simondi, S.; Casaretto, E.; Quero, G.; Ceretta, S.; Bonnecarrère, V.; Borsani, O. A Simple and Accurate Method Based on a Water-Consumption Model for Phenotyping Soybean Genotypes under Hydric Deficit Conditions. *Agronomy* **2022**, *12*, 575. [CrossRef]
8. Berriel, V.; Perdomo, C.H.; Signorelli, S.; Monza, J. Crop Performance Indexes Applied to Legume Used as Summer Cover Crops under Water Deficit Conditions. *Agronomy* **2022**, *12*, 443. [CrossRef]
9. Kulczycki, G.; Sacała, E.; Chohura, P.; Załuska, J. Maize and Wheat Response to Drought Stress under Varied Sulphur Fertilisation. *Agronomy* **2022**, *12*, 1076. [CrossRef]
10. Signorelli, S.; Dewi, J.R.; Considine, M.J. Soil Water Content Directly Affects Bud Burst Rate in Single-Node Cuttings of Perennial Plants. *Agronomy* **2022**, *12*, 360. [CrossRef]
11. Conde-Innamorato, P.; García, C.; Villamil, J.J.; Ibáñez, F.; Zoppolo, R.; Arias-Sibillotte, M.; De León, I.P.; Borsani, O.; García-Inza, G.P. The Impact of Irrigation on Olive Fruit Yield and Oil Quality in a Humid Climate. *Agronomy* **2022**, *12*, 313. [CrossRef]
12. Severino, V.; Arias-Sibillotte, M.; Dogliotti, S.; Frins, E.; Yuri, J.A.; González-Talice, J. Pre- and Postharvest Management of Sunburn in 'Granny Smith' Apples (Malus × domestica Borkh) under Neotropical Climate Conditions. *Agronomy* **2021**, *11*, 1618. [CrossRef]
13. Rasouli, F.; Amini, T.; Asadi, M.; Hassanpouraghdam, M.B.; Aazami, M.A.; Ercisli, S.; Skrovankova, S.; Mlcek, J. Growth and Antioxidant Responses of Lettuce (*Lactuca sativa* L.) to Arbuscular Mycorrhiza Inoculation and Seaweed Extract Foliar Application. *Agronomy* **2022**, *12*, 401. [CrossRef]
14. Medic, A.; Zamljen, T.; Grohar, M.C.; Slatnar, A.; Hudina, M.; Veberic, R. Using HPLC–MS/MS to Assess the Quality of Beet, Mizuna, Lettuce and Corn Salad after Juglone and Walnut Leaf Extract Treatments. *Agronomy* **2022**, *12*, 347. [CrossRef]

agronomy

Article
The Effect of Night Low Temperature on Agronomical Traits of Thirty-Nine Pepper Accessions (*Capsicum annuum* L.)

Sherzod Nigmatullayevich Rajametov †, Kwanuk Lee †, Hyo-Bong Jeong, Myeong-Cheoul Cho, Chun-Woo Nam and Eun-Young Yang *

National Institute of Horticultural & Herbal Science, Rural Development Administration, Wanju 55365, Korea; sherzod_2004@list.ru (S.N.R.); kwanuk01@korea.kr (K.L.); bong9846@korea.kr (H.-B.J.); chomc@korea.kr (M.-C.C.); cwsky1004@daum.net (C.-W.N.)
* Correspondence: yangyang2@korea.kr; Tel.: +82-(0)63-238-6613; Fax: +82-(0)63-238-6605
† Equal contribution.

Abstract: Pepper plants are subject to complex environmental factors including abiotic and biotic stresses in fields, as well as the significant effects of climate changes, including low and high temperatures. Low temperature stress in the growth and development of pepper plants is one of the most critical issues, and directly impacts the crop yield and productivity of pepper plants. Therefore, it is essential to select and breed low temperature-(LT) tolerant pepper (*Capsicum annuum* L.) cultivars. This research was conducted to assess the agronomical traits of 39 pepper accessions belonging to the chili and bell fruit varieties which were cultivated under two different night temperature set-points: at 15 °C for a suboptimal temperature (CT) and at 10 °C for a low temperature (LT). The plant heights (PH) of most pepper accessions in a LT were significantly decreased compared to those in a CT. The stem diameter (SD) and the length of main axis (LMA) varied depending on the genotypes under LT. Moreover, the number of flowers (NFL), total number of fruits (NFR), fruit yield (FY), fruit fresh weight (FFW), fruit length (FL), fruit diameter (FD), and number of seeds in a fruit (NSF) remarkably declined in a LT compared to in a CT. The evaluated agronomical traits between LT and CT were further applied for the correlation analysis, principal component analysis (PCA), and hierarchical cluster analysis. Notably, the FY trait was correlated with other reproductive traits including NFR, FFW, FD, and FL on the positive directions and thirty-nine LT-treated pepper accessions were clustered into seven groups by the hierarchical clustering analysis. The selected accessions were primarily involved in the positive trends with the reproductive index including NFR, FL, FD, and FFW traits and could be used for pepper breeding programs to develop LT-tolerant cultivars.

Keywords: chili and bell pepper; low temperature stress; vegetative and reproductive traits; pepper breeding; PCA; hierarchical cluster analysis

Citation: Rajametov, S.N.; Lee, K.; Jeong, H.-B.; Cho, M.-C.; Nam, C.-W.; Yang, E.-Y. The Effect of Night Low Temperature on Agronomical Traits of Thirty-Nine Pepper Accessions (*Capsicum annuum* L.). *Agronomy* **2021**, *11*, 1986. https://doi.org/10.3390/agronomy11101986

Academic Editor: Santiago Signorelli

Received: 3 September 2021
Accepted: 29 September 2021
Published: 30 September 2021

1. Introduction

Pepper plants (*Capsicum annuum* L.) originate from the American tropics, are classified into the Solanaceae family, and are considered as an essential horticultural crop. Among the 30 species in *Capsicum*, around five species, including *C. annum*, *C. baccatum*, *C. chinense*, *C. frutescens mill*, and *C. pubescens*, are broadly domesticated by plant breeders and farmed in agricultural areas [1]. The pepper fruits of chili and bell peppers are used in diverse cuisines as a source of basic ingredients, contributing a variety of vitamins, phytochemicals, minerals, food colors, and capsaicin [2–4]. The importance of peppers in agriculture and the cultivation area and production of peppers is increasingly worldwide. According to the Food and Agriculture Organization (FAO) and the Korean Statistical Information Service (KOSIS) in 2019, the cultivation area and the production of peppers accounted for approximately 4.5 million ha and 61 million tonnes of the total green and dried peppers grown worldwide (http://www.fao.org/faostat/), as well as around 36,600 ha and 334,280 tonnes of green and dried peppers of both chili and bell varieties in Korea (https://kosis.kr/eng (accessed on 9 August 2021), respectively.

Climate changes, including low and high temperatures, strong wind, drought, flooding, and heavy rainfall can severely influence crop yield and productivity [2,5]. Particularly, reports demonstrated the impact of the night low temperature (NLT) with different temperature regimes during the period of entire growth and development [6,7]. The NLT remarkably affected seed germination, seedling growth, leaf morphology, stem diameter (SD), and plant height (PH) during the vegetative stages of tomato, cucumber, and pepper plants [7–14], causing poor growth and development. The temperature stress also affected the reproductive index, including the agronomical traits of the number of flowers (NFL), the number of fruits (NFR), fruit set (FS), and fruit yield (FY) during reproductive stages of pepper plants [15–17]. Since the flower development was highly associated with NFR, FS, and FY in response to LT, many studies focused on the development of the floral organs. For example, the plants grown in temperatures below 15 °C resulted in abnormal flower shapes, reduced pollen activity and quality [18–20], parthenocarpic fruits [21], a decrease in the number of seeds in the fruit (NSF) [21–23]. Moreover, in comparison with those grown in normal and low temperatures, flower development was shown in the malformation of unexpanded petals, stunted stamens containing a few of pollen, and a reduced pollen germination activity in androecium [19,22]. Flower development was also impaired by the swollen ovary and shorter styles in the gynoecium, resulting in the poor quality of fruit development; small, flat, and irregular shaped fruits [15–22]. Although the effects of NLT on the aforementioned vegetative and reproductive parameters were determined in pepper plants, the impact of NLT was only characterized in limited accessions and mainly in sweet peppers.

The temperature regulation is one of the most important factors for pepper growth and is preferentially considered for cultivation in a winter greenhouse. According to the Korea Meteorological Administration (KMA) (https://www.weather.go.kr/w/index.do (accessed on 9 August 2021)) and our previous publication [19], in Wanju, where this study was performed in Korea, the minimum and maximum temperatures fluctuated from −6.3 to 8.7 °C. Additionally, the average temperature was maintained from −1.7 to 3.9 °C for the winter season, as has been the case since 1970, indicating that the climate of Wanju fluctuated from −7 to 10 °C during the period of winter season. Generally, the heating demand for greenhouse cultivation significantly increased at night time in winter [24] and the heating accounted for around 19–23% of the operating costs in Korea from 2017–2019 [25]. Several studies showed that sub-optimal temperatures in greenhouses could decrease energy costs by approximately 16% with a 2 °C temperature reduction [26], suggesting that the sub-optimal temperature ranging from 15–20 °C could be considered as the minimal low temperature, which avoided serious damage to plant growth and development [27,28]. Interestingly, it was shown that LT led to a higher fruit yield than the optimal temperature (>20 °C) [22], whereas LT (<10 °C) resulted in the severe defective growth and development of the plants [29,30]. However, a few studies recently elucidated the development of breeding systems for selecting low temperature (LT)-tolerant pepper accessions. There is still a possibility of developing breeding programs with a large number of different fruit types for pepper plants, economically regulating NLT at 10 °C with the optimal temperature set-points at 15–20 °C in winter greenhouse.

In this work, we analyzed the agronomical traits of the pepper plants, including chili peppers (n = 27) and bell peppers (n = 12), in response to night low temperatures in greenhouse conditions. The vegetative parameters of PH, SD, and LMA and the reproductive parameters of NFL, FY, FW, FD, NFR, and NSF were investigated among 39 pepper genotypes between 15 °C and 10 °C in a greenhouse. On the basis of this correlation, as well as PCA and cluster analysis, together with ten agronomical traits, we selected four genotypes of chili peppers and four genotypes of bell peppers exhibiting low or high performances of vegetative and reproductive parameters, which were associated with high FY traits between LT and CT. The characterized and identified pepper genotypes in the present study will be used as a good resource for pepper breeders to breed LT stress-tolerant cultivars in winter greenhouses, with the consideration of a high FY index.

2. Materials and Methods

2.1. Plant Material and Growth Conditions

A total of 39 pepper breeding lines including chili (n = 27) and bell peppers (n = 12) from National Institute of Horticultural and Herbal Science (NIHHS, Korea, 35°83′ N, 127°03′ E) were utilized in the experiments (Table S1). The seeds of the 39 pepper lines were sown on 28 September 2020 in plastic trays which were 52 cm × 26 cm in pot size and 6 cm × 6 cm in cell size. The trays were transferred into a glasshouse that maintained 26/18 °C (day/night) and relative humidity from 65–70% and given one liter of water, daily. 14-day-old seedlings were transplanted on 13 November 2020 into two plastic film greenhouses. Pepper seedlings of ten plants per accession were planted with a distance of 1.5 m × 35 cm between plants in both LT and CT greenhouses. To help the pepper seedlings adapt new environment conditions, night temperature set-point was initially maintained at 15 °C for 2 weeks in both greenhouses and then was modulated for LT and CT, respectively.

2.2. Soil Preparations

The soil preparations in the trays and two greenhouses were completed as previously described in [16]. The commercial media (Bio Sangto, Seoul, Korea) which consisted of coco peat (47.2%), peat moss (35%), zeolite (7%), vermiculite (10.0%), dolomite (0.6%), humectant (0.006%), and fertilizers (0.194%), which were made of 270 mg kg^{-1} of N, P and K, respectively. The soil preparations in two greenhouses were prepared, following the recommendations of the Korea Soil Information System (KSIS) (https://soil.rda.go.kr (accessed on 28 August 2021)), equally with pre-plant broadcast manure at a dose of 1 kg m^{-2} and basal fertilizer containing 16 g m^{-2} N, 8 g m^{-2} K_2O, and 16 g m^{-2} P_2O_5, which was regularly fertigated with the mixture of solution A (5.5% nitrogen, 4.5% potassium, 4.5% calcium, 0.00014% boron, 0.05% iron, 0.0001% zinc, and 0.0002% molybdenum) and B (6% nitrogen, 2% phosphorus, 4% potassium, 1% magnesium, 0.05% boron, 0.01% manganese, 0.005% zinc, and 0.0015% copper) in 1200 L water (Mulpure, Daeyu Co., Ltd., Gyeongsan, Korea).

2.3. Temperature Regulations

The winter climate where this experiment was conducted was shown in our previous report [16]. Briefly, the temperature was monitored and recorded in LT and CT greenhouses during the period of the pepper cultivation using data logger (Figure S1) (WatchDog 1450, Spectrum Technologies Inc., Aurora, IL, USA). Nighttime temperature was maintained by heating machine (Model TKP-800, Tae Kwang Machine Co., LTD., Daegu, Korea) whenever the temperature decreased below the 10 °C and 15 °C set-points. The relative humidity (RH) was kept within approximately 40–60% in both greenhouses, respectively.

2.4. Diseases and Pest Controls

Diseases and pest controls were conducted as previously described in [16]. Briefly, 20% of Spiromesifen (Farmhannong, Seoul, Korea) was diluted with the ratio of 1:2000 for controlling whitefly; 5% of Rampage (Hankooksamgong, Seoul, Korea) was diluted with the ratio of 1:1000 for controlling thrips. In addition to this, 30% of Iminoctadine tris (Farmhannong, Seoul, Korea) was diluted with the ratio of 1:1000 for controlling leaf and gray molds; 50% of Polyoxin B (Farmhannong, Seoul, Korea) was diluted with 1 to 5000 for controlling powdery mildew and fungi.

2.5. Data Collections

The pepper accessions were planted with the same arrangement in LT and CT greenhouses from three independent biological plants, which were randomly selected among ten plants for measurement the reproductive parameters. The number of flowers (NFL) was determined from the second to fifth internodes and the vegetative parameters, including plant height (PH), stem diameter (SD), and length of main axis (LMA), were measured at 120 days after transplanting (DAT) of seedlings. Total number of fruits per plant (NFR) and fruit yield per plant (FY) were measured from three individual plants, randomly. Five fruits of each accession were collected for measurement of fruit fresh weight (FFW), fruit length (FL), fruit diameter (FD), and the number of seeds in a fruit (NSF) using a digital electron Micro Weighing Scale MW-II (CAS), a ruler, and a caliper, respectively.

2.6. Data Analysis

The significant difference in vegetative parameters of PH, SD, and LMA, and reproductive parameters of NFL, NFR, FY, FD, FL, FFW and NSF were assessed as described in the figure legends with Student's t-test using EXCEL 2016 software (Microsoft Co., Ltd., Albuquerque, NM, USA). To figure out the effects of LT on the evaluated traits, the score of agnomical traits was calculated by dividing LT by CT and multiplying by 100 (%). The analysis of correlation coefficients was performed among the total populations ($n = 39$) using EXCEL 2016 (Microsoft, Redmond, WA, USA). The multivariate analysis, including correlation analysis, principal components analysis (PCA), and hierarchical cluster analysis were assessed using SPSS program (IBM SPSS v27.0, Chicago, IL, USA). The adequacy of the samples was carried out by the Kasier-Meyer-Olkin (KMO > 0.5), and Bartlett's test of Sphericity (BTS < 0.001) was utilized as an indicator in the proper construct of the PCA model to evaluate the relationship between variables.

3. Results

3.1. The Vegetative Traits with Chili and Bell Peppers

In order to understand the response of pepper plants to the night low temperature (NLT), the vegetative parameters including PH, SD, and LMA were investigated among 39 pepper accessions at 120 DAT in LT and CT greenhouses. The growth rate of most of the accessions in PH significantly reduced in LT compared to that in CT (Figure 1A). Only one accession C22 of chili pepper was observed with no significant difference in both LT and CT. The lowest significant difference ($p \leq 0.05$) was identified in C17 of chili pepper and P06 of bell pepper.

Next, in order to evaluate the effect of LT on the stem growth, the plant stem diameter (SD) was measured among 39 pepper accessions. Sixteen accessions of chili pepper and seven accessions of bell pepper were observed to have no significant difference between LT and CT (Figure 1B). The significant differences ($p \leq 0.01$) were identified with C02, C05, C17, and C20 of chili peppers and P01 of bell pepper. Interestingly, the SD of C05 accession was higher in LT than in CT. In addition to this, the growth of LMA was investigated and thirteen accessions of chili peppers and two accessions of bell peppers were observed with no remarkable differences between LT and CT (Figure 1C). The significant difference ($p \leq 0.001$) was determined in C13, C16, C21, and C23 of chili peppers and P01 and P02 of bell peppers. Taken together, the observations indicated that the effect of LT on vegetative traits was widely varied from genotypes among 39 pepper accessions.

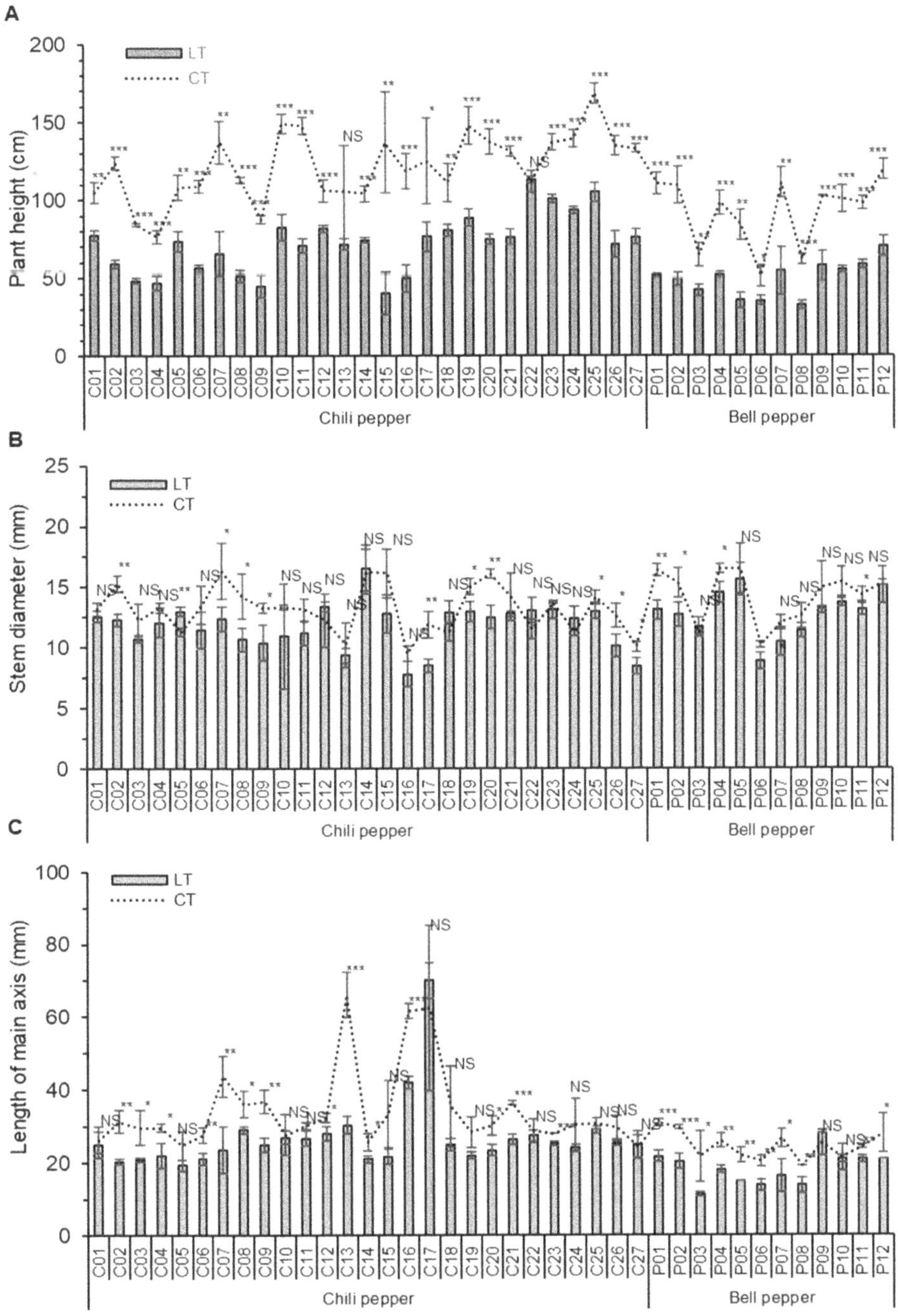

Figure 1. The evaluation of vegetative traits on (**A**) plant height, (**B**) plant stem diameter, and (**C**) length of main axis among 39 pepper accessions in LT and CT greenhouses. Plant height, stem diameter, and length of main axis were measured at 120 days after transplanting. Significant differences were evaluated with Student's *t*-test with $p \leq 0.05$, $p \leq 0.01$, and $p \leq 0.001$ and denoted by *, ** and ***, respectively. NS means not significant and bars indicate ± standard deviation (n = 3).

3.2. The Reproductive Traits of NFL, NFR, and FY with Chili and Bell Peppers

In order to determine the response of pepper accessions to NLT regarding the reproductive traits, the reproductive parameters including NFL, NFR, and FY were investigated among 39 pepper accessions at 120 DAT in LT and CT greenhouses. The effect of LT on NFL was different depending on the pepper accessions (Figure 2A). The floral organs were not developed in C04, C13, and C16 accessions in both LT and CT conditions, and the flowers of C03, C17, and P11 were developed in CT but not in LT. No significant difference in NFL was observed except for the P06 and P10 accessions between both LT and CT. In addition, the effect of LT on NFR per plant at 120 DAT was subsequently evaluated and observed in the significant reduction in most pepper accessions in LT (Figure 2B). Notably, C04, C08, C13, and C16 of chili peppers were not shown in any fruit in both LT and CT, whereas C22, P08, and P09 were not different between LT and CT. Because high fruit yield was one of the most important parameters to determine LT-tolerant pepper cultivars in breeding programs, the evaluation of FY was implemented and it was drastically decreased in LT compared to CT, except for P08 and P09 of bell peppers (Figure 2C). Interestingly, the highest index of FY over 500 g was identified in C01, C02, C05, C18, C19, C23, and C24 of chili peppers and in P01, P02, P03, and P04 of bell peppers in CT, whereas it was observed in C22 (226.7 g) of chili pepper and P04 (215.0 g) of bell pepper in LT.

3.3. The Fruit Traits of FD, FL, FFW, and NSF with Chili and Bell Peppers

In order to explore the impact of fruit traits in response to LT, the fruit traits-related FD, FL, FFW, and NSF were evaluated among 39 pepper accessions at 120 DAT in LT and CT greenhouses. FD dramatically decreased in most pepper accessions in LT in comparison with CT (Figure 3A). However, no appreciable difference in FD was found in C23 of chili peppers and P01 and P04 of bell peppers between LT and CT. In detail, the widest FD over 25 mm and 60 mm was found in C02, C05, C14, C15, C21, and C22 of chili peppers and P02, P03, P5, and P10 of bell peppers in CT, respectively. The widest FD over 15 mm and 50 mm was observed in C02, C05, C14, C22, and C23 accessions, and in P01, P02, P05, P08, and P09 accessions in LT.

A previous study reported that FD was highly associated with FL [31,32]. In order to validate the effect of LT on FL, together with FD, FL was measured among 39 pepper accessions and FL was noticeably reduced in most pepper genotypes in LT compared to CT (Figure 3B). However, no appreciable difference in FL was observed in P06 and P08 of bell peppers between LT and CT. In detail, the longest FL over 15 cm and 10 cm in CT was determined in C02, C05, C06, C14, C18, C19, C22, C23, and C24 of chili peppers, and in P01 and P02 of bell peppers. Moreover, the longest FL over 10 cm and 5 cm in LT was observed in C18, C19, C22, and C23 of chili peppers and P02, P04, and P09 of bell peppers, respectively. FFW was also investigated and decreased in most pepper accessions in LT compared to CT, except for P08 and P09 of bell pepper which exhibited no significant difference between LT and CT conditions (Figure 3C). One study determined the effect of LT on the seed development in the pepper fruit, which caused the growth of seedless fruit (referred to as parthenocarphy) and a reduced fruit marketability [18]. To further confirm the effect of LT on seed development in a fruit, the number of seeds per fruit was counted. The results showed that NSF was reduced in a variety of pepper accessions and even all fruits of bell peppers did not develop any seeds in LT condition (Figure 3D). Intriguingly, the NSF of C12 accession increased in LT compared to CT and no appreciable difference in NSF was found in C25 and P05 accessions between LT and CT. In addition, the highest NSF over 80, 70, and 40 seeds in CT was identified in C20, C21, and C24 of chili peppers and P01, P02, and P09 of bell peppers and the highest NSF over 40 in LT was observed in C12, C24 and C25 of chili peppers.

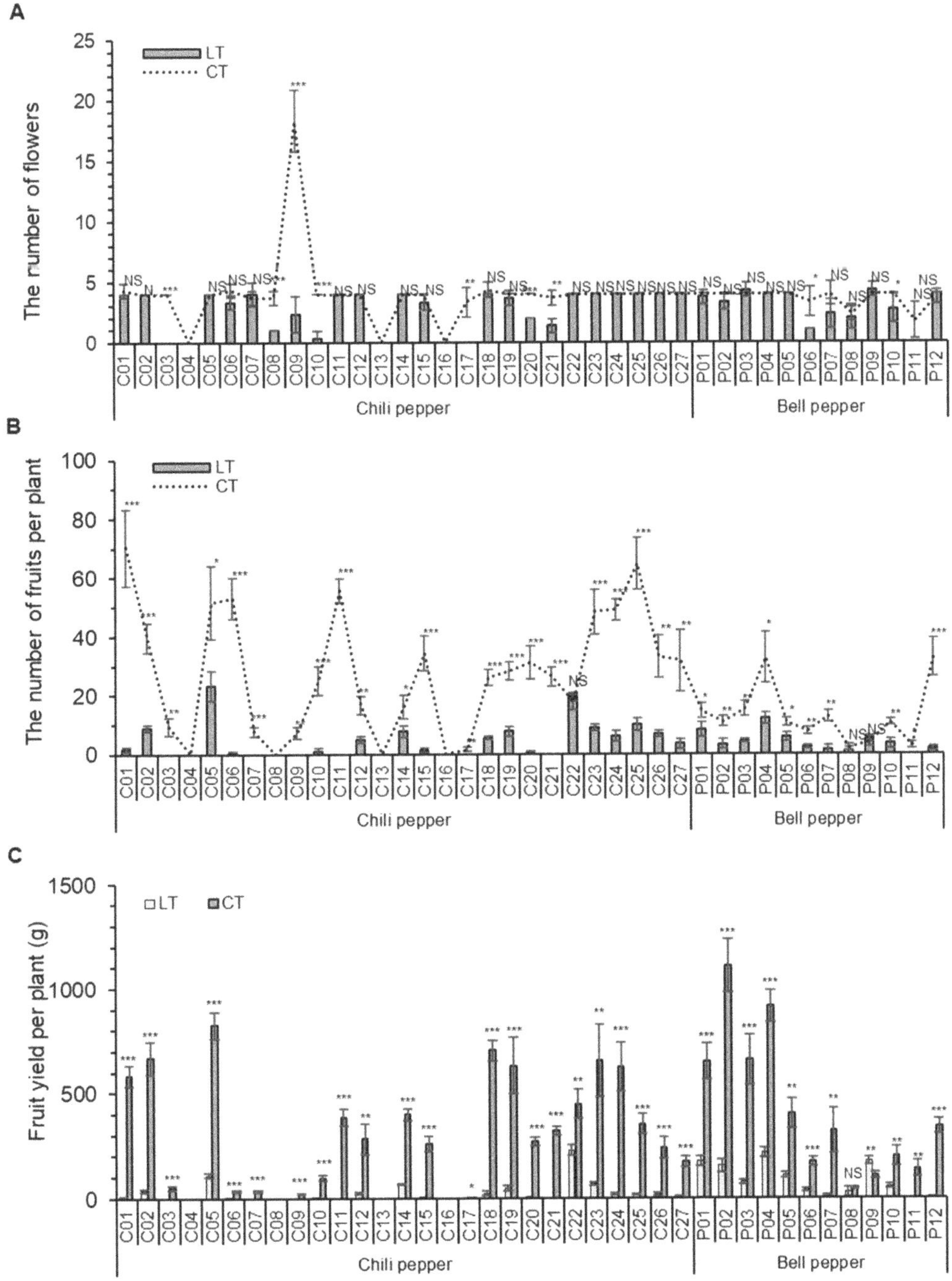

Figure 2. The evaluation of reproductive traits on (**A**) the number of flowers, (**B**) the number of fruits, and (**C**) fruit yield among 39 pepper accessions in LT and CT greenhouses. The reproductive parameters were measured at 120 days after transplanting. Significant differences were evaluated with Student's t-test with $p \leq 0.05$, $p \leq 0.01$, and $p \leq 0.001$ and denoted by *, **, and ***, respectively. NS means not significant and bars indicate ± standard deviation. (n = 3).

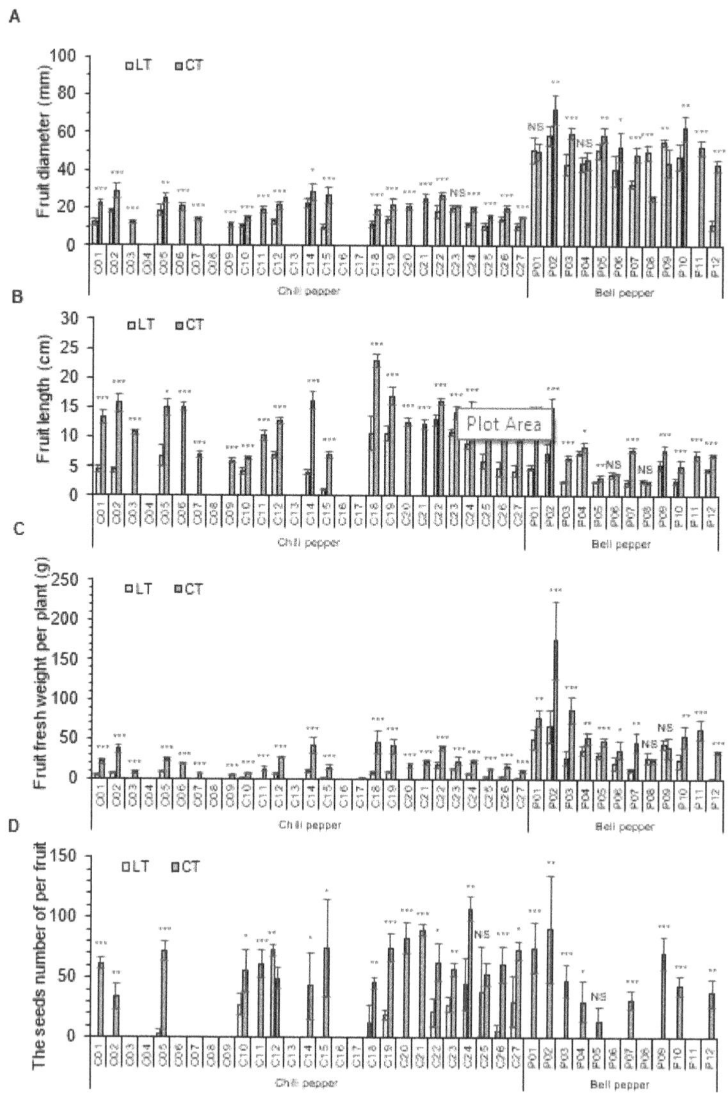

Figure 3. The evaluation of reproductive traits on (**A**) fruit diameter, (**B**) fruit length (cm), (**C**) fruit fresh weight, and (**D**) the number of seeds among 39 pepper accessions in LT and CT greenhouses. The reproductive parameters were measured at 120 days after transplanting. Significant differences were evaluated with Student's t-test with $p \leq 0.05$, $p \leq 0.01$, and $p \leq 0.001$ and denoted by *, **, and ***, respectively. NS means not significant and bars indicate ± standard deviation (n = 3).

3.4. The Principal Component Analysis (PCA) of Agronomical Traits

In order to understand the relationship between the multiple variables, including the aforementioned parameters, a principal component analysis (PCA) was conducted (Figure 4). A correlation matrix of 10 variables was produced in LT and CT conditions (Figure S2). The Kaiser-Meyer-Olkin (KMO) was calculated for the adequacy of measured samples on vegetative and reproductive traits, and the score was 0.720. Bartlett's Test of Sphericity (BTS) was significantly lower than 0.001, indicating that the samples and

PCA construct were appropriate for PCA analysis. Five PCs were extracted from ten studied agronomical traits and the eigenvalues were greater than one from the first three (Table S2). To reduce the dimensions of the data space, the correlation matrix with the first two components of PCA was applied. The total variance of the acquired data was explained with the 65.76%, which represented 46.63% from component factor 1 (PC1) and 19.13% from component factor 2 (PC2), respectively. The traits (scores > 0.30) were loaded onto PC1 and PC2 (Table S2A). The first component PC1 contributed to multiple traits, concluding that FL had a major contribution towards positive loading vectors (0.885), followed by FD (0.878) and FW (0.791) within the first component, and PC2 exhibited FY (0.914) and NFR (0.815). This indicated that three major and two minor variables had a strong correlation with the first PC and second PC, respectively.

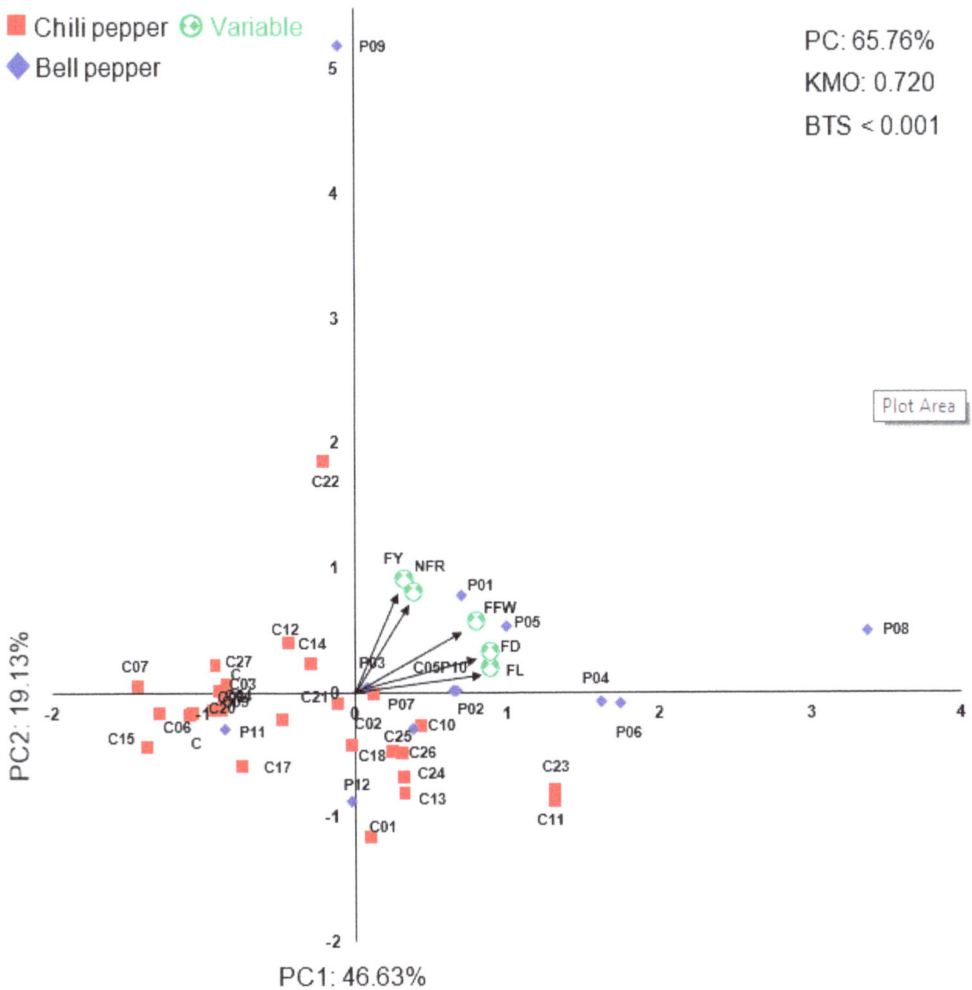

Figure 4. Biplot for first two principal components was analyzed using the principal component analysis (PCA) for 10 agronomical traits among 39 pepper accessions possessing chili (red square) and bell (blue diamond) fruit varieties between LT and CT conditions. The green round circles with multiple diamonds and the arrows represent the variables and the direction of vectors, respectively.

3.5. Clustering Analysis

In order to analyze the association with the PCs in 39 pepper accessions, a scatter plot matrix was drawn using the component factor 1 (PC1) and factor 2 (PC2), exhibiting no clear pattern for the grouping of pepper fruit varieties (Figure 4). The PC1 and PC2 factors were further justified to agglomerative hierarchical clustering utilizing the Euclidean distance matrix through Ward's method and the dendrogram was produced on the basis of the results. Seven major groups were distinctly clustered (Figure 5A). The results showed that group 1 primarily had a very low value of LMA and PH; group 2 had a moderate low value of SD and NFL; group 3 had a low value of both vegetative and reproductive traits; and group 4 had a moderate value of SD in vegetative traits, and a high value of FL and FD in reproductive traits. Additionally, it displayed that the group 5 had a high value of PH and SD in vegetative traits and NFR and FL in reproductive traits; group 6 had a high value of FW, FL, and FD in reproductive traits; and group 7 had a high value of LMA in vegetative traits, and a high value of NFL, NFR, FY, FFW, and FD in reproductive traits. Depending on the clustering and the studied agronomical traits, the plants displaying a low performance of the agronomical traits were selected in two accessions (C01 and C02) of chili peppers and two accessions (P01 and P02) of bell peppers, and the plants showing a high performance of agronomical traits were selected in two accessions (C22 and C23) of chili peppers and two accessions (P08 and P09) of bell peppers (Figure 5B).

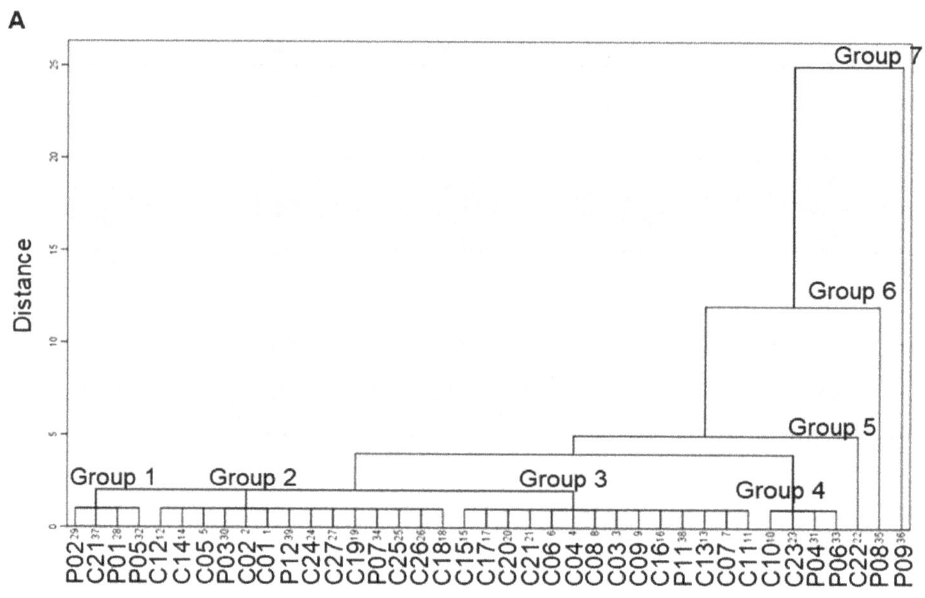

Figure 5. *Cont.*

B

Figure 5. (**A**) Dendrogram was performed by the cluster analysis of 39 pepper accessions on the basis of the Euclidean distance using components factors 1 and factors 2. (**B**) Four pepper accessions in chili and bell pepper were selected and the pictures were shown in the fruits of pepper plants taken at 120 days between LT and CT greenhouse conditions.

4. Discussion and Conclusions

Pepper plants were naturally exposed to harsh cold stress during winter season in agriculture and possessed the cellular and molecular mechanisms to acclimate and overcome low temperature stress [33–36]. Previous reports demonstrated a response to LT with limited accessions, mainly in reproductive traits, including flower morphologies, fruit shapes, fruit varieties, and fruit yield [6,18,19]. Additionally, the effect of LT on the agronomical traits was determined primarily in sweet peppers. In this study, we assessed 39 pepper genotypes including chili (n = 27) and bell fruit varieties (n = 12) and evaluated the vegetative and reproductive traits during the entire period of the pepper growth and development stages under LT, which could economically reduce the heating demand of the pepper cultivation in a winter greenhouse.

The effect of LT showed that the PH decreased significantly in almost all accessions, except for one accession, C32, which was not significantly different among the chili peppers grown between the LT and CT greenhouses (Figure 1A). In line with our current results, previous studies determined that the effect of LT on PH influences the retarded growth of plant height in pepper and cucumber plants with different low temperature regimes [7,9,37]. In addition to this, previous studies also reported that the reduction in tomato plant height in LT could be correlated to the number of leaves, which influenced the relative growth rate and net assimilation rate with the modulated photosynthetic ability [9]. Next, it was in our interest to further investigate how the number of pepper leaves, leaf length, and leaf width, as well as the photosynthetic parameters were involved in PH in LT.

The effect of LT on SD exhibited a reduction in twenty-one accessions (40.8%) of chili peppers and five accessions (41.6%) of bell peppers, respectively (Figure 1B), whereas other accessions in LT did not significantly decrease in SD, except for C05 which showed a higher growth of SD in LT than in CT. This result was in agreement with a previous study which determined the difference in the SD of tomato plants under LT [15,16]. The finding showed that the SDs varied depending on the tomato genotypes without fruit types [16]. Moreover, the effects of LT on LMA displayed a reduction in 13 accessions (48.2%) of chili peppers and 10 accessions (83.3%) of bell peppers, respectively (Figure 1C), indicating that the effect of LT on SD and LMA might be also significant among pepper genotypes, irrelevant of fruit types (Figure 1), and that the bell peppers would be very sensitive to LT in the vegetative traits compared to those in CT, as previous studies mentioned [18]. In the current study, the positive high temperature (day temperature–night temperature > 0) in tomato plants led to an increased stem thickness and enlogation, as well as the number of xylem vessels via the modulation of genes involved in the cell wall, GA, and auxin biosynthesis [38]. Furthermore, our studies were essential to provide factors affecting the pepper stem diameter and enlogation with the phytohormones under diverse LT regimes. Collectively, we did not completely understand how LT affected the vegetative index, such as SD and LMA, in pepper plants. Nevertheless, our current results clearly suggested that LT influenced pepper growth and development during the vegetative stages.

We conducted the investigation of a reproductive index including NFL, NFR, and FY. A study reported that NFL, in the first truss of the tomato plant, displayed no significant difference under a low temperature in the atmosphere, while the NFL was promoted by a low temperature in the root area [39]. In addition, our previous research found that the LT did not affect the NFL of the tomato plant and LT influenced the NFL with a genotype-specific interaction [15,16]. In the agreement with our previous studies, our current finding also showed that the NFL of most pepper accessions showed no remarkable difference between the LT and CT, although NFL was reduced in several accessions of chili and bell peppers (Figure 2). Despite that, we did not understand why the effect of LT on NFL in most accessions was not influenced. Our further studies need to focus on elucidating the mechanistic role of the impact of LT on NFL in pepper plants with a consideration of the air and root low temperature.

Previous studies showed that NFR was closely associated with FY in LT [13,39] and that NFR and fruit setting were key determinants to select LT-tolerant tomatoes and peppers with high fruit yields [16,40]. In line with a previous study, the effect of LT on NFR and FY resulted in the drastic decrease compared to those in CT (Figure 2B,C), suggesting that NFR and FY were closely correlated in LT. Moreover, LT affected the reduction in FD, FL, and NSF (Figure 3), resulting in irregular fruit shapes (Figure 5B). These findings were concordant with previous effects of LT on the flower morphology and the fruit development [18]. The studies assessed that the impact of LT on anther and ovary shapes caused the malformation of floral organs including the stunted stamen, the decreased number of pollen, and a reduced pollen activity by hindering pollination and fertilization [18,22], which further led to the abnormal formations of fruit development, irregular fruit shapes, and parthenocarpic fruits, together with reduced NSF [18,41]. The previous studies determined that the malformation of fruit shapes in rice, mango, and pepper plants were produced by the swollen ovary and shortened style in LT [2,6,42,43], indicating that the development of floral organs with a stamen and an ovary was very sensitive to LT. Furthermore, the parthenocarpic fruits and the declined NSF were associated with the balance of plant hormones, including auxin, gibberillin, and cytokinin due to the lack of fertilization, but might not be a defect of pollination [44,45]. Next, our important endeavor was to dissect the involvement of the plant hormones and the anatomic structures of floral organs during the period of flower development under different low temperature regimes.

We utilized the correlation matrix to perform a PCA analysis determining the crucial factors with 10 variables of agronomical traits between LT and CT, and 5 traits (scores > 0.30) were loaded into the plot. The first two PCA explained 65.76% of the agronomical variables

(Figure 4). The angle between the vectors of traits, including FFW, FD, and FL in PC1 and NFR in PC2 with FY, were less than around 80°, indicating that the positive correlations were exhibited in the variables which were in line with previously published statistics [16]. Our finding showed that FFW, FD, and FL were closely correlated in LT effect. Studies were not conducted on the effects of LT and the LTN condition on fruit shapes such FFW, FD, and FL. On the one hand, a study reported that the FFW was positively associated with FD and pericarp thickness, but not in FL [46,47] under normal conditions. Moreover, some studies determined that the quantative trait locus (QTL) governing FFW and FD was linked on the choromosome P12, indicating that the positive trends between FFW and FD could be described from our PCA result [48]. Additionally, studies assumed that the mechanism of fruit development factors, including FD, FL, and FFW, were most likely shared [31,32]. Notably, in our current study, the effect of LT on the parameters of fruit development uncovered that FL was highly correlated with FD. Given that one study mentioned that the FL was governed by between 3 and 10 pairs of genes and was most likely affected by environmental conditions [46], further studies should investigate the mechanistic role of how the fruit developmental factors are governed by the gene clusters.

As our previous publication mentioned [27–29], the decreased NFR of sweet pepper in LT was associated with FY, showing the close correlation between the traits. In line with our current results, the previous PCA analysis of tomato plants also exhibited a strong correlation with FY and NFR in LT effect [16], suggesting that the NFR plays an important role in determining the yield-related parameters such as fruit set and fruit yield in pepper plants for breeding programs when selecting LT-tolerant peppers. A biplot analysis was further conducted to understand the multivariate relationships with 39 pepper accessions containing 27 chili and 12 bell peppers accessions (Figure 4). The evaluated agronomical traits and correlation matrix were applied to the cluster analysis and classified into seven groups (Figure 5A). Our findings showed that group 1 to group 3 tended to have negative trends with regard to the vegetative and reproductive parameters, but some accessions still exhibited moderate high and low values for these parameters. On the one hand, group 4 to group 7 tended to have positive trends, mainly with the reproductive parameters and with one or two of the vegetative parameter(s). Our previous study determined, in the selection criteria for LT-tolerant tomatoes, that tomato plants displayed different vegetative or reproductive indexes depending on fruit types [16]. Given that the clusters grouped with the reproductive index still exhibited one or two vegetative parameters in some accessions, or vice versa in different clusters, it is probable that the screening of the pepper accessions tolerant to LT were also taken into consideration with a different selection index among pepper accessions in the greenhouse. On the basis of our clustering of 39 pepper accessions (Figure 5A), we selected two LT-sensitive two chili and bell pepper accessions (C01 and C02 in group 2, P01 and P02 in group 1) which showed the low values of the vegetative and reproductive indexes, respectively. Additionally, it was observed that two LT-tolerant chili and bell peppers (C23 in group 4 and C22 in group 5; P08 in group 6 and P09 in group 7) mainly showed a high value of the reproductive index, such as FY, FFW, FL, and FD. Intriguingly, the fruit shapes of LT-sensitive chili and bell pepper genotypes were much smaller than those in CT, whereas the the fruit shapes of LT-tolerant genotypes were almost similar, or smaller to some exent, in comparison with those in CT (Figure 5B). Our results indicated that FL and FD played a crucial role in selecting LT-tolerant cultivars. Taken together, we assume that the results from our PCs and cluster analysis can be importantly considered as a measure for the selection of LT-tolerant pepper cultivars with reproductive and fruit shape-related traits, although the left traits are shown to have a minimal contribution to positive or negative directions. The selected accessions could be further used for pepper breeding programs to develop LT-tolerant chili and bell pepper cultivars with the selection criteria in winter greenhouses.

Supplementary Materials: The following are available online at https://www.mdpi.com/article/10.3390/agronomy11101986/s1, Figure S1: Air temperature was measured in LT and CT greenhouse during the period of pepper growth and development, respectively. Figure S2: The correlations coefficients between vegetative and reproductive traits in total population of pepper between LT and CT, Table S1: The information of chili and bell pepper accessions for evaluating agronomical traits under night low temperature in winter 2020–2021. Table S2: Loading matrix associated with the principal components analysis (PCA) for 10 agronomical traits.

Author Contributions: Conceptualization, M.-C.C. and E.-Y.Y.; methodology, S.N.R., M.-C.C., C.-W.N. and E.-Y.Y.; investigation, S.N.R., K.L., H.-B.J. and E.-Y.Y.; writing—original draft preparation, S.N.R. and K.L.; writing—review and editing, S.N.R., K.L., E.-Y.Y. and M.-C.C.; visualization, S.N.R. and K.L.; supervision, M.-C.C. All authors have read and agreed to the published version of the manuscript.

Funding: This study was supported by a grant (Project No: PJ01267102 "Study on the physiological mechanism of temperature adaptable pepper lines") from the National Institute of Horticultural and Herbal Science, Rural Development Administration.

Institutional Review Board Statement: Not applicable.

Informed Consent Statement: Not applicable.

Data Availability Statement: The datasets presented in this study are available upon request to the corresponding author.

Conflicts of Interest: The authors declare no conflict of interest.

References

1. Kothari, S.; Joshi, A.; Kachhwaha, S.; Ochoa-Alejo, N. Chilli peppers—A review on tissue culture and transgenesis. *Biotechnol. Adv.* **2010**, *28*, 35–48. [CrossRef]
2. Bhutia, K.; Khanna, V.; Meetei, T.; Bhutia, N. Effects of climate change on growth and development of chilli. *Agrotechnology* **2018**, *7*, 2. [CrossRef]
3. Prohens, J.; Nuez, F. *Handbook of Plant Breeding. Vegetables II: Fabaceae, Liliaceae, Solanaceae and Umbelliferae*; Springer: New York, NY, USA, 2008; Volume 3, pp. 30–40.
4. Sun, T.; Xu, Z.; Wu, C.T.; Janes, M.; Prinyawiwatkul, W.; No, H. Antioxidant activities of different colored sweet bell peppers (*Capsicum annuum* L.). *J. Food Sci.* **2007**, *72*, 98–102. [CrossRef]
5. Sarada, C.; Ratnam, M.; Naidu, L.; Ramana, C.; Rajani, A.; Vijaya, T. Chilli production and productivity in relation to Seasonal weather conditions in Guntur District of Andhra Pradesh. *Int. J. Pure Appl. Biosci.* **2015**, *3*, 207–213.
6. Cruz-Huerta, N.; Williamson, J.G.; Darnell, R.L. Low night temperature increases ovary size in sweet pepper cultivars. *HortScience* **2011**, *46*, 396–401. [CrossRef]
7. Toki, T.; Ogiwara, S.; Aoki, H. Effect of varying night temperature on the growth and yields in cucumber. *Acta Hortic.* **1978**, *87*, 233–238. [CrossRef]
8. Horie, T.; de Wit, C.T.; Goudriaan, J.; Bensink, J. A formal template for the development of cucumber in its vegetative stage (I, II and III). In *Proceedings of the Koninklijke Nederlandse Akademie Van Wetenschappen. Serie C: Biological and Medical Sciences*; Wageningen University: Wageningen, The Netherlands, 1979; Volume 82, pp. 433–479.
9. Nilwik, H. Growth analysis of sweet pepper (*Capsicum annuum* L.) 1. The influence of irradiance and temperature under glasshouse conditions in winter. *Ann. Bot.* **1981**, *48*, 129–136. [CrossRef]
10. Ji, L.; Li, P.; Su, Z.; Li, M.; Guo, S. Cold-tolerant introgression line construction and low-temperature stress response analysis for bell pepper. *Plant Signal. Behav.* **2020**, *15*, 1773097. [CrossRef] [PubMed]
11. Foolad, M.; Lin, G. Relationship between cold tolerance during seed germination and vegetative growth in tomato: Germplasm evaluation. *J. Am. Soc. Hortic. Sci.* **2000**, *125*, 679–683. [CrossRef]
12. O'SULLIVAN, J.; Bouw, W. Pepper seed treatment for low-temperature germination. *Can. J. Plant Sci.* **1984**, *64*, 387–393. [CrossRef]
13. Seo, J.-U.; Hwang, J.-M.; Oh, S.-M. Effects of night temperature treatment of raising seedlings before transplanting on growth and development of pepper. *J. Bio-Env. Con.* **2006**, *15*, 149–155.
14. Bhatt, R.; Srinivasa Rao, N. Response of bell-pepper (*Capsicum annuum*) photosynthesis, growth, and flower and fruit setting to night temperature. *Photosynthetica* **1994**, *28*, 127–132.
15. Sherzod, R.; Yang, E.Y.; Cho, M.C.; Chae, S.Y.; Kim, J.H.; Nam, C.W.; Chae, W.B. Traits affecting low temperature tolerance in tomato and its application to breeding program. *Plant Breed. Biotechnol.* **2019**, *7*, 350–359. [CrossRef]
16. Rajametov, S.N.; Lee, K.; Jeong, H.-B.; Cho, M.-C.; Nam, C.-W.; Yang, E.-Y. Physiological Traits of Thirty-Five Tomato Accessions in Response to Low Temperature. *Agriculture* **2021**, *11*, 792. [CrossRef]
17. Oh, S.-Y.; Koh, S.C. Fruit Development and Quality of Hot Pepper (*Capsicum annuum* L.) under Various Temperature Regimes. *Hortic. Sci. Technol.* **2019**, *37*, 313–321.

18. Mercado, J.; Mar Trigo, M.; Reid, M.; Valpuesta, V.; Quesada, M. Effects of low temperature on pepper pollen morphology and fertility: Evidence of cold induced exine alterations. *J. Hortic. Sci.* **1997**, *72*, 317–326. [CrossRef]
19. Pressman, E.; Moshkovitch, H.; Rosenfeld, K.; Shaked, R.; Gamliel, B.; Aloni, B. Influence of low night temperatures on sweet pepper flower quality and the effect of repeated pollinations, with viable pollen, on fruit setting. *J. Hortic. Sci. Biotechnol.* **1998**, *73*, 131–136. [CrossRef]
20. Shaked, R.; Rosenfeld, K.; Pressman, E. The effect of low night temperatures on carbohydrates metabolism in developing pollen grains of pepper in relation to their number and functioning. *Sci. Hortic.* **2004**, *102*, 29–36. [CrossRef]
21. Kato, K. Flowering and fertility of forced green peppers at lower temperatures. *J. Jpn. Soc. Hortic. Sci.* **1989**, *58*, 113–121. [CrossRef]
22. Rylski, I. Effect of night temperature on shape and size of sweet pepper (*Capsicum annuum* L.). *Amer. Soc. Hort. Sci. J.* **1973**, *98*, 149–152.
23. Aloni, B.; Pressman, E.; Karni, L. The effect of fruit load, defoliation and night temperature on the morphology of pepper flowers and on fruit shape. *Ann. Bot.* **1999**, *83*, 529–534. [CrossRef]
24. De Koning, A. The effect of different day/night temperature regimes on growth, development and yield of glasshouse tomatoes. *J. Hortic. Sci.* **1988**, *63*, 465–471. [CrossRef]
25. Rural Development Administration (RDA). *Data Book of Agricultural Products Income for the Improvement of Agricultural Management in 2019*; Rural Development Administration: Jeonju, Korea, 2020.
26. Elings, A.; Kempkes, F.; Kaarsemaker, R.; Ruijs, M.; Van De Braak, N.; Dueck, T. The Energy Balance and Energy-Saving Measures in Greenhouse Tomato Cultivation. *Acta Hortic.* **2005**, *691*, 67–74. [CrossRef]
27. Rylski, I.; Spigelman, M. Effects of different diurnal temperature combinations on fruit set of sweet pepper. *Sci. Hortic.* **1982**, *17*, 101–106. [CrossRef]
28. Rylski, I. Investigations on the Influence of Suboptimal Temperatures on the Flowering, Fruit Setting and Development of Sweet Pepper (Capsicum annum L.). Ph. D. Thesis, Hebrew University of Jerusalem, Jerusalem, Israel, 1971; pp. 1–96.
29. Rylski, E.; Kempler, H. Fruit set of sweet pepper (*Capsicum annuum L.*) under plastic covers. *HortScience* **1972**, *7*, 422–423.
30. Wang, C. Alleviation of chilling injury in tropical and subtropical fruits. In *III International Symposium on Tropical and Subtropical Fruits*; International Society for Horticultural Science: Bierbeek, Belgium, 2004; Volume 864, pp. 267–273.
31. Barchi, L.; Lefebvre, V.; Sage-Palloix, A.-M.; Lanteri, S.; Palloix, A. QTL analysis of plant development and fruit traits in pepper and performance of selective phenotyping. *Theor. Appl. Genet.* **2009**, *118*, 1157–1171. [CrossRef] [PubMed]
32. Yarnes, S.C.; Ashrafi, H.; Reyes-Chin-Wo, S.; Hill, T.A.; Stoffel, K.M.; Van Deynze, A. Identification of QTLs for capsaicinoids, fruit quality, and plant architecture-related traits in an interspecific Capsicum RIL population. *Genome* **2013**, *56*, 61–74. [CrossRef] [PubMed]
33. Yang, S.; Tang, X.-F.; Ma, N.-N.; Wang, L.-Y.; Meng, Q.-W. Heterology expression of the sweet pepper CBF3 gene confers elevated tolerance to chilling stress in transgenic tobacco. *J. Plant Physiol.* **2011**, *168*, 1804–1812. [CrossRef] [PubMed]
34. Hou, X.-M.; Zhang, H.-F.; Liu, S.-Y.; Wang, X.-K.; Zhang, Y.-M.; Meng, Y.-C.; Luo, D.; Chen, R.-G. The NAC transcription factor CaNAC064 is a regulator of cold stress tolerance in peppers. *Plant Sci.* **2020**, *291*, 110346. [CrossRef]
35. Kong, X.-M.; Zhou, Q.; Zhou, X.; Wei, B.-D.; Ji, S.-J. Transcription factor CaNAC1 regulates low-temperature-induced phospholipid degradation in green bell pepper. *J. Exp. Bot.* **2020**, *71*, 1078–1091. [CrossRef]
36. Chinnusamy, V.; Zhu, J.-K.; Sunkar, R. Gene Regulation During Cold Stress Acclimation in Plants. *Methods Mol. Biol.* **2010**, *639*, 39–55. [PubMed]
37. Bakker, J.; Van Uffelen, J. The effects of diurnal temperature regimes on growth and yield of glasshouse sweet pepper. *Neth. J. Agri. Sci.* **1988**, *36*, 201–208.
38. Ohtaka, K.; Yoshida, A.; Kakei, Y.; Fukui, K.; Kojima, M.; Takebayashi, Y.; Yano, K.; Imanishi, S.; Sakakibara, H. Difference Between Day and Night Temperatures Affects Stem Elongation in Tomato (*Solanum lycopersicum*) Seedlings via Regulation of Gibberellin and Auxin Synthesis. *Front. Plant Sci.* **2020**, *11*, 1947. [CrossRef] [PubMed]
39. Phatak, S. Top and root temperature effects on tomato flowering. *J. Am. Soc. Hortic. Sci.* **1966**, *88*, 527–531.
40. Goodstal, F.J.; Kohler, G.R.; Randall, L.B.; Bloom, A.J.; Clair, D.A.S. A major QTL introgressed from wild Lycopersicon hirsutum confers chilling tolerance to cultivated tomato (*Lycopersicon esculentum*). *Theor. Appl. Genet.* **2005**, *111*, 898–905. [CrossRef]
41. Patterson, B.D.; Reid, M.S. Genetic and environmental influences on the expression of chilling injury. In *Chilling Injury of Horticultural Crops*; CRC Press: Boca Raton, FL, USA, 1990; pp. 87–112.
42. Issarakraisila, M.; Considine, J. Effects of temperature on pollen viability in mango cv. 'Kensington'. *Ann. Bot.* **1994**, *73*, 231–240. [CrossRef]
43. Satake, T.; Hayase, H. Male sterility caused by cooling treatment at the young microspore stage in rice plants: V. Estimations of pollen developmental stage and the most sensitive stage to coolness. *Jpn. J. Crop Sci.* **1970**, *39*, 468–473. [CrossRef]
44. Polowick, P.; Sawhney, V. Temperature effects on male fertility and flower and fruit development in *Capsicum annuum* L. *Sci. Hortic.* **1985**, *25*, 117–127. [CrossRef]
45. Sawhney, V.K.; Shukla, A. Male sterility in flowering plants: Are plant growth substances involved? *Am. J. Bot.* **1994**, *81*, 1640–1647. [CrossRef]
46. Zhigila, D.A.; AbdulRahaman, A.A.; Kolawole, O.S.; Oladele, F.A. Fruit morphology as taxonomic features in five varieties of *Capsicum annuum* L. Solanaceae. *J. Bot.* **2014**, *2014*, 1–6. [CrossRef]

47. Paran, I.; Van Der Knaap, E. Genetic and molecular regulation of fruit and plant domestication traits in tomato and pepper. *J. Exp. Bot.* **2007**, *58*, 3841–3852. [CrossRef] [PubMed]
48. Vilarinho, L.B.O.; da Silva, D.J.H.; Greene, A.; Salazar, K.D.; Alves, C.; Eveleth, M.; Nichols, B.; Tehseen, S.; Khoury, J.K.; Johnson, J.V. Inheritance of fruit traits in Capsicum annuum: Heirloom cultivars as sources of quality parameters relating to pericarp shape, color, thickness, and total soluble solids. *J. Am. Soc. Hortic. Sci.* **2015**, *140*, 597–604. [CrossRef]

agronomy

MDPI

Article

Transcriptomic Analysis of Short-Term Salt-Stress Response in Mega Hybrid Rice Seedlings

Noushin Jahan [1,2,†], Yang Lv [1,†], Mengqiu Song [1,†], Yu Zhang [1,3], Lianguang Shang [4], Ying Lu [5], Guoyou Ye [4], Qian Qian [1], Zhenyu Gao [1] and Longbiao Guo [1,*]

1 State Key Lab for Rice Biology, China National Rice Research Institute, Hangzhou 310006, China; nowshin012@gmail.com (N.J.); lvyang0725@gmail.com (Y.L.); songmengqiu718@gmail.com (M.S.); zhangyu08@caas.cn (Y.Z.); qianqian188@hotmail.com (Q.Q.); gaozhenyu@caas.cn (Z.G.)
2 Department of Agronomy, Khulna Agricultural University, Khulna 9100, Bangladesh
3 Tobacco Research Institute of Chinese Academy of Agricultural Sciences, Qingdao 266101, China
4 Shenzhen Agricultural Genome Research Institute of Chinese Academy of Agricultural Sciences, Shenzhen 518120, China; shanglianguang@caas.cn (L.S.); g.ye@irri.org (G.Y.)
5 CAAS Channel (ShenZhen) Biological Breeding Technology Co., Ltd., Shenzhen 518120, China; ying_lulu@outlook.com
* Correspondence: guolongbiao@caas.cn; Tel.: +86-571-63370537
† These authors contributed equally to this work.

Citation: Jahan, N.; Lv, Y.; Song, M.; Zhang, Y.; Shang, L.; Lu, Y.; Ye, G.; Qian, Q.; Gao, Z.; Guo, L. Transcriptomic Analysis of Short-Term Salt-Stress Response in Mega Hybrid Rice Seedlings. *Agronomy* **2021**, *11*, 1328. https://doi.org/10.3390/agronomy11071328

Academic Editor: Santiago Signorelli

Received: 25 May 2021
Accepted: 28 June 2021
Published: 29 June 2021

Abstract: Salinity is a major abiotic stressor that leads to productivity losses in rice (*Oryza sativa* L.). In this study, transcriptome profiling and heterosis-related genes were analyzed by ribonucleic acid sequencing (RNA-Seq) in seedlings of a mega rice hybrid, Liang-You-Pei-Jiu (LYP9), and its two parents 93–11 and Pei-ai64s (PA64s), under control and two different salinity levels, where we found 8292, 8037, and 631 salt-induced differentially expressed genes (DEGs), respectively. Heterosis-related DEGs were obtained higher after 14 days of salt treatment than after 7 days. There were 631 and 4237 salt-induced DEGs related to heterosis under 7-day and 14-day salt stresses, respectively. Gene functional classification showed the expression of genes involved in photosynthesis activity after 7-day stress treatment, and in metabolic and catabolic activity after 14 days. In addition, we correlated the concurrence of an expression of DEGs for the bHLH transcription factor and a shoot length/salinity-related quantitative trait locus qSL7 that we fine-mapped previously, providing a confirmed case of heterosis-related genes. This experiment reveals the transcriptomic divergence of the rice F1 hybrid and its parental lines under control and salt stress state, and enlightens about the significant molecular mechanisms developed over time in response to salt stress.

Keywords: expression profiling; heterosis; salinity stress; seedlings; rice

1. Introduction

Heterosis is an event whereby the heterozygous first filial (F1) hybrid displays growth or fertility superiority over its homozygous parents, and has been widely used in rice-breeding practices for increasing grain yield. Heterosis breeding is a powerful tool to secure global food demands, because of its 10–20 percent higher yield than its two parental lines [1]. The underlying molecular mechanisms of heterosis are broadly exploited by plant breeders in modern breeding programs, however, as yet complex heterosis yet remains poorly understood [2]. Before the development of molecular genetics tools, heterosis focused mainly on three non-mutually exclusive hypotheses, of dominance, single-locus overdominance, and pseudo-overdominance [3]. Evidence for each hypothesis has been presented [4–7]. Recently, the yield heterosis of super hybrid rice was studied by integrating phenomic, genomic, and transcriptomic data. The comprehensive mapping and analysis of heterosis QTLs with multi-omics tools provide valuable data for both testing heterosis hypothesis and purposely manipulating heterosis for breeding new cultivars.

However, research reported that differential expression of genes could be responsible for heterosis in parental lines and hybrids [8,9]. The present advancement of molecular techniques facilitates expression profiling analysis at the RNA level, and transcriptome profiling can be used to identify the underlying molecular mechanisms of heterosis between a hybrid and its parents. For example, with the RNA-seq analysis, a total of 536 and 269 differentially expressed genes (DEGs) were identified between a hybrid and its parents at the seedling stage for roots' response under potassium deficiency, respectively [10]. Transcriptome analysis of divergent tissues of the hybrid rice and parents revealed that 10% of genes were differentially expressed among parents and hybrid, and about 0.5–1.4% of expressed genes were detected as non-additive genes [11]. Transcriptome-based heterosis-associated gene analysis has been carried out for a wide range of traits in rice, maize, cotton and Arabidopsis [10–14]. Heterosis contributes to superior phenotypic performance for a broad range of attributes, such as plant height, biomass, grain and grain yield and environmental stress adaptation, but salt stress response transcriptomic expression in hybrid rice is still not clearly understood.

Salinity stress is one of the most consequential abiotic stresses that hamper the plant lifecycle. The world climate change is seriously contributing to increases of soil salinity and reducing crop productivity. Soil salinity is one of the critical abiotic stresses significantly limiting crop production all over the world. Approximately 20% of irrigated and 8% of rainfed agricultural land are affected by salinity [15,16]. Several transcriptome analyses have revealed that thousands of transcription factor genes are altered in response to salt stress [17,18]. However, limited statistics are available on the underlying mechinisms of salt stress and heterosis-related genes. Transcriptomic analysis has been used to recognize DEGs of the rice inbred lines 93-11 and PA64s and their heterozygous F1 (LYP9) hybrid in response to grain quality traits and root traits under potassium stress [10,11], but salt stress response DEGs for heterosis are not widely reported. Therefore, finding out if genes are associated to heterosis is crucial to unveil candidate genes with important functions in the response to salinity stress and the underlying mechanism of heterosis.

Salinity is a major problem for rice (*Oryza sativa* L.)-based farming systems in coastal areas. About one third of the irrigated rice-growing area is affected by salinity [19]. Soil salinity effects on rice may vary depending upon its different development stages and stress duration period [20]. The effect of salinity stress arises as a result of the relationship between the physiological and molecular responses of plants [21]. Understanding the salt stress-tolerance mechanism of rice is crucial for identifying the responsible genetic material and introducing hybrid rice in salinity stress-prone areas. To investigate the molecular mechanisms in the response to salt stress, the expression profiles of LYP9 and its parental lines were compared in seedlings under 7- and 14-day salinity stress treatment by RNA-seq, and heterosis-associated genes were analyzed to explore the underlying mechanism of heterosis. The results lay an important foundation for a better understanding of salt-tolerance mechanisms in hybrids, and allow insight into the underlying molecular basis of heterosis under salt stress.

2. Materials and Methods

2.1. Plant Growth Condition and Salinity Stress Treatment

Rice cultivars 93–11, PA64s and their hybrid LYP9 were cultured hydroponically in a growth chamber under the following conditions: 14 h photoperiod, under a temperature regime of 30 °C/25 °C and a relative humidity level of ~70%. To grow seedlings under hydroponics culture, the grains were surface-sterilized and allowed to germinate in water for 3 days. The germinated seedlings were then transferred to nutrient solution containing 1.425 mM NH_4NO_3, 0.42 mM NaH_2PO_4, 0.510 mM K_2SO_4, 0.998 mM $CaCl_2$, 1.643 mM $MgSO_4$, 0.168 mM Na_2SiO_3, 0.125 mM Fe-EDTA, 0.019 mM H_3BO_3, 0.009 mM $MnCl_2$, 0.155 mM $CuSO_4$, 0.152 mM $ZnSO_4$ and 0.075 mM Na_2MoO_4 [22]. The pH was adjusted to 5.5–5.8. The nutrient solution was replaced every 2 days to adjust the volume and pH of the nutrient solution. Seedlings were cultured in hydroponic solution for two weeks, and after that, half of the seedlings were transferred to NaCl-containing solution under 100 mM NaCl/L salt treatment as the salt stress treatment (S), and half of the seedlings were cultured in normal solution as a control (CK) for a further 2 weeks. Shoots of control seedlings were collected at 14 days and treated at 7 and 14 days of salinization for further assays.

2.2. Phenotype Evaluation of Seedlings under Salt Stress

Ten plants per line of uniform growth were evaluated for traits related to salinity tolerance, such as shoot length (SL), root length (RL), shoot fresh (SFW) and dry weight (SDW) and root fresh (RFW) and dry weight (RDW). Shoot lengths were measured from the base of the culm to the tip of the tallest leaf, expressed in centimeters, and fresh weights were taken of the same leaves in milligrams. Root lengths were measured from the base of the culm to the tip of the longest root, and fresh weights were also measured of the same roots. Root lengths were measured in centimeters and fresh weights in milligrams. For dry weights, five plants per line per replication were collected and dried at 65 °C in an oven for 5 days prior to weighing and expressed in milligrams. For transcriptomic analysis, we collected 5 shoot samples from each of 2 biological replicates of each group after 7 and 14 days of treatment. The shoot samples were frozen in liquid nitrogen and stored at −80 °C until they were used for transcriptome sequencing.

2.3. RNA Extraction and cDNA Library Preparation and Sequencing

Total RNA was isolated from both the parents 93-11, PA64s and LYP9 under control and salt treatment by using the TruSeq RNA Sample Preparation Kit (Illumina, San Diego, CA, USA) and purified using poly-T oligo-attached magnetic beads (Illumina, San Diego, CA, USA). RNA was stored at −80 °C until it was used for transcriptome sequencing and real-time fluorescent quantitative PCR validation. The purified RNA was used to construct the cDNA library using the AMPure XP system (Beckman Coulter, Beverly, CA, USA). Fragmentation was carried out using divalent cations under elevated temperature in an Illumina proprietary fragmentation buffer. First-strand cDNA was synthesized using random oligonucleotides and SuperScript II. Second-strand cDNA synthesis was subsequently performed using DNA polymerase I and RNase HExtracted. After the library construction was completed, the library fragments were enriched by PCR amplification, and then the library selection was performed according to the fragment size, and the library size was 300–400 bp. Next, the library was subjected to a quality test by an Agilent 2100 Bioanalyzer, and the total library concentration and the effective concentration of the library were examined. After RNA extraction, purification and library construction, the library was subjected to paired-end (PE) sequencing based on the Illumina HiSeq sequencing platform using next-generation sequencing (NGS). Subsequently, removing adaptor sequences and low-quality reads, the high-quality paired-end reads were mapped to the 93–11 reference genome (Oryza_indica.ASM465v1.dna.toplevel.fa) using the spliced read mapper TopHat version 2.0.12 [23]. The prcomp function in R software with default settings was used to perform principal component analysis (PCA) on each sample based on the amount of expression and interpret the relatedness among all replicas in each genotype.

2.4. Data Filtering and Assembly

Before assembly of the transcriptome data, a stringent filtering process was carried out. Data filtering was performed using Cutadapt. For filtering, the 3′ end of the adaptor was cut with Cutadapt, and the removed part and the known linker have at least 10 bp overlap. The adapter sequences of the raw reads were removed, and low-quality sequences (reads with ambiguous bases 'N') and reads with more than 20% $Q < 20$ bases were removed.

2.5. Analysis of Differentially Expressed Genes and Functional Annotation

Differentially expressed genes (DEGs) were analyzed using the DESeq method by comparing the read counts of the transcripts of the control and salt-treated samples. The expression of all genes was determined with FPKM (fragments per kilobase of exon per million fragments mapped) values using the software Cufflinks. Significant differentially expressed genes were determined based on a threshold of two-fold expression change and false discovery rate (FDR) < 0.05. Differentially expressed genes were annotated using the "Ensembl" database (http://www.ensembl.org/, accessed on 1 October 2019) data version ASM465v1.40 and were subjected to GO enrichment analysis using topGO.

2.6. RT-qPCR Analysis

Total RNA was extracted from fresh leaf samples of the super rice hybrid LYP9, and its parents PA64s and 93–11, under control and two salt stresses using a Micro RNA Extraction kit (Axygen) and reverse transcribed into cDNA using a Rever Tra Ace qPCR-RT kit (TOYOBA, Japan). qPCR for the qSL7 gene was conducted on an ABI PRISM 7900HT Sequence Detector (Applied Biosystems) according to the manufacturer's instructions. Primers for RT-qPCR are listed in Supplementary Table S3. The rice histone gene was used as an internal control. Two biological replicates of each sample were prepared.

3. Results

3.1. Transcriptome Sequencing of the Hybrid LYP9 and Its Parents under Two Salinity Stresses

To investigate the transcriptional changes in rice under salinity stress, we grew LYP9, 93-11 and PA64s plants hydroponically under 0 and 100 mM NaCl and evaluated the shoot length (SL), shoot fresh weight (SFW), shoot dry weight (SDW), root length (RL), root fresh weight (RFW) and root dry weight (RDW) (Supplementary Table S1). However, the salt injury symptoms appeared in the seedlings after one day of NaCl treatment. Visual symptoms showed up as brownish leaf tips, yellowing of leaves, drying leaves, reduction in shoot growth and stunted height in both parents. Similar damage symptoms appeared in LYP9, but in less number of leaves and increased in time of stress imposed (Supplementary Figure S1).

An Illumina HiSeq™ 2500 was used to conduct high-throughput transcriptome analysis of control and salt-treated rice seedling samples of super rice LYP9 (l), and its parents 93-11 (y) and PA64s (p) at three time points (Supplementary Figure S1). To evaluate the potential molecular mechanism underlying salt tolerance, we compared LYP9 and both parents at the transcriptional level. DESeq software was used to compare the expression levels of all transcripts, followed by topGO for enrichment analysis of the GO categories. We used two biological replicates of super hybrid LYP9 and two parents under control and stress conditions, thereby constructing 18 cDNA libraries. After removing short and poor-quality reads, clean reads were obtained at three time points, and accounted for over 88% of the total sequences (Table 1). The Q20 (~97%) and Q30 (~93%) values indicated that the quality of the sequencing data was ample to carry out further transcriptome analysis (Table 1).

Approximately 88% of the clean reads were aligned with the reference genome. We examined the density plot based on the distributions of FPKM scores and correlations among the experimental samples using Pearson's correlation coefficient (Supplementary Figure S2A,B). As shown in Supplementary Figure S3, PCA and the r-index values among the sample ranged from 0.69 to 0.99, revealing high correlation among the two biological replicates of eighteen LYP9, 93-11 and PA64s samples under CK and salt treatments. All of the tested samples were efficient, the experiment was repeatable and the transcriptome data were dependable for further assessment of the DEGs.

Table 1. Sequencing statistics of 18 RNA libraries.

Sample	Raw Data (bp)	Clean Data (bp)	Q20 (%)	Q30 (%)	Total Mapped
CKl_1	7,923,985,022	7,518,704,344	97.32	93.36	46,372,093 (93.13%)
CKl_2	6,480,020,342	6,146,625,630	97.58	93.36	37,908,455 (93.13%)
CKy_1	7,446,201,694	7,064,073,846	97.44	93.61	44,311,077 (94.72%)
CKy_2	7,585,693,380	7,207,629,244	97.58	93.89	44,852,039 (93.97%)
CKp_1	7,190,156,100	6,797,538,000	97.47	93.56	43,303,322 (95.56%)
CKp_2	7,312,592,400	6,912,223,200	97.51	93.64	43,166,663 (93.67%)
S-l_7_1	7,253,315,100	6,839,539,800	97.53	93.67	42,824,053 (93.92%)
S-l_7_2	7,295,699,100	6,890,722,800	97.35	93.24	43,026,950 (93.66%)
S-y_7_1	7,274,557,500	6,870,173,400	97.54	93.66	43,685,297 (95.38%)
S-y_7_2	8,194,719,300	7,745,438,400	97.38	93.35	49,557,272 (95.97%)
S-p_7_1	7,487,282,700	7,087,584,600	97.56	93.74	45,026,497 (95.29%)
S-p_7_2	7,109,772,900	6,710,789,400	97.51	93.64	42,096,782 (94.10%)
S-l_14_1	8,286,700,612	7,863,880,378	97.30	93.34	48,575,755 (93.27%)
S-l_14_2	6,374,416,680	6,056,998,372	97.24	93.25	36,757,809 (91.64%)
S-y_14_1	7,336,768,974	6,948,243,860	97.49	93.72	43,152,182 (93.78%)
S-y_14_2	7,461,388,972	7,071,076,924	97.46	93.64	44,220,646 (94.43%)
S-p_14_1	8,189,553,600	7,718,297,100	98.01	94.62	48,529,365 (94.31%)
S-p_14_2	6,494,686,800	6,111,858,300	97.97	94.51	38,808,204 (95.24%)

3.2. Identification of Salt Stress Response Differentially Expressed Genes (DEGs) by RNA-Seq

We compared the levels of all transcripts among 7-day and 14-day salt-treated samples using DESeq software. Significant DEGs were screened with the criteria of fold change ≥ 1 and FDR $\leq 5\%$ among different samples. Hierarchical cluster analysis showed that all the samples were in the immediate vicinity of each other, which reflected the high efficiency of reproducible results of RNA-Seq (Figure 1, Supplementary Table S2).

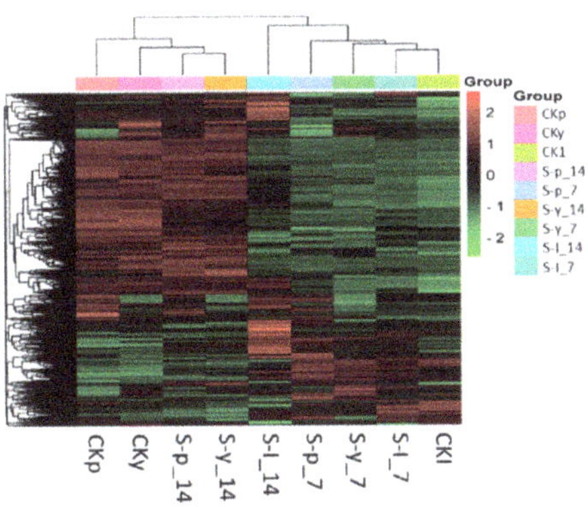

Figure 1. Hierarchical cluster analysis of 18 samples of three rice varieties, LYP9 (l), PA64s (p) and 93-11 (y), under control (CK) and salt stress (S) conditions at 7 and 14 days.

To identify the DEGs in the three genotypes, six pairwise comparisons: A1 (CKp vs. S-p_7), B1 (CKp vs. S-p_14), A2 (CKy vs. S-y_7), B2 (CKy vs. S-y_14), A3 (CKl vs. S-l_7) and B3 (CKl vs. S-l_14), were performed. Comparison among genotypes shows that, in A1—8292 (3562 up- and 4730 down-regulated), A2—8037 (3585 up- and 4452 down-regulated) and A3—631 (440 up- and 191 down-regulated) DEGs were identified (Figure 2, Supplementary Figure S4). The number of expressed DEGs was lower in LYP9 at 7 days than that in its parents. Whereas, at 14 days, the expressed DEGs were significantly higher. More gene activity was found after 7 days of salt stress acclimation in the parents, but the hybrid showed gene activity after 14 days of acclimation.

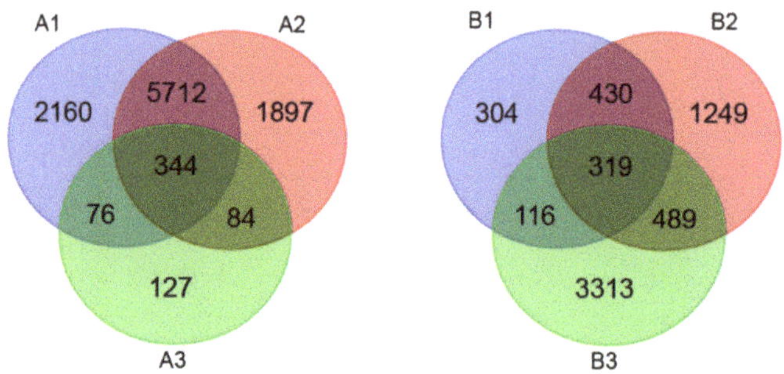

Figure 2. Venn diagram for differentially expressed genes of three rice varieties, LYP9, PA64s and 93-11, between the control (CK) and two salt (S) treatments. Where, A1 (CKp vs. S-p_7), B1 (CKp vs. S-p_14), A2 (CKy vs. S-y_7), B2 (CKy vs. S-y_14), A3 (CKl vs. S-l_7), B3 (CKl vs. S-l_14).

Additionally, the DEGs were divided into different types according to the heterosis characteristics among the hybrid rice LYP9, and its parents 93-11 and PA64s, under two stress levels. We defined DEG related with heterosis because of the differences in expression among F1 (LYP9) and its parents: the DEGs between parental lines (93-11 and PA64s) were defined as DEGpp, and the DEGs between the hybrid rice LYP9 and its parents were defined as DEGhp. DEGhp was further divided into two types, i.e., DEGhpu denotes the unique portion of DEGhp and DEGo denotes the overlap between DEGpp and DEGhp. H2P, B2P and L2P represent higher than both parents, between both parents and lower than both parents, respectively. In heterosis analysis, 1727, 1729 and 342 DEGs related with heterosis were observed in control (CK), and after 7 and 14 days of salt stress treatments, respectively. However, the heterosis-related DEGs were found higher in number when compared to the LYP9 and its parents, i.e., in control condition, the LYP9 vs. 93-11 (L/Y) showed 9724 DEGs, whereas LYP9 vs. PA64s (L/P) showed 9827 DEGs. In the salt stress condition (L/Y, L/P), less heterosis-related DEGs were found in 7-day than in 14-day acclimation (Table 2). In the stress condition, the highest number of heterosis-related genes overlapped between DEGpp and DEGhp at 7 days of salt acclimation. Additionally, the DEGs of the hybrid were found to be higher than both parents (3251) at 14 days after salt treatment, whereas the common DEGs of the hybrid and its parents were found to be lower in number at 14 days (Table 2). This suggested that the hybrid performed better than its parents under prolonged salt stress conditions. To further understand the function of DEGs, we classified these genes according to their functional categories and relatedness.

Table 2. Number and classification of DEGs according heterosis analysis.

Traits	DEGpp	DEGhp						
		L/Y	L/P	DEGhpu	DEGo	H2P	B2P	L2P
CK	1727	9724	9827	10,139	470	3430	1523	6505
7-day	1729	1397	1952	1592	142	1412	811	501
14-day	342	7581	5473	7993	65	3251	192	4834

Note: Y, L and P refer to 93-11, LYP9 and PA64s, respectively. CK refers to the control condition. DEGpp refers to DEGs between both parents, DEGhp refers to DEGs between the hybrid and parents. L/Y refers to DEGs between LYP9 and 93-11, and L/P refers to DEGs between LYP9 and PA64s under salt stress. DEGhpu denotes the unique portion of DEGhp, and DEGo denotes the overlap between DEGpp and DEGhp. H2P, B2P and L2P represent higher than both parents, between both parents and lower than both parents, respectively.

3.3. Functional Classification of Common DEGs

To assess the functional and biological process categories of the DEGs in LYP9 and both parents under stress conditions, we performed GO enrichment analysis. Comparative GO analysis between the three genotypes revealed that the salt-stress-related DEGs are related with various molecular, cellular and biological processes (Figure 3). A total of 27,956 GO terms were matched with all predicted genes. Among the molecular functions, genes that encoded proteins with chlorophyll binding and protein kinase binding activities accounted for a large portion under 7-day and 14-day salt stress acclimation, respectively. In the 7-day stress treatment, most of the DEGs were involved in photosynthesis activities in cellular and biological processes (Figure 3A). Whereas, after the 14-day stress treatment, DEGs were found to be involved in metabolic and catabolic activities (Figure 3B).

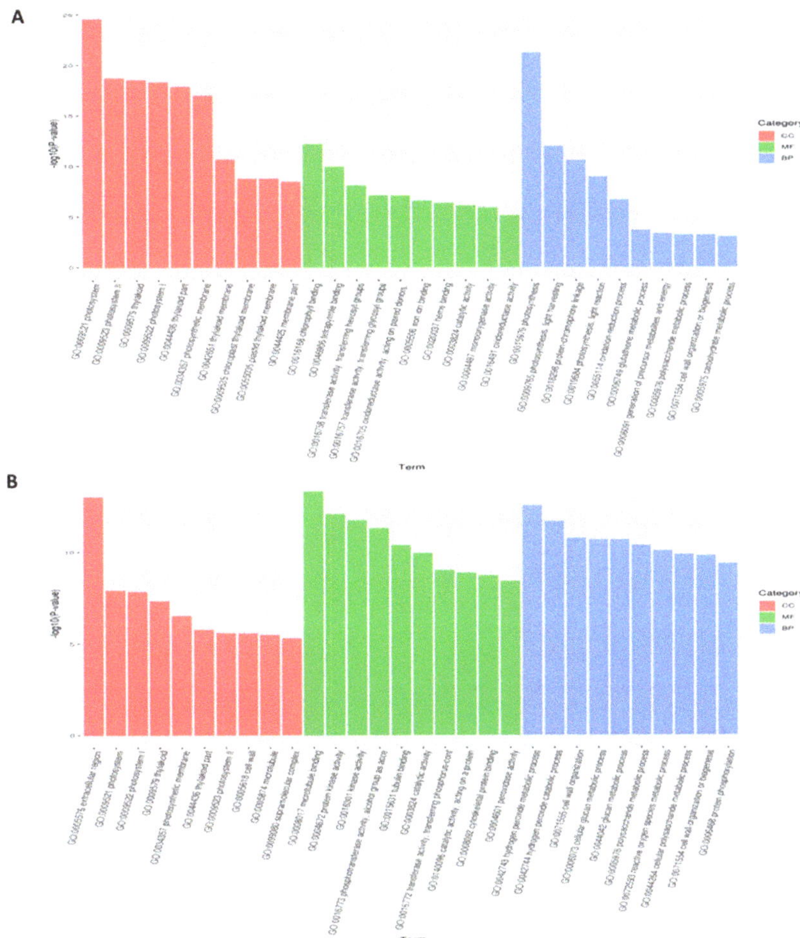

Figure 3. GO classification of differentially expressed genes of three rice varieties, LYP9, PA64s and 93-11, under two salt stresses. Here, (**A**) GO classification at 7 days, (**B**) GO classification at 14 days. CC—cellular component, MF—molecular function and BP—biological process.

Additionally, protein encoding genes confined in organelles were also considered for a large number of those that were differentially expressed under salt stresses. Lots of these genes take part in photosynthesis and their expression was stimulated by adverse environmental conditions. GO classification analysis of the DEGs revealed that these genes perform important roles in response to salt stress.

To gain a better understanding of metabolic processes, we performed Kyoto Encyclopedia of Genes and Genomes (KEGG) pathway enrichment analysis of the identified DEGs (Figure 4). The KEGG pathway analysis showed the top 20 most enriched pathway subcategories, including the most profuse group represented as the phenylpropanoid biosynthesis, plant hormone signaling and sugar metabolism. Enrichment analysis also revealed that the DEGs involved in biosynthesis of metabolites were strongly influenced by salt treatment, indicating that salt stress also affects the normal biological functions of the plant (Figure 4). In the 7-day stress treatment, DEGs were involved in stress sensing and processing pathways, such as plants' hormone signal transduction and MAPK signaling pathways and plant pathogen interaction pathways (Figure 4A). In the 14-day stress treatment, DEGs were also found to be active in the ABC transporter pathway along with other stress sensing pathways (Figure 4B).

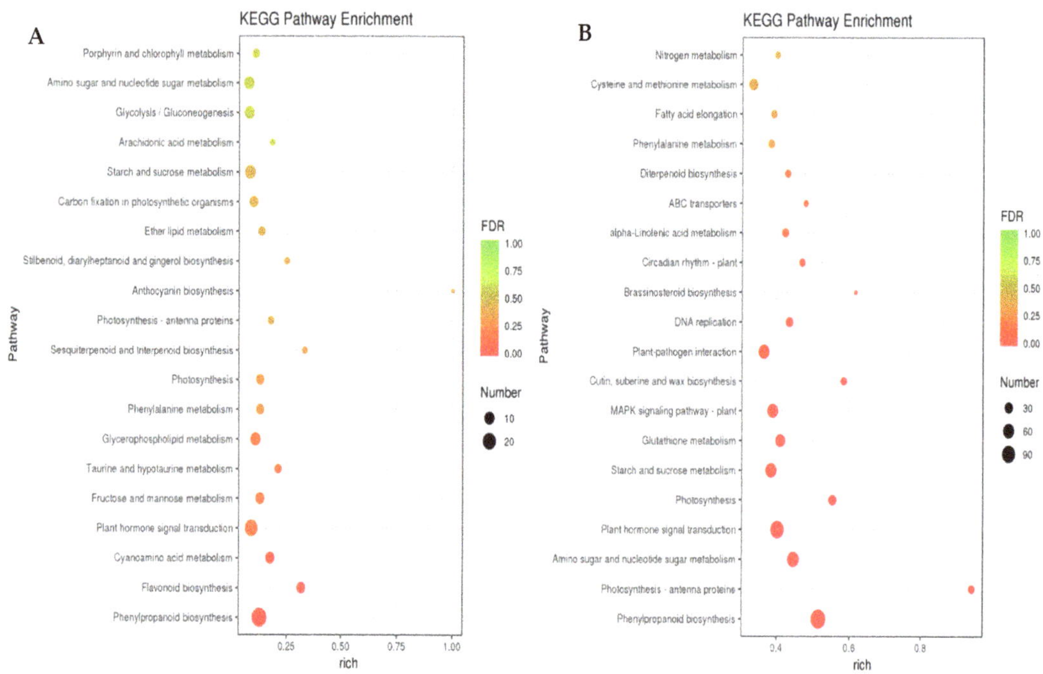

Figure 4. Top enriched KEGG pathways of the DEGs in PA64s 93-11 and LYP9 at 7- and 14-day stress treatments. Here, (**A**) KEGG pathway of DEGs at 7 days, (**B**) KEGG pathway of DEGs at 14 days. The corresponding pathways are listed on the Y-axis, and the rich factor values are listed on the X-axis. Different sized circles represent different numbers of DEGs, whereas their color (FDR) represents the adjusted *p*-value.

3.4. Identification of Transcription Factors That Respond to Salt Stress

High-throughput sequencing is one of the efficient assays for gene expression identification in compared to microarrays and less abundant gene expressions are also detected in this method, such as transcription factors. Further analysis of differentially expressed genes (DEGs) in rice shoots showed that there were 18,992 transcription factors belonging to 59 transcription factor families (Figure 5). Of these, salt stress transcription factors and bHLH DNA binding factor family members were the most numerous. The bHLH transcription factor concurrence correlated to a shoot length/salinity-related quantitative trait locus qSL7 that we fine-mapped previously [16].

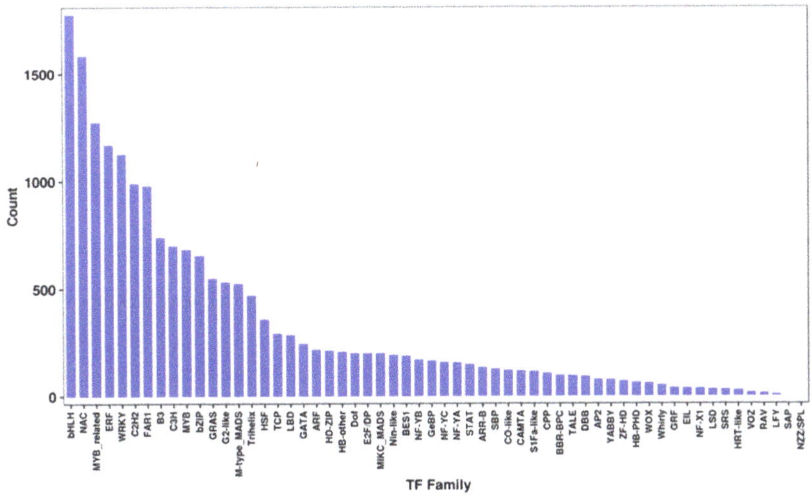

Figure 5. The summary of TFs' gene families represented in differentially expressed genes.

Then, we used qPCR to detect the target gene and found that the expression of LYP9 was significantly higher than that of 93-11 and PA64s after 7 days of salt acclimation. Whereas, no significant differences were found among LYP9, 93-11 and PA64s expression under the control and 14-day salt stress treatment (Figure 6 and Supplementary Table S3). The salt stress transcription factor family regulates different stress responses in plants, and the ethylene response factor family are primarily involved in DNA binding. Other transcription factors are included in ion homeostasis and biological processes, including signal transduction and the adverse environmental conditions guarding mechanism.

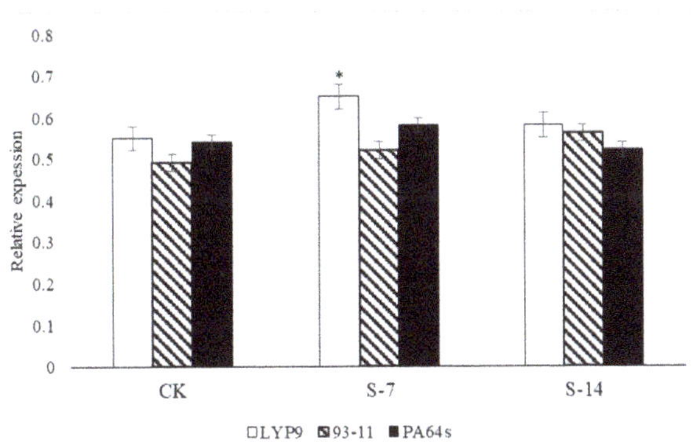

Figure 6. qPCR analysis of qSL7 for expression analysis of three rice varieties, LYP9 (l), PA64s (p) and 93-11 (y), CK refers to control condition, S-7 refers to salt stress treatment at 7 days, S-14 refers to salt stress treatment at 14 days. * Indicates the 5% level of significance.4. Discussion.

This experiment engaged in high-throughput sequencing using three materials with two different time point for the analysis of the rice shoot transcriptome under control and salt stress conditions. The findings of the present study revealed that a large number of genes exhibit clear expression changes in response to different salt stress conditions. Among the expressed genes, a large number of them had unknown functions, implying that research on the transcriptional changes in rice caused by salt stress is an abundant area for future investigation.

Furthermore, due to the higher performance of hybrids in comparison with their parental lines, heterosis has been widely studied in plant breeding for years. Three main basic models of classical quantitative genetics could explain heterosis, including comprising dominance, overdominance and epistasis hypotheses [24]. Different study results have revealed that gene expression level dissimilarity between hybrids and their parents may be responsible for heterosis. However, the differences could also be because of the presence of the multi-genetics background in F1. Furthermore, it is well-known that during meiosis division, genes organized in tandem undergo to recombination events that produce new genes in tandem duplicate, which causes dissimilarities between the hybrid and its parents [25]. On the other hand, soil salinity is a major abiotic stress that has huge room for improvement. Despite the complexity of abiotic stress responses, there are several hundred salt-responsive genes that have been identified. Among those, a few genes could be effectively utilized in breeding programs because of their complex biological responses [26]. However, heterosis could be a great genetic tool to improve salt tolerance in plants. In the present study, RNA-seq was used to find out the expression patterns and heterosis in seedlings of the rice hybrid LYP9 and its parental lines 93-11 and PA64s grown under control and salt stress conditions.

The significant difference in gene expression between 9311 and PA64s may be an important genetic element that causes the heterosis of LYP9. We also observed that the number of DEGs increased between the F1 hybrid and its parental line. Further, the heterosis-related DEGs were found to be higher in number when compared to LYP9 and its parents, i.e., in control condition, the LYP9 vs. 93-11 (L/Y) showed 9724 DEGs, whereas in LYP9 vs. PA64s (L/P), it showed 9827 DEGs. In the salt stress condition (L/Y, L/P), less heterosis-related DEGs were found in 7-day than in 14-day acclimations. In the stress condition, the highest number of heterosis-related genes overlapped between DEGpp and DEGhp at 7 days of salt acclimation. Additionally, the DEGs of the hybrid were found to be higher than both parents (3251) at 14 days after salt treatment, whereas the common DEGs of the hybrid and its parents were found to be lower in number at 14 days. These results suggested that the hybrid developed a coping mechanism under the salt stress condition over time.

Among the six pairwise comparisons, the total number of DEGs in control vs. salt stress samples was significantly higher than most of the other salt-treated samples: a total of 8292 (3562 up- and 4730 down-regulated), 8037 (3585 up- and 4452 down-regulated) and 631 (440 up- and 191 down-regulated). However, in tolerant cultivar LYP9, there were more upregulated genes at the 14-day than the 7-day salt treatment, which were different from those of genes showing changes in expression. This suggested that the genes showed transcriptional changes at time periods under salt stress, which supports a previous experiment [27]. The total number of DEGs increased under salt stress, and similar results were obtained for other crop species in previous studies [28,29]. More gene activity was found after 14 days of salt stress acclimation.

Functional classification of DEGs was further assorted by GO enrichment analysis. Many genes were overrepresented in biological processes associated with diverse stress response activities. Both GO and KEGG pathway analysis revealed that most of the DEGs were involved in photosynthesis activity, light reaction and stress responses. Photosynthesis is a major physiological mechanism that produces the energy required for growth and development in plants. Salt stress usually restricts growth or directly leads to metabolic dysfunction in plants [13,30], with strong effects on plant physiology and biochemistry [31–33]. Salt stress directly affects photosynthesis by reducing cellular CO_2 levels in leaves due to limited diffusion through the stomata [34]. The current findings reveal that salt stress in rice affects the expression of photosynthesis-related genes, including genes in the categories: photosystem I, photosystem II, Rubisco, chloroplast, curvature thylakoid protein, chlorophyll a-b binding protein and PsbP-like protein 1. We uncovered how a hybrid responds under salt stress. These results provide an overview of how heterosis improves the salt-tolerance critical process of plants.

Furthermore, TFs play vital roles in the salt tolerance mechanism in crops via transcriptional regulation. Particular TF families were identified to control and attune salt stress adaptive pathways, such as bZIP, WRKY, ERF/AP2, HB, MYB and bHLH [35–38]. The HB family TFs with basic helix loop domain are transcriptionally regulated in an ABA-dependent manner, and also act in a signal transduction pathway which arbitrates the abiotic stress response. Several transcription factors of the bHLH gene family were discovered for regulating plant growth and plant responses to salt stress, such as OsbHLH094, OsbHLH062 and OsbHLH035, already reported to take part in salt tolerance by DNA-binding in the promoter region of some ion transporter genes [39–41]. In the present study, the highest number of bHLH DETFs were identified. This result is also similar to our previous study, where we identified salt tolerance QTL in a unique position [16]. The temporal gene expression patterns of these TFs involved in RNA processing in the hybrid and its parents could be helpful in understanding the transcriptional regulation in rice under salt stress.

4. Conclusions

Here, we provided a global view of the transcriptomic differences of the elite super hybrid rice LYP9 and its parental lines under two salt stress treatments using RNA-seq. This study suggested that the hybrid has an important role in the reaction to salt stress, and provided new insights into the mechanisms of heterosis in salt tolerance. Additionally, the hybrid LYP9 can tolerate longer periods of salt stress at the seedling stage. Finally, the hybrid LYP9 and its parents PA64s and 93-11 will provide genetic resources for future advancement of rice salinity tolerance.

Supplementary Materials: The following are available online at https://www.mdpi.com/article/10.3390/agronomy11071328/s1.

Author Contributions: Conceptualization, N.J. and L.G.; Data curation, N.J., Y.L. (Yang Lv) and M.S.; Formal analysis, N.J., L.S., Y.L. (Ying Lu) and M.S.; Funding acquisition, L.G.; Investigation, N.J. and Y.L. (Yang Lv); Methodology, N.J., Y.L. (Yang Lv) and M.S.; Project administration, L.G.; Software, N.J., Y.L. (Yang Lv) and M.S.; Supervision, N.J., G.Y., Z.G. and L.G.; Validation, N.J., Y.L. (Yang Lv) and M.S.; Visualization, Y.Z.; Writing—original draft, N.J., Y.L. (Yang Lv) and M.S.; Writing—revew & editing, N.J., Z.G., Q.Q. and Y.Z. All authors have read and agreed to the published version of the manuscript.

Funding: This work was supported by Shenzhen S&T Project: (GJHZ20190821163601707) and Shandong Agricultural Elite Variety Project (2019LZGC003), The National Key Research and Development Program of China (Grant No. 2016YFD0100902-07).

Institutional Review Board Statement: Not applicable.

Informed Consent Statement: Not applicable.

Data Availability Statement: Not applicable.

Conflicts of Interest: The authors declare no conflict of interest.

References

1. Huang, X.; Yang, S.; Gong, J.; Zhao, Q.; Feng, Q.; Zhan, Q.; Zhao, Y.; Li, W.; Cheng, B.; Xia, J.; et al. Genomic architecture of heterosis for yield traits in rice. *Nature* **2016**, *537*, 629–633. [CrossRef]
2. Kaeppler, S. Heterosis: Many Genes, Many Mechanisms-End the Search for an Undiscovered Unifying Theory. *ISRN Bot.* **2012**, *2012*, 682824. [CrossRef]
3. Zhao, C.; Liu, C.; Zhang, Y.; Cui, Y.; Hu, H.; Jahan, N.; Lv, Y.; Qian, Q.; Guo, L. A 3-bp deletion of WLS5 gene leads to weak growth and early leaf senescence in rice. *Rice* **2019**, *12*, 26. [CrossRef]
4. Bao, J.; Lee, S.; Chen, C.; Zhang, X.; Zhang, Y.; Liu, S.; Clark, T.; Wang, J.; Cao, M.; Yang, H.; et al. Serial Analysis of Gene Expression Study of a Hybrid Rice Strain (LYP9) and Its Parental Cultivars. *Plant Physiol.* **2005**, *138*, 1216. [CrossRef]
5. Krieger, U.; Lippman, Z.B.; Zamir, D. The flowering gene SINGLE FLOWER TRUSS drives heterosis for yield in tomato. *Nat. Genet.* **2010**, *42*, 459–463. [CrossRef]
6. Shang, L.; Wang, Y.; Cai, S.; Wang, X.; Li, Y.; Abduweli, A.; Hua, J. Partial Dominance, Overdominance, Epistasis and QTL by Environment Interactions Contribute to Heterosis in Two Upland Cotton Hybrids. *G3* **2015**, *6*, 499–507. [CrossRef]

7. Swanson-Wagner, R.A.; DeCook, R.; Jia, Y.; Bancroft, T.; Ji, T.; Zhao, X.; Nettleton, D.; Schnable, P.S. Paternal dominance of trans-eQTL influences gene expression patterns in maize hybrids. *Science* **2009**, *326*, 1118–1120. [CrossRef]
8. Paschold, A.; Jia, Y.; Marcon, C.; Lund, S.; Larson, N.B.; Yeh, C.-T.; Ossowski, S.; Lanz, C.; Nettleton, D.; Schnable, P.S.; et al. Complementation contributes to transcriptome complexity in maize (*Zea mays* L.) hybrids relative to their inbred parents. *Genome Res.* **2012**, *22*, 2445–2454. [CrossRef] [PubMed]
9. Song, R.; Messing, J. Gene expression of a gene family in maize based on noncollinear haplotypes. *Proc. Natl. Acad. Sci. USA* **2003**, *100*, 9055–9060. [CrossRef] [PubMed]
10. Zhang, X.; Jiang, H.; Wang, H.; Cui, J.; Wang, J.; Hu, J.; Guo, L.; Qian, Q.; Xue, D. Transcriptome Analysis of Rice Seedling Roots in Response to Potassium Deficiency. *Sci. Rep.* **2017**, *7*, 5523. [CrossRef] [PubMed]
11. Wei, G.; Tao, Y.; Liu, G.; Chen, C.; Luo, R.; Xia, H.; Gan, Q.; Zeng, H.; Lu, Z.; Han, Y.; et al. A transcriptomic analysis of superhybrid rice LYP9 and its parents. *Proc. Natl. Acad. Sci. USA* **2009**, *106*, 7695. [CrossRef]
12. Long, L.; Yang, W.; Liao, P.; Guo, Y.; Kumar, A.; Gao, W. Transcriptome analysis reveals differentially expressed ERF transcription factors associated with salt response in cotton. *Plant Sci.* **2019**, *281*, 72–81. [CrossRef] [PubMed]
13. Li, P.; Wang, M.; Lu, Q.; Ge, Q.; Rashid, M.H.o.; Liu, A.; Gong, J.; Shang, H.; Gong, W.; Li, J.; et al. Comparative transcriptome analysis of cotton fiber development of Upland cotton (*Gossypium hirsutum*) and Chromosome Segment Substitution Lines from *G. hirsutum* × *G. barbadense*. *BMC Genom.* **2017**, *18*, 705. [CrossRef]
14. Li, X.; Wei, Y.; Nettleton, D.; Brummer, E.C. Comparative gene expression profiles between heterotic and non-heterotic hybrids of tetraploid Medicago sativa. *BMC Plant Biol.* **2009**, *9*, 107. [CrossRef]
15. FAO Soils Portal. Available online: http://www.fao.org/soils-portal/soil-management/management-of-some-problem-soils/salt-affected-soils/more-information-on-salt-affected-soils/en/ (accessed on 21 June 2021).
16. Jahan, N.; Zhang, Y.; Lv, Y.; Song, M.; Zhao, C.; Hu, H.; Cui, Y.; Wang, Z.; Yang, S.; Zhang, A.; et al. QTL analysis for rice salinity tolerance and fine mapping of a candidate locus qSL7 for shoot length under salt stress. *Plant Growth Regul.* **2020**, *90*, 307–319. [CrossRef]
17. Zhou, Y.; Yang, P.; Cui, F.; Zhang, F.; Luo, X.; Xie, J. Transcriptome Analysis of Salt Stress Responsiveness in the Seedlings of Dongxiang Wild Rice (*Oryza rufipogon* Griff.). *PLoS ONE* **2016**, *11*, e0146242. [CrossRef] [PubMed]
18. Krishnamurthy, P.; Mohanty, B.; Wijaya, E.; Lee, D.-Y.; Lim, T.-M.; Lin, Q.; Xu, J.; Loh, C.-S.; Kumar, P.P. Transcriptomics analysis of salt stress tolerance in the roots of the mangrove *Avicennia officinalis*. *Sci. Rep.* **2017**, *7*, 10031. [CrossRef] [PubMed]
19. Prasad, S.R.; Bagali, P.G.; Hittalmani, S.; Shashidhar, H.E. Molecular mapping of quantitative trait loci associated with seedling tolerance to salt stress in rice (*Oryza sativa* L.). *Curr. Sci.* **2000**, *78*, 162–164.
20. Yamaguchi, T.; Blumwald, E. Developing salt-tolerant crop plants: Challenges and opportunities. *Trends Plant Sci.* **2005**, *10*, 615–620. [CrossRef]
21. Zhang, Y.; Lv, Y.; Jahan, N.; Chen, G.; Ren, D.; Guo, L. Sensing of Abiotic Stress and Ionic Stress Responses in Plants. *Int. J. Mol. Sci.* **2018**, *19*, 3298. [CrossRef]
22. Li, Q.; Yang, A.; Zhang, W. Efficient acquisition of iron confers greater tolerance to saline-alkaline stress in rice (*Oryza sativa* L.). *J. Exp. Bot.* **2016**, *67*, 6431–6444. [CrossRef] [PubMed]
23. Kim, D.; Pertea, G.; Trapnell, C.; Pimentel, H.; Kelley, R.; Salzberg, S.L. TopHat2: Accurate alignment of transcriptomes in the presence of insertions, deletions and gene fusions. *Genome Biol.* **2013**, *14*, R36. [CrossRef] [PubMed]
24. Shao, L.; Xing, F.; Xu, C.; Zhang, Q.; Che, J.; Wang, X.; Song, J.; Li, X.; Xiao, J.; Chen, L.-L.; et al. Patterns of genome-wide allele-specific expression in hybrid rice and the implications on the genetic basis of heterosis. *Proc. Natl. Acad. Sci. USA* **2019**, *116*, 5653. [CrossRef] [PubMed]
25. Freeling, M. Bias in plant gene content following different sorts of duplication: Tandem, whole-genome, segmental, or by transposition. *Annu. Rev. Plant Biol.* **2009**, *60*, 433–453. [CrossRef]
26. Kim, D.; Ju, H.; Kwon, T.; Oh, C.; Ahn, S. Mapping QTLs for salt tolerance in an introgression line population between Japonica cultivars in rice. *J. Crop Sci. Biotechnol.* **2009**, *12*, 121. [CrossRef]
27. Ma, T.; Wu, W.; Wang, Y. Transcriptome analysis of rice root responses to potassium deficiency. *BMC Plant Biol.* **2012**, *12*, 161. [CrossRef]
28. Gao, S.; Zhang, H.; Tian, Y.; Li, F.; Zhang, Z.; Lu, X.; Chen, X.; Huang, R. Expression of TERF1 in rice regulates expression of stress-responsive genes and enhances tolerance to drought and high-salinity. *Plant Cell Rep.* **2008**, *27*, 1787–1795. [CrossRef] [PubMed]
29. Li, Q.; Yang, A.; Zhang, W. Comparative studies on tolerance of rice genotypes differing in their tolerance to moderate salt stress. *BMC Plant Biol.* **2017**, *17*, 141. [CrossRef]
30. Gundel, P.E.; Garibaldi, L.A.; Martínez-Ghersa, M.A.; Ghersa, C.M. Neotyphodium endophyte transmission to *Lolium multiflorum* seeds depends on the host plant fitness. *Environ. Exp. Bot.* **2011**, *71*, 359–366. [CrossRef]
31. Chaves, M.M.; Flexas, J.; Pinheiro, C. Photosynthesis under drought and salt stress: Regulation mechanisms from whole plant to cell. *Ann. Bot.* **2009**, *103*, 551–560. [CrossRef]
32. Luo, J.; Huang, C.h.; Peng, F.; Xue, X.; Wang, T. Effect of salt stress on photosynthesis and related physiological characteristics of *Lycium ruthenicum* Murr. *Acta Agric. Scand. Sect. B Soil Plant Sci.* **2017**, *67*, 680–692. [CrossRef]
33. Wang, N.; Qian, Z.; Luo, M.; Fan, S.; Zhang, X.; Zhang, L. Identification of Salt Stress Responding Genes Using Transcriptome Analysis in Green Alga *Chlamydomonas reinhardtii*. *Int. J. Mol. Sci.* **2018**, *19*, 3359. [CrossRef]

34. Moolna, A.; Bowsher, C.G. The physiological importance of photosynthetic ferredoxin $NADP^+$ oxidoreductase (FNR) isoforms in wheat. *J. Exp. Bot.* **2010**, *61*, 2669–2681. [CrossRef]
35. Jaradat, M.; Feurtado, J.; Huang, D.; Lu, Y.; Cutler, A.J. Multiple roles of the transcription factor AtMYBR1/AtMYB44 in ABA signaling, stress responses, and leaf senescence. *BMC Plant Biol.* **2013**, *13*, 192. [CrossRef] [PubMed]
36. Golldack, D.; Lüking, O.; Yang, O. Plant tolerance to drought and salinity: Stress regulating transcription factors and their functional significance in the cellular transcriptional network. *Plant Cell Rep.* **2011**, *30*, 1383–1391. [CrossRef] [PubMed]
37. Zhu, J. Salt and drought stress signal transduction in plants. *Annu. Rev. Plant Biol.* **2002**, *53*, 247–273. [CrossRef]
38. Hsieh, T.-H.; Li, C.-W.; Su, R.-C.; Cheng, C.-P.; Sanjaya; Tsai, Y.-C.; Chan, M.-T. A tomato bZIP transcription factor, SlAREB, is involved in water deficit and salt stress response. *Planta* **2010**, *231*, 1459–1473. [CrossRef]
39. Blumwald, E. Sodium transport and salt tolerance in plants. *Curr. Opin. Cell Biol.* **2000**, *12*, 431–434. [CrossRef]
40. Chen, H.-C.; Cheng, W.-H.; Hong, C.-Y.; Chang, Y.-S.; Chang, M.-C. The transcription factor OsbHLH035 mediates seed germination and enables seedling recovery from salt stress through ABA-dependent and ABA-independent pathways, respectively. *Rice* **2018**, *11*, 50. [CrossRef]
41. Toda, Y.; Tanaka, M.; Ogawa, D.; Kurata, K.; Kurotani, K.-i.; Habu, Y.; Ando, T.; Sugimoto, K.; Mitsuda, N.; Katoh, E.; et al. RICE SALT SENSITIVE3 Forms a Ternary Complex with JAZ and Class-C bHLH Factors and Regulates Jasmonate-Induced Gene Expression and Root Cell Elongation. *Plant Cell* **2013**, *25*, 1709. [CrossRef]

agronomy

MDPI

Article

Identification of Drought Tolerance on the Main Agronomic Traits for Rice (*Oryza sativa* L. ssp. *japonica*) Germplasm in China

Muhammad Shafiq Ahmad, Bingrui Wu, Huaqi Wang * and Dingming Kang *

College of Agronomy and Biotechnology, China Agricultural University, Beijing 100193, China; ahmad@cau.edu.cn (M.S.A.), wubingrui@mengniu.cn (B.W.)

* Correspondence: wanghuaqi@cau.edu.cn (H.W.); kdm@pku.edu.cn (D.K.); Tel.: +86-10-62733982 (H.W.); +86-10-62756091 (D.K.)

Abstract: Drought is a major abiotic factor restricting rice yield; therefore, to cope with this stress, 2030 *japonica* rice accessions from China and other countries were evaluated in Beijing in 2017 and 2018. This was the first time six agronomic traits in the large-scale germplasm of rice under lowland and upland conditions with an augmented randomized complete block design (ARCBD) were analysed. The genotypes revealing drought resistant grade (DRG) scores of 1, 1–3, 3 and 3–5 were considered drought-tolerant and comprised 10% of the assessed germplasm. These findings were consistent with the agglomerative hierarchical cluster (AHC) analysis that classified germplasm in nine clusters. The generated clusters were further grouped in A, B, C and D classes based on the stress response. Approximately half of the genotypes with an upland ecotype were distributed in drought-resistant class A (cluster VII and VI) and moderately resistant class B (VIII and IX). The majority of the genotypes from China, Korea and Japan fall in drought-susceptible classes C and D. Genotypes of DRG 1, 1–3 and 3 belonged to the clusters VII and VI. Finally, we screened out 42 elite genotypes including seven improved upland rice lines (D78, LB37-13, NSU77, Handao 385, Handao 306, SF83 and HF6-65-119), three upland released varieties (Liaogeng 27, Hanfeng 8 and IRAT109) and three traditional lowland cultivars (Hongmaodao, Weiguo 7 and Xiaohongbandao). These genotypes might be used as priority parents in drought-tolerant rice breeding programmes and some of them could be recommended directly to farmers in water-deficient rice areas of China.

Keywords: field identification; drought resistance; *japonica* rice; germplasm; agronomic trait

Citation: Ahmad, M.S.; Wu, B.; Wang, H.; Kang, D. Identification of Drought Tolerance on the Main Agronomic Traits for Rice (*Oryza sativa* L. ssp. *japonica*) Germplasm in China. *Agronomy* **2021**, *11*, 1740. https://doi.org/10.3390/agronomy11091740

Academic Editor: Santiago Signorelli

Received: 2 July 2021
Accepted: 16 August 2021
Published: 30 August 2021

1. Introduction

Drought is a direct result of water scarcity, in particular the long-term lack of precipitation that could result in low soil moisture content, reduced crop yields, and increased water demand for irrigation (low runoff) and groundwater extraction [1]. Although China accounts for 6% of the world's freshwater, a total volume of 2.8×10^{12} m^3 per capita water availability is just 25% of the global average and the lowest among Asian countries [2]. Rice (*Oryza sativa* L.), a major food crop for over half of the world's population, is cultivated worldwide in over 95 countries, with around 90% of the world's rice being produced and consumed in Asia, providing 35% to 60% of the dietary calories for more than 3 billion people [3]. Yet, due to industrial and urban development, declining supply because of resource depletion, pollution and population growth, water is becoming readily shorter in supply [4], and its supply for rice production is becoming limited. Therefore, there is a need for another production system for rice, especially for the areas with reduced labour resources and supply of water.

Germplasm banks are warehouses of the crop plants, and its pools characterize a large capability for resources of stress tolerance. The Institute of Crop Sciences at the Chinese Academy of Agricultural Sciences (CAAS), Beijing, China has maintained a genetic pool of rice genotypes—mostly landraces or breeding materials of *Oryza sativa, Oryza glaberrima*, representative species of the genus *Oryza* and other wild species. The future improvement of crops is based on the genetic variability from traditional varieties and concerned wild species to manage many abiotic and biotic stresses that impacted the production of rice in the whole world [5]. The vast array of rice genotypic variation in the national germplasm of China presents an outstanding prospect to identify and screen genotypes for stress adaptability and to research the genetic characteristics which allow those accessions to withstand the negative impact of stress on yield and growth of rice. To mitigate the increasing food crisis, it is imperative to recognize or cultivate rice varieties that could endure the stress of water scarcity and generate cost-effective yields. The maximum amounts of improved varieties that are cultivated in water-deficient rain-fed lowlands were grown under watered conditions and were not chosen for resistance to drought [6]. While breeding attempts for lowland, aerobic, and upland rice make considerable developments of new rice varieties for drought-prone conditions [7–11], more sources of drought tolerance are required to tackle reduction problems in yield under a variety of drought conditions. Additionally, yield assessment under normal environments is critical for the recognition of accessions that would be valuable for agriculturists tackling water scarcity in some years and adequate rainwater in others [6].

In the previous decade, a lot of strategies have been adopted to assess the drought resistance of crops [12–14]. Even utilizing transgenic methods, several genes have been found and bred for enhancing water-deficiency tolerance; yet, the majority of the research was carried out in the controlled conditions, e.g., in greenhouses [15]. Accurately and effectively examining the tolerance against drought in the rice field has turned out to be a crucial goal for breeding against drought-tolerance. For breeding resources, screening based on drought adaptive and physio-morphological traits, phenotyping is considered main tool [16,17]. Therefore, experiments were conducted to check heat and drought interactive effects during flowering and grain filling phases in divergent rice cultivars sown in the field environments [18]. Further, screening based on the sole grain yield trait turned out to be ineffective [19]; nevertheless, Torres et al. (2013) utilized the same method for screening the germplasm of 988 accessions in a trial in IRRI. Similarly, sole agglomerative hierarchical cluster (AHC) analysis resulted in poor performance due to the merging errors [20]. On the other hand, AHC proved to be reliable for the identification of drought-tolerant genotypes when the selection was made on multiple traits [21].

It is appropriate to develop a methodology to breed drought-resistant upland rice which is accurate and rapid to differentiate drought resistance among different rice cultivars displaying a good combination of important agronomic attributes cumulatively imparting to enhanced yields [22]. Normally, selection would aim at accessions with comparatively increased yields under both optimal and stressed conditions to adapt them better to climate change. Thus, it is necessary to determine the relative drought stress susceptibility (RDS) of test genotypes using AHC. For this, core yield attributes of rice are pertinent for drought identification and screening consist of the following: plant height to leaf and panicle (PHL and PHP), aboveground biomass plant^{-1} (ABP^{-1}), grain yield plant^{-1} (GYP^{-1}) and harvest index (HI). Days to flowering (DF) are imperative as well, while breeding for drought stress at the terminal phase as this assist drought escape [23]. This is the first time an integrated selection based on combined results from RDS, drought resistance grade (DRG) and AHC from large-scale rice germplasm under lowland and upland conditions applying augmented randomized complete block design (ARCBD) has been made. Controlled water application-based selection, with the support of RDS and AHC, presents efficient yield-based genotype screening, permitting for identification of better yielding genotypes under both stressed and watered environments.

The study aimed to uncover the genetic variability for drought resistance among miscellaneous rice accessions based on agronomic characters and to sort out improved promising lines for breeding. In this work, 2030 *Oryza sativa* ssp. *japonica* accessions from various regions of China, Japan, South Korea, the Ivory Coast and Brazil were subjected to field trials conducted to find out the drought-tolerant and -susceptible genotypes which can be used in future breeding programs, and also if a few of them can be grown directly in farmers' field.

2. Materials and Methods

2.1. Experimental Site and Plant Material

The field trials were carried out from May to October at Shangzhuang Agricultural Research Station (40°08′13.4″ N, 116°11′06.6″ E), China Agricultural University, Beijing, China in 2017 and 2018 to identify drought-resistant genotypes. For this study, 2000 japonica rice accessions were supplied by the Institute of Crop Sciences from the Chinese Academy of Agricultural Sciences (CAAS) and the Rice Research Centre, CAU provided 30 genotypes. The evaluated germplasm was distributed among China (1900), South Korea (70), Japan (57), the Ivory Coast (2) and Brazil (1). Concerning China, the maximum number of genotypes originated from Liaoning (336), Jilin (330), Jiangsu (328) and Heilongjiang (244) (Table 1). Soil observations including weather conditions for two consecutive years of experiments are shown in Figure 1 and Table S1. Various types of genotypes were included like traditional cultivars, several exotic varieties, promising lines and released varieties with two main ecotypes of lowland and upland (Table S2).

Table 1. Origin, ecotype and type of germplasm evaluated in the study.

Ecotype / Origin	Lowland				Upland			Total
	Exotic	Line	Variety	Cultivar	Line	Variety	Cultivar	
China		195	1351	203	31	28	92	1900
Anhui			8					8
Beijing			37		31	8		76
Guizhou			6					6
Hebei			31	20			3	54
Heilongjiang			192	37		2	13	244
Henan			67			4		71
Inner Mongolia				4			4	8
Jiangsu		176	150	1		1		328
Jilin			263	51		1	15	330
Liaoning			247	41		12	36	336
Ningxia			85	13				98
Shandong		19	60	25			16	120
Tianjin				2			2	4
Xinjiang			17	9			3	29
Yunnan			188					188
Brazil						1		1
Ivory Coast						2		2
Japan	56		1					57
South Korea	70							70
Total	126	195	1352	203	30	31	92	2030

Note. Exotic = exotic variety with unknown type; line = promising line; variety = released variety; cultivar = traditional cultivar.

These accessions accompanied by three control genotypes namely B_1 [24] and Handao 277 [25] as tolerant controls and 297-28 as a susceptible control [24] were assessed in lowland and upland environments. The Beijing-approved upland variety, Handao 277 is widely cultivated by farmers in Huang (Yellow) and in the Huai River basins of China. From 2004 to 2019, National Upland Rice Variety Regional Trials of China utilized this genotype as a control in the same river basin areas of the country. B_1 was developed from IAPAR9 through mutation by γ-ray radiation and characterized as a semi-dwarf mutant with high drought resistance. However, 297-28 is a drought-susceptible mutant derived from Handao 297 by the same exposure to the same radiations. This mutant was characterized by slender leaves, culms, roots and more tiller plants^{-1} from its wild-type under lowland conditions but leaves wilt readily under drought stress, leading to a decline in yield. Agronomic traits were solely considered for the screening of drought-tolerant accessions for two years of the experiments.

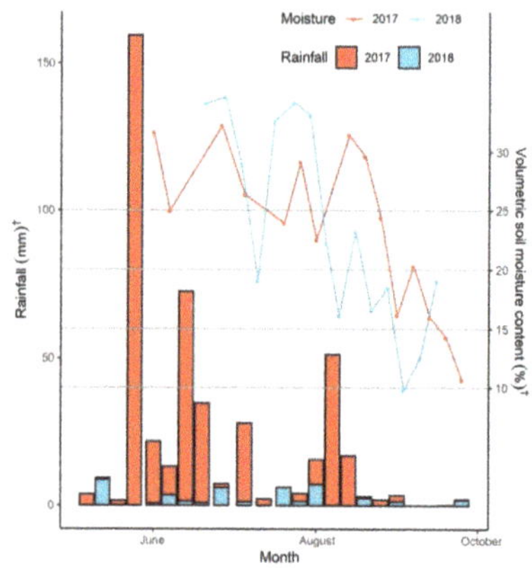

Figure 1. Rainfall (mm) and volumetric soil moisture content (%) situations at Shangzhuang Agricultural Research Station, Beijing, China for the study carried out consecutively in 2017 and 2018 from May to October. † Left and right axes have different scales.

2.2. *Field Management*

Before conducting the trials, the experimental field was levelled. The lowland and upland experiments were established under irrigated non-puddled and rain-fed conditions, respectively. Before sowing, the seeds of all germplasm were handled with the fungicide-suggested dosage. Both experiments were manually established by seeding directly into the dry soil at a depth of 2–3 cm per genotype in two shallow groves spread out at a distance of 28 cm with approximately 80 grains per genotype for lowland conditions and 25 cm spaced two rows with 100 seeds per accession for the upland conditions.

The row lengths for lowland and upland were kept 1.45 m and 0.80 m, respectively, with no specific plant-to-plant distance. The wet condition was maintained in non-stress experiments from the date of sowing until maturity by either natural rainfall or supplementary irrigation as required. General cultural procedures were performed in two trials. The applied basal fertilizers contained 48 kg·ha^{-1} of nitrogen (N), 120 kg·ha^{-1} of phosphorus pentaoxide (P_2O_5), 100 kg·ha^{-1} of potassium oxide (K_2O), 22.5 kg·ha^{-1} of zinc sulphate ($ZnSO_4$) and 30 kg·ha^{-1} of iron sulphate ($FeSO_4$) and were applied in both conditions. Later, at the seedling stage, nitrogen was placed in the amount of 45 kg·ha^{-1}. Additionally, dried sheep manure was supplied at a rate of 3.75 t·ha^{-1} consisting of 5% N. The chemical properties of the 0 cm to 30 cm soil layer of the study site in 2017 were as follows: extracted mineral N (Nmin) 7.0 mg·kg^{-1}, pH (H_2O) 7.7, Olsen-P 21.2 mg·kg^{-1}, NH_4OAc-K 106.1 mg and organic matter 15.2 g·kg^{-1} [26]. In upland experiments, unwanted plants and weeds were restricted by 2–3 times of cultural hoeing in each season, by uprooting, and along borders by spraying herbicide. According to the requirement of the experimental layout, the upland-released variety Handao 297 was utilized as a border crop, which protected the experimental area of 0.24 ha in lowland and 0.81 ha in upland conditions.

2.3. *Drought Trials and Investigations in the Field*

Drought experiments were carried out in upland conditions to avoid rainfall-associated water accumulation during the stress period. Under drought conditions, water was supplied only once just after sowing for the sake of germination. Even after the onset of drought, the germplasm was not irrigated again despite the severity of the stress, except for the upland condition in 2017, when it was irrigated a second time after herbicide spraying. For the lowland condition, irrigation was applied for the entire growth period to maintain a shallow layer of water on the surface and a wet state in the experimental field. Investigation of the characters was conducted at maturity just before harvesting with an exception for phenology. We investigated 28 plant and soil traits in total, including visually scored parameters (Table S3), but for selection, six core plant traits were used. For the whole experiment, the investigated traits' detailed information is shown in Table 2. TRIME-PICO32 instrument (IMKO Micromodultechnik GmbH, Ettlingen, Germany) was utilized to record volumetric soil moisture content with a soil depth of 15 cm in three random locations of the upland field. The drought was categorized as slight, mild, moderate, severe and extremely severe based on the volumetric soil moisture content of 30% to 25%, 25% to 20%, 20% to 15%, 15% to 10% and less than 10% in the topsoil.

Table 2. Selected traits of study, their abbreviations and measurement details.

Trait	Description
DF	Days to flowering was recorded as the number of days from sowing to the time when inflorescences had emerged above the flag leaf sheath for more than half of the individuals of a landrace.
PHL (cm)	Measured height from ground to the highest leaf tip with a meter rod.
PHP (cm)	Measured height from ground to panicle tip with a meter rod.
ABP^{-1} (g)	Shoot dry weight, including the grain yield and straw, were weighed for each plant after being dried in an oven at 105 °C for 30 min and at 80 °C for 23.5 h.
GYP^{-1} (g)	Total grain weight $plant^{-1}$ was weighed after drying at 105 °C for 30 min and then 80 °C for 23.5 h in an oven.
HI	The harvest index was computed as the ratio of filled spikelet weight to total aboveground biomass.

Note. DF = days to flowering; PHL = plant height to leaf; PHP = plant height to panicle; ABP^{-1} = aboveground biomass $plant^{-1}$; GYP^{-1} = grain yield $plant^{-1}$; HI = harvest index.

2.4. Large-Scale Germplasm Adjustments for Screening

Augmented randomized complete block design (ARCBD), widely used to phenotype large populations [27–30], was implemented to lowland and upland field trials. The study was amended for large-scale germplasm to manage the capital, land and resources. Under ARCBD experimental design, all the control genotypes were sown adjacently and on average, there was one control group (B_1, Handao 277 and 297-28) for 127 and 120 genotypes under lowland conditions in 2017 and 2018, respectively. While under upland conditions, on average each control group was assigned to 79 and 83 genotypes in 2017 and 2018, correspondingly. The number of control groups was 26 in 2017 and 24 in 2018 for upland conditions, while there were 16 groups in total for the lowland condition in each season. In ARCBD, the number of blocks must be the same as that of control groups. For the reduction in environmental errors in these uni-replicate experiments, accessions were planted in two rows, and means were computed from measured values of traits from three selected sample $genotypes^{-1}$. ABP^{-1}, GYP^{-1} and HI were measured in 2018 only to adjust labour and capital. The year factor was pooled for graphical representation of population distribution for all the traits. To calculate the relative drought stress susceptibility for traits (RDS_T; Eq. [5]), we marked 1118 genotypes whose flowering and filling phases across both conditions in each season were not affected by the cool temperature of Beijing. Shared variation by RDS_T was utilized for trait statistical ranking based on SD/s^2. We also compared relative drought stress susceptibility (RDS) for specific traits in both seasons, as well as in pooled situation owing to same-scaled data for all the traits.

2.5. Statistical Analysis

2.5.1. Phenotypic Data Analysis

Following the previously described procedure to minimize the effects of environmental variation [27–30], the 2 years of phenotypic data were fitted with a linear mixed model that included the effects of genotypes, conditions, years, blocks, genotypes × conditions, genotypes × years, conditions × years and genotypes × conditions × years. R version

3.6.2 [31] was run for statistical and graphical analysis of data. Further, agricolae [32] and augmentedRCBD [33] libraries were loaded to compare treatment means by Tukey's honestly significant difference (HSD) at 5% probability. Four standard errors (*SE*) of the mean difference for ARCBD were calculated in the following equations:

$$Sc = \sqrt{\frac{2MSe}{r}} \tag{1}$$

$$Sb = \sqrt{2MSe} \tag{2}$$

$$Sv = \sqrt{2MSe\left(1 + \frac{1}{c}\right)} \tag{3}$$

$$Svc = \sqrt{MSe(1 + \frac{1}{r} + \frac{1}{c} + \frac{1}{rc})} \tag{4}$$

Here, *Sc* is the dissimilarity between the control genotypes, *Sb* and *Sv* are the dissimilarities concerning evaluated genotypes in the same and different blocks, respectively, and *Svc* is the dissimilarity about the control and evaluated genotypes. *c* and *r* represent the number of controls and the number of blocks in the experiment. The first three *SE* were multiplied by $q(t, v; \alpha)$ to calculate Tukey's HSD, where q was the studentized range statistic for the total number of group means (t) with degrees of freedom for error (v) at a specific significance level (α), whereas $q(t, v; \alpha)$ was multiplied by $\sqrt{\frac{MSe}{H'}}$ to compute the value of HSD for *Svc* and here, H' was the harmonic mean of the coefficients for standard errors of a difference. In our study, we used the relative drought stress susceptibility index for a particular trait (RDS_T) already utilized by Huang et al. (2018).

$$RDS_T\ (\%) = \frac{T_p - T_s}{T_p} \times 100 \tag{5}$$

$$RDS\ (\%) = \frac{1}{n}\sum_{i=1}^{n} RDS_T\ (\%) \tag{6}$$

$$RDS_{Tw}\ (\%) = \frac{RDS_T\ (\%) \times w\ (\%)}{100} \tag{7}$$

$$RDS - I\ (\%) = \sum_{i=1}^{n} RDS_{Tw}\ (\%) \tag{8}$$

WTp is the performance of a particular trait in the lowland condition, *Ts* is the performance of a particular trait in the upland condition, *RDS* is the average of RDS_T for all the traits under investigation, *n* is the total number of traits and w is the weighted factor. Weighted *RDS* for a particular trait (RDS_{Tw}) was estimated directly from RDS_T by multiplying with w and then aggregated to compute the integrated relative drought stress susceptibility (*RDS-I*). While summing up RDS_T and RDS_{Tw} to *RDS* and *RDS-I* correspondingly, the respective value of DF was subtracted due to its adverse direction from the breeding point of view. *RDS* and *RDS-I* were utilized to compute drought-resistance ranking of genotypes and to obtain the drought resistance grade (DRG). *RDS* and *RDS-I* values were directly proportional to the susceptibility rate, i.e., the higher the value of *RDS* and *RDS-I*, the more susceptible was the genotype and the lower the value, the more resistant was the investigated genotype. DRG was derived by dividing the total range of phenotypic expression under *RDS-I* into nine defined classes that were consistent with visual score practice of standard evaluation system (SES) in rice [34]. DRG values ranged from one to nine responsible for drought resistance level, with grades "1" and "9" revealing the "strongest" and "weakest" response against stress, respectively. Descriptive and genetic variability statistics were estimated by loading augmentedRCBD library [33] in R.

2.5.2. Graphical Representation

Package ggplot2 [35] accompanied by additional packages were loaded in R software for all graphical visualization of data with the exception of the agglomerative hierarchical cluster (AHC) plot. R function cor with the Pearson method was used to calculate trait correlations, corr.test from the psych package utilized to determine *p*-values and the package ggcorrplot was used for its graphical interpretation. To draw AHC, we loaded dendextend and heatmaply packages in R, and for scatter graph visualization, a scales package along with the previously mentioned libraries were used. The map in this article was created using ArcGIS® software by Esri. ArcGIS® and ArcMap™ are the intellectual property of Esri and are used herein under license. For more information about Esri® software, please visit www.esri.com (accessed on 16 August 2021).

3. Results

3.1. Meteorological and Soil Observations

During the rice-growing season from May to September, total rainfall in trial fields was 430.6 mm and 235.8 mm in 2017 and 2018, respectively (Figure 1).

The entire water availability to crops involving rainfall and irrigation was 565.5 mm to 595.5 mm and 310.8 mm to 340.8 mm, respectively. That water supply was lower than the required range of 750 mm to 1400 mm for aerobic rice [36]. The range of mean minimum and maximum temperatures (T_{min} and T_{max}) from May to October was 15.0 °C to 27.7 °C and 15.2 °C to 27.7 °C in 2017 and 2018, respectively. As a whole, in 2017, the drought stress trend under upland conditions was mild to moderate. Though a little rainfall deficit was observed at the seedling stage in May and June, the soil was not dry because the field was irrigated a second time to control weeds by application of herbicide. The drought stress from the tillering (late June) to filling phase (early September) was slight to mild as there was a 30% to 20% volumetric soil moisture content at 15 cm depth. Yet, stress became moderate to severe from mid-September to early October until harvesting (moisture decreased from 20% dropping close to 10%). However, in 2018, stress was generally moderate to severe. Seedlings in May and June suffered from drought stress caused by more than one month of high temperature without rain. Furthermore, from late August to the whole of September, volumetric soil moisture decreased from 20% to below 10% (moderate and severe drought). Only in July to August, the stress was slight to mild with moisture contents ranging from 35% to 20%.

3.2. Variation of Main Agronomic Traits among Controls and Populations

Table 3 represents the abridged outcomes of the statistical assessment of the effects of genotype, condition and year for the measured traits. All the main effects exhibited significant differences excluding plant height to leaf (PHL), which was not affected by year. The mean values of days to flowering (DF) and plant height to panicle (PHP) were reduced in 2018 as compared to 2017. Genotype and year interactions expressed significance for DF, PHP and PHL. With respect to condition × year effects, those were highly significant for all the observed parameters. Interaction of genotype, condition and year affected PHP only. Table 4 shows pooled traits including DF, PHL, PHP, aboveground biomass plant^{-1} (ABP^{-1}), grain yield plant^{-1} (GYP^{-1}) and harvest index (HI) for genotypes and controls with post hoc mean comparison analysis (HSD) that was performed after ARCBD analysis. Entries and controls denoted significant differences among considered traits. Control 297-28 expressed higher values of DF and PHL and lower ABP^{-1}, GYP^{-1} and HI in comparison with other controls under both conditions with an exception for high HI under lowland conditions while comparing to control B_1. For plant height, B_1 was the shortest control, whereas Handao 277, being a released upland rice cultivar, performed best in GYP^{-1}, ABP^{-1} and HI. Control B_1 revealed a more consistent measurement of the same trait despite being measured in lowland and upland conditions. Contrary to this, control 297-28 exhibited obvious differences for lowland and upland environments. After summarizing all assessments, Handao 277 proved to be the best control, control B_1 was the runner up and 297-28 obtained the third rank regarding yield and yield attributes. While focusing on core trait for genotypes, GYP^{-1} in the lowland condition was 103.4% higher on average than that of the upland condition. That could be attributed to ABP^{-1} accumulation in the lowland condition. The HI was also significantly different between the two cultivated conditions.

Table 3. Summary of the analysis of variance (ANOVA) for the effects of genotype (G), condition (C), year (Y), and their possible interactions on rice phenotypic traits.

Trait	G	C	Y	G × C	G × Y	C × Y	G × C × Y
DF	0.000	0.000	0.000	0.000	0.012	0.000	0.979
PHL	0.000	0.000	0.997	0.002	0.015	0.000	0.535
PHP	0.000	0.000	0.000	0.000	0.000	0.000	0.004
ABP^{-1}	0.000	0.000		0.006			
GYP^{-1}	0.000	0.000		0.001			
HI	0.000	0.000		0.221			

Note: above-mentioned data are depicted as *p*-values. DF = days to flowering; PHL = plant height to leaf; PHP = plant height to panicle; ABP^{-1} = aboveground biomass $plant^{-1}$; GYP^{-1} = grain yield $plant^{-1}$; HI = harvest index.

Table 4. Performance of measured six traits for genotypes and controls under lowland and upland conditions in 2017 and 2018 along with condition × year.

Factor	DF	PHL (cm)	PHP (cm)	ABP^{-1} (g)	GYP^{-1} (g)	HI
Genotype	100.07 ± 20.88	91.52 ± 19.10	88.21 ± 18.28	11.09 ± 5.79	4.17 ± 2.61	0.372 ± 0.11
Control	98.69 ± 8.47	93.44 ± 18.37	89.15 ± 19.61	11.09 ± 4.66	4.06 ± 2.3	0.348 ± 0.12
B_1	96.04 ± 6.53	75.34 ± 7.86	68.12 ± 7.61	9.48 ± 2.84	3.49 ± 1.4	0.36 ± 0.07
HD-277	94.39 ± 5.67	96.41 ± 10.37	94.68 ± 10.33	12.88 ± 4.44	5.52 ± 1.97	0.43 ± 0.07
297-28	106.2 ± 7.82	109.24 ± 15.83	105.3 ± 15.89	11 ± 5.53	3.24 ± 2.6	0.26 ± 0.12
Condition (C)						
Lowland	99.17 ± 17.94	98.62 ± 16.82	95.63 ± 15.93	13.93 ± 6.04	5.39 ± 2.82	0.39 ± 0.11
Upland	100.27 ± 21.02	84.29 ± 14.24	80.77 ± 13.09	7.99 ± 3.77	2.83 ± 1.55	0.36 ± 0.11
Year (Y)						
2017	103.8 ± 20.8	91.02 ± 19.96	88.57 ± 19.3			
2018	96.05 ± 19.63	92.18 ± 18.13	87.94 ± 17.35	11.09 ± 5.75	4.16 ± 2.6	0.371 ± 0.11
$HSD_{0.05}$						
Sc	7.00	7.11	7.21	4.82	2.17	0.11
Sb	55.96	64.45	65.29	30.63	13.77	0.60
Sv	68.54	74.43	75.43	35.35	15.94	0.65
Svc	34.53	37.44	37.94	17.90	8.05	0.45
C	0.276	0.409	0.379	0.267	0.118	0.0055
Y	0.276	NS	0.378			
C × Y	0.553	0.821	0.758			

DF = days to flowering; PHL = plant height to leaf; PHP = plant height to panicle; ABP^{-1} = aboveground biomass $plant^{-1}$; GYP^{-1} = grain yield $plant^{-1}$; HI = harvest index; HSD = Tukey's honestly significant difference; *Sc* = difference between two control genotypes; *Sb* = difference between two test genotypes in the same block; *Sv* = difference between two test genotypes in different blocks; *Svc* = difference between a control and test entry; NS = non-significant.

Regarding population distribution, the six examined traits of germplasm depicted vast phenotypic variability as shown in Figure 2. Most of the traits revealed normal distribution to skewness and kurtosis and that was endorsed by q-q plot (Figure S1). Differences between conditions were clear in the comparative population (lowland and upland) histograms. Among the measured traits, the spread of genotypes in days to flowering skewed to earlier in lowland condition in contrast to upland condition. That was accredited to North China germplasm and traditional upland cultivars. Comparing both conditions, germplasm expressed higher plant height, dry biomass and harvest index under the lowland situation. Plant height distributed among genotypes was somewhat similar in both ecotypes with a little left skewness in the lowland environment.

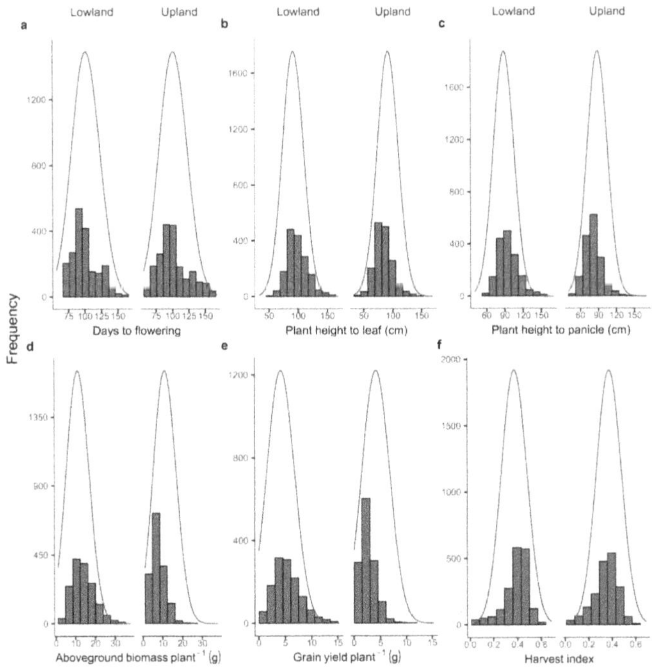

Figure 2. Frequency distribution of the six traits in lowland and upland conditions. Note: (**a**) Days to flowering. (**b**) Plant height to leaf (cm). (**c**) Plant height to panicle (cm). (**d**) Aboveground biomass plant^{-1} (g). (**e**) Grain yield plant^{-1} (g). (**f**) Harvest index.

3.3. Determination of Genetic Variation

A wide range of variation among the 2030 rice genotypes was observed for all the traits (Table 5). The phenotypic coefficient of variation (PCV) in 2017 and 2018 of all traits was higher than the genotypic coefficient of variation (GCV) in both trials. The PCV and GCV for all the traits recorded under drought were higher compared to lowland conditions, with some exception as GCV for PHP 2018, GYP^{-1} and HI and PCV for PHP 2018 and GYP^{-1} were higher in the lowland condition. In the upland condition, the highest PCV (49.58%) was recorded for the GYP^{-1} followed by ABP^{-1} (42.28%). The lowest PCV was recorded for PHP 2018 (16.07%) under drought stress. Similarly, the highest GCV was recorded for GYP^{-1} (34.52%) under the lowland situation, followed by GYP^{-1} (33.08%) under the upland situation. GCV ranged from 16.36% to 34.52% and 13.17% to 33.08% in lowland and upland conditions, respectively, with PHP 2018 possessing the lowest and GYP^{-1} possessing the highest values in drought condition. Broad sense heritability (H^2) estimates showed that traits such as DF, PHL, and PHP were highly heritable for both years, but the H^2 of HI was high only in lowland situation. The rest of the traits showed medium heritability. The highest H^2 (95.72%) was recorded for DF 2018 in lowland condition followed by DF 2017 (94.52%), and DF 2018 (92.29%) in the upland condition.

Table 5. Descriptive statistics of days to flowering, plant height to leaf, plant height to panicle, above-ground biomass plant^{-1}, grain yield plant^{-1} and harvest index under lowland and upland conditions in 2017 and 2018.

Trait	Year	C	Mean	Min	Max	*CV* (%)	SE_M	*SD*(±)	*GCV* (%)	*PCV* (%)	H^2 (%)	*GA* (%)	$HSD_{0.05}$
Days to flowering	2017	L	104.1	67	174	9.23	0.44	19.73	16.36	18.78	75.91	30.63	50.16
		U	101.9	61	170	4.96	0.51	21.84	20.76	21.36	94.52	42.43	31.41
	2018	L	94.24	67	138	3.93	0.37	16.16	18.6	19.01	95.72	35.38	18.90
		U	96.92	55	155	6.19	0.47	19.64	21.63	22.51	92.29	41.54	29.37
Plant height to leaf (cm)	2017	L	99.67	43.8	187	6.82	0.5	19.43	17.45	18.74	86.64	33.39	35.90
		U	81	33	149.33	8.77	0.43	16.47	19.41	21.35	82.63	29.48	35.28
	2018	L	97.2	42	178.33	7.09	0.41	16.94	18.6	19.9	87.36	34.85	34.95
		U	85.84	45	139.67	9.56	0.42	15	27.03	31.76	72.43	2.27	39.15
Plant height to panicle (cm)	2017	L	97.62	44	191	6.55	0.48	18.59	16.92	18.15	86.93	31.77	33.68
		U	78.2	25.6	147.67	8.68	0.4	15.57	18.91	20.86	82.17	27.66	33.70
		L	93.25	54	185	6.84	0.4	16.39	18.58	19.79	88.15	33.56	32.30
		U	82.41	45	139.67	9.27	0.38	14.1	13.17	16.07	67.12	18.34	36.36
Aboveground biomass plant^{-1} (g)	2018	L	13.93	2.50	48.66	33.58	0.15	6.04	24.81	41.83	35.18	4.23	24.05
		U	7.99	1.22	26.47	28.19	0.1	3.77	32.36	42.28	58.57	4.08	10.48
Grain yield plant^{-1} (g)		L	5.39	0.1	17.36	36.1	0.07	2.82	34.52	50.22	47.26	2.64	10.07
		U	2.83	0.05	10.65	39.18	0.04	1.55	33.08	49.58	44.5	1.29	5.03
Harvest index		L	0.39	0.01	0.69	15.36	0.002	0.11	22.23	27.07	67.43	0.15	0.31
		U	0.36	0.01	0.65	24.6	0.003	0.11	16.22	28.99	31.28	0.07	0.48

Note. C = conditions; *CV* = coefficient of variation; SE_M = standard error of the mean; *SD* = standard deviation; *GCV* = genotypic coefficient of variation; *PCV* = phenotypic coefficient of variation; H^2 = broad sense heritability; *GA* = genetic advance; HSD = Tukey's honestly significant difference; L = lowland; U = upland.

3.4. Correlational Studies of Traits in Lowland and Upland Conditions along with Their RDS Index

Traits observed in lowland and upland conditions with their respective *RDS* were subjected to correlational studies. To determine the association of days to flowering, plant height, dry biomass and harvest index, a correlation matrix was implemented with significance based on *p*-values (Figure 3). There was a mostly high significant relationship in both conditions including RDS. Lowland conditions revealed a strong association among measured traits, followed by upland conditions and RDS. Accessions that had high grain yield under lowland and upland conditions and RDS also had high ABP^{-1}, PHP and HI. PHL showed a high correlation with PHP and DF in lowland and upland conditions, and under RDS except for its negative correlation to DF in RDS. DF, PHL, PHP and ABP^{-1} denoted a negative relationship with HI in both cultivated conditions indicating that less growth duration imparted better yield. Opposite to this, under RDS, higher HI is attributed to taller plant height, higher ABP^{-1}, and earlier days to flowering.

3.5. Relative Drought Stress Susceptibility (RDS) and Drought Resistance Grade (DRG) of Genotypes

Traits under *RDS* were significant at the 5% level of significance from each other, except for PHL, PHP and HI so representing the same letters when applied Tukey's honestly significant difference (HSD; Table 6). GYP^{-1} ranked first due to the highest difference of performance in lowland to upland (37.44%), while DF ranked last having the least value of *RDS* (0.21%). Then, these traits were ranked again according to their practical importance for final selection and accuracy in field condition. Pooled variation, such as variance and standard deviation for *RDS*, was in harmony to those of shared variation by *RDS*, and here GYP^{-1} showed maximum variability (576.29), followed by ABP^{-1} (509.97) and HI (361.97). On the other hand, days to flowering was accredited with the least variation (38.9). As shown in Equation (6), *RDS* was derived by taking an average of RDS_T for each trait to obtain *RDS* ranking for all the genotypes (Table S2). In that ranking, the

North Chinese germplasm was mostly top ranked, followed by Central Chinese and South Chinese germplasms. While the trend of *RDS-I* ranking was somewhat similar to the *RDS* ranking, there were some exceptions as well. Some germplasms of *RDS* ranking degraded in subsequent rankings due to either less dry biomass (ABP^{-1} and GYP^{-1}) or a lower plant height (PHL and PHP). Conversely, some lower-grade genotypes were upgraded because they possessed higher dry biomass or plant height. Nonetheless, on average, the germplasm distribution pattern was similar to the *RDS* ranking.

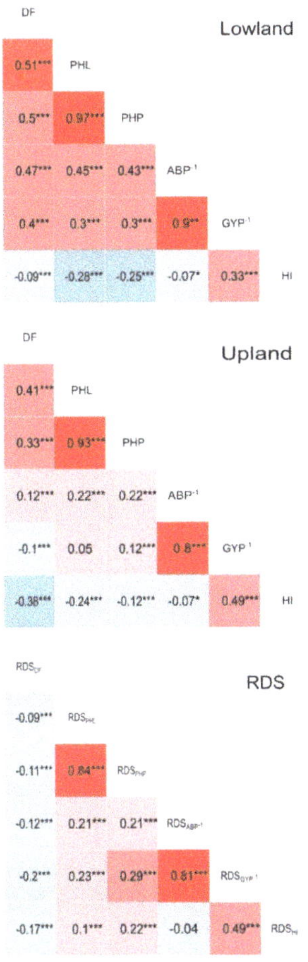

Figure 3. The relationship among days to flowering (DF), plant height to leaf (PHL; cm), plant height to panicle (PHP; cm), aboveground biomass plant^{-1} (ABP^{-1}; g), grain yield plant^{-1} (GYP^{-1}; g) and harvest index (HI) under lowland condition, upland condition and respective relative drought stress susceptibility (*RDS*) index. Note: * $p < 0.05$. ** $p < 0.01$. *** $p < 0.001$.

Drought resistance grade (DRG) was computed from *RDS-I* from the mean of DRG 1 being −28.92, followed by the means of DRGs 1–3 and 3 that were −18.36 and −8.81 individually (Table 7). Each DRG was distinct from each other whether they were compared by concerning means or discrete trait values of *RDS-I*. Tables S4 and S5 indicated that drought resistance of the majority of germplasms (73.79%) from the designated 1118 genotypes was somewhat weak, with DRG scores of 7 (weak), 5–7 (relatively weak) and 7–9 (weaker) corresponding to 29.34%, 22.36% and 22.09% of the population proportion under study.

Table 6. Statistical analysis of *RDS*, derivation of weighted factor (w), range and variation of *RDS* for six traits in 2017, 2018 as well as pooled data.

Variable	Factor	DF	PHL	PHP	ABP^{-1}	GYP^{-1}	HI
RDS	2017	2.12 ± 7.89 ^{NS}	17.52 ± 12.69 ***	18.32 ± 13.31 ***			
	2018	−3.32 ± 8.59 ***	9.53 ± 12.82 ***	10.3 ± 13.21 ***	42.72 ± 27.91 ***	52.45 ± 26.57 ***	16.56 ± 26.44 ***
	Pooled	−0.3 ± 8.25 *** d	13.82 ± 12.76 *** c	14.39 ± 13.26 *** c	42.78 ± 27.91 *** b	52.34 ± 26.57 *** a	16.19 ± 26.44 *** c
Weighted Factor Calculation	Shared *RDS* variation (%)	0.21	9.89	10.29	30.60	37.44	11.58
	Original ranking	6	5	4	2	1	3
	Practical ranking	5	2	4	1	3	6
	Weighted factor (w; %)	9.89	30.60	10.29	37.44	11.58	0.21
Minimum	2017	−56.16	−49.46	−65.50			
	2018	−60.22	−61.90	−80.69	−66.06	−120.60	−111.98
	Pooled	−60.22	−61.90	−80.69	−66.06	−120.60	−110.98
Maximum	2017	53.59	67.28	74.54			
	2018	29.63	46.84	52.42	91.75	98.73	98.60
	Pooled	30.67	57.00	49.96	91.75	98.73	98.60
Range	2017	109.76	116.74	140.03			
	2018	89.84	108.75	133.11	157.81	219.33	210.57
	Pooled	90.88	118.90	130.65	157.81	219.33	209.57
Variance (s^2)	2017	62.21	161.13	177.07			
	2018	73.87	164.43	174.54	779.19	705.76	699.27
	Pooled	68.04	162.78	175.81	779.19	705.76	699.27
Standard Deviation (*SD*)	2017	7.89	12.69	13.31			
	2018	8.59	12.82	13.21	27.91	26.57	26.44
	Pooled	8.25	12.76	13.26	27.91	26.57	26.44

Note: *RDS* = relative drought stress susceptibility; DF = days to flowering; PHL = plant height to leaf; PHP = plant height to panicle; ABP^{-1} = aboveground biomass $plant^{-1}$; GYP^{-1} = grain yield $plant^{-1}$; HI = harvest index; NS = non-significant. Pooled traits having the same letter are not different from each other according to Tukey's honestly significant difference ($HSD_{0.05}$). *** $p < 0.001$. *Shared RDS variation* (%) = $(RDS_T\ SD / \sum RDS_T\ SD) \times 100$, where RDST SD is the standard deviation of RDS for a particular trait.

Table 7. Drought resistance grade (DRG), number of genotypes in each grade, mean of six traits under integrated relative drought stress susceptibility (RDS-I), mean of each grade, clusters, groups, ecotypes and origin of genotypes in each grade.

DRG	G. N.	DF	PHL	PHP	ABP^{-1}	GYP^{-1}	HI	Mean	Cluster	Class	Ecotype	Origin [a]
1	2	0.91	2.08	1.16	−22.53	−10.50	−0.037	−28.92	VII	A	RU	North, Ivory Coast
1–3	5	0.06	−1.78	0.03	−14.44	−2.25	0.025	−18.36	VI, VII	A		North
3	18	0.05	1.49	0.59	−10.72	−0.27	0.035	−8.81	VI, VII, VIII, IX	A, B	TL, IU, TU	North, Centre, South
3–5	67	0.20	2.25	0.96	−2.76	1.73	0.039	2.42	V, VI, VII, VIII, IX	A, B, C	TL, IU, RU, TU	North, Centre, South, South Korea, Japan
5	141	0.12	3.12	1.09	4.70	2.93	0.025	11.98	V, VI, VII, VIII, IX	A, B, C	TL, IU, RU, TU	North, Centre, South, South Korea, Japan
5–7	250	0.04	3.45	1.18	12.64	4.93	0.023	22.27	I, II, III, V, VI, VII, VIII, IX	A, B, C, D	TL, IU, RU, TU	North, Centre, South, South Korea, Japan
7	328	−0.07	4.39	1.48	19.54	7.00	0.030	32.36	I, II, III, IV, V, VI, VII, VIII, IX	A, B, C, D	TL, IU, RU, TU	North, Centre, South, South Korea, Japan
7–9	247	−0.14	5.46	1.93	25.95	8.82	0.041	42.06	I, II, III, IV, V, VIII	B, C, D	TL, IU, RU, TU	North, Centre, South, South Korea, Japan, Brazil
9	60	−0.29	7.90	2.87	29.69	10.02	0.064	50.26	I, IV, V	C, D	TL, RU, TU	North, Centre, South, South Korea, Japan
SD		0.32	2.55	0.76	17.51	5.99	0.03					
SE_M		0.11	0.85	0.25	5.84	2.00	0.01					

Note. G. N. = number of genotypes; DF = days to flowering; PHL = plant height to leaf; PHP = plant height to panicle; ABP^{-1} = aboveground biomass $plant^{-1}$; GYP^{-1} = grain yield $plant^{-1}$; HI = harvest index; *SD* = standard deviation; SE_M = standard error of means; RU = released upland rice variety; TL = traditional lowland rice cultivar; IU = improved upland rice breeding line; TU = traditional upland rice cultivar. [a] North = North China, Centre = Central China, South = South China.

Most of the genotypes from Yunnan and Jiangsu seemed weaker (7–9). In comparison with major "weak" type germplasms, the "strong" genotypes in the top four DRGs (1, 1–3, 3 and 3–5) were less than 10% and the number was even lower than 3% for the top three DRGs particularly. However, there were some selectable genotypes in DRG 5 (moderate) that was 12.61% of the whole population. The potential genotypes with better drought resistance were mainly from Northmost (Heilongjiang, and Jilin) and North Central China (Liaoning, Ningxia and CAU, Beijing). Remarkably, the elite genotypes in DRG 1, 1–3 and 3 were not only identified in upland rice ecotypes, but also lowland rice ecotypes. There were the top 25 genotypes (Table S2) in which we obtained four upland rice genotypes (IRAT109 with DRG 1; originated from the Ivory Coast and released in Jiangsu, Hupiheimangdao with DRG 3; a traditional cultivar in Xinjiang and two promising lines D78 and LB37-13 from CAU, Beijing with DRG 3) and 21 lowland genotypes. Concerning those lowland genotypes, nine were distributed in Heilongjiang with DRG 1 (Wunongqiyi), DRG 1–3 (Hejiang 21, Songjing 17 and Puzhan 6) and DRG 3 (Zaoshuqingsen, Longjing 4, Songzhan 1, Puxuan 10 and Hejiang 20), five originated from Jilin with DRG 1–3 (Jijing 101 and Yanjing 22) and DRG 3 (Jiudao 31, Jijing 88 and Ji 89–60), four from Liaoning with DRG 3 (Fuyou 33, Liaoxing 4, Xinyu 4 and Tiejing 5) and one of each with DRG 3 from Jiangsu (Huainuo 9702) and Yunnan (Yunjing 32). The last one was the traditional lowland cultivar Fanlongdao from Heilongjiang.

3.6. *Agglomerative Hierarchical Clustering (AHC)*

A distance matrix was calculated applying the "Canberra" method using *RDS-I* of the investigated traits, and cluster analysis was performed to identify grouping among genotypes using the "Ward.D2" [37] hierarchical agglomerative method (Figure 4 and Figure S2). The summary of this analysis is shown in Table 8, while detailed information is provided as Table S2. AHC resulted in nine clusters that were categorized subsequently in four classes, A, B, C and D, regarding drought-resistant performance. The nine clusters and four classes revealed different characteristics responsible for drought stress in different germplasms. The *RDS-I* of dry biomass and plant height in clusters VII (ABP^{-1} = 7.43, GYP^{-1} = 2.20, PHL = 1.76 and PHP = 0.58) and VI (ABP^{-1} = 6.18, GYP^{-1} = 5.57, PHL = 2.12 and PHP = 0.81) was lower as compared to the other seven clusters. We concatenated those two clusters in class A (drought-resistant) that possessed drought-resistant genotypes and contained all the genotypes with DRG 1 and 1–3 and the majority of genotypes from DRG 3, 3–5 and 5. On the contrary, class D consisted of clusters III, IV and V whose *RDS-I* of dry biomass (ABP^{-1}; 19.44 to 26.68 and GYP^{-1}; 5.6 to 8.87) and plant height (PHL; 6.22 to 6.42 and PHP 1.96 to 2.35) was higher in comparison with the other six clusters, and therefore attributed as a drought-susceptible class. Between class A and D, class B covered clusters VIII and IX that were just opposite to class C (clusters I and II). The *RDS-I* for dry biomass was lower, and that of plant height was higher in class B, hence drought resistance ranged from moderate (DRG 5) to strong (DRG 3–5 and 3), whereas *RDS-I* for dry biomass was higher, and that of plant height was lower in class C, which resulted in weak drought resistance among most of the genotypes in that class, with mostly DRGs 7, 5–7 and 9.

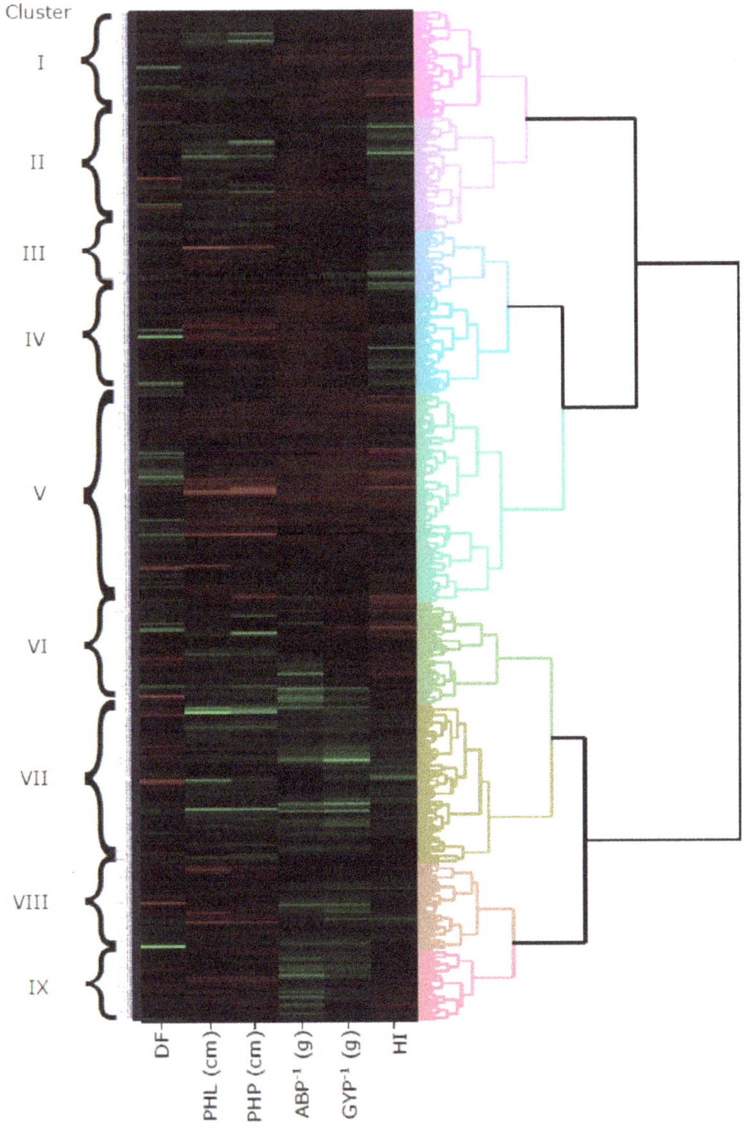

Figure 4. Agglomerative hierarchical cluster analysis (AHC) and heat map of agronomic traits of rice genotypes regarding integrated relative drought stress susceptibility. Note: comparative variable levels correspond to the colour temperature. The colour temperature scheme indicates relative variable levels ranging from minimum (green) to maximum (red) contents of the respective variable. DF = days to flowering; PHL = plant height to leaf; PHP = plant height to panicle; ABP^{-1} = above-ground biomass $plant^{-1}$; GYP^{-1} = grain yield $plant^{-1}$. For high resolution illustration, please refer to Figure S2.

Table 8. Clusters with integrated relative drought stress susceptibility (*RDS-I*) for six traits and their relation to drought resistance grade (DRG).

Cluster	G. N.	Statistic	DF	PHL	PHP	ABP^{-1}	GYP^{-1}	HI	Total
VI	113	Mean	−0.14	2.12	0.81	6.18	5.57	0.08	14.62
		Min	−2.42	−6.51	−2.85	−23.28	−1.91	0.03	−22.77
		Max	2.6	5.5	2.14	16.3	11	0.18	31.66
VII	176	Mean	0.21	1.76	0.58	7.42	2.20	0.00	12.18
		Min	−1.04	−13.18	−2.43	−24.71	−13.99	−0.11	−35.85
		Max	1.9	8.42	2.58	17.39	6.07	0.09	32.91
VIII	97	Mean	0.00	6.13	2.07	8.88	3.05	0.01	20.13
		Min	−3.41	2.41	0.86	−10.79	−2.29	−0.07	−5.2
		Max	1.93	17.44	4.36	15.88	5.87	0.03	37.87
IX	76	Mean	0.10	5.83	2.02	1.56	3.85	0.06	13.42
		Min	−1.15	3.64	0.86	−19.81	−0.09	0.02	−10.82
		Max	3.04	12.81	4.71	12.92	6.62	0.13	30.25
I	117	Mean	−0.05	2.69	1.04	22.90	8.51	0.06	35.16
		Min	−1.96	−2.58	−1.69	16.36	6.28	−0.19	20.16
		Max	2.33	5.91	2.01	33.22	11.36	0.18	48.45
II	126	Mean	0.06	2.00	0.65	22.31	6.60	−0.01	31.61
		Min	−1.55	−3.94	−1.90	16.00	1.99	−0.17	20.18
		Max	2.09	6.91	1.96	31.61	10.00	0.03	45.33
III	69	Mean	−0.08	6.42	1.96	19.44	5.60	−0.02	33.33
		Min	−1.47	4.44	0.4	16.03	0.95	−0.14	24.14
		Max	1.58	16.44	4.02	24.27	7.22	0.03	47.79
IV	113	Mean	−0.08	6.23	2.03	26.68	8.24	0.00	43.10
		Min	−3.3	4.21	−0.14	20.85	6.33	−0.12	34.06
		Max	2.44	10.89	3.8	34.31	10.6	0.05	58.52
V	231	Mean	−0.22	6.22	2.35	22.07	8.87	0.08	39.38
		Min	−4.11	2.69	1.19	−16.46	6.17	0.04	4.99
		Max	2.07	14.2	5.15	33.48	11.45	0.2	55.38
SD			0.13	2.14	0.71	9.18	2.44	0.04	
SE_M			0.04	0.71	0.24	3.06	0.81	0.01	

RDS-I Performance	Class	DRG	with G. (%)
Dry biomass (ABP^{-1} and GYP^{-1}) and plant height (PHL and PHP) were lowest or lower, thus genotypes have strong drought resistance (all the genotypes in top DRGs 1 and 1–3 and the majority of them from DRGs 3, 3–5 and 5 fall in these clusters)	A	1	100
		1–3	100
		3	77.78
		3–5	68.66
		5	65.25
		5–7 [a]	47.2
Comparatively, dry biomass was lower but plant height was higher. Hence, drought resistance of genotypes varied with moderate (DRG 5) to somewhat strong resistance (DRG 3–5 and 3)	B	3	22.22
		3–5	29.85
		5	34.04
		5–7	33.2
		7	5.18
		7–9	0.4
Contrary to class B above, dry biomass and plant height were higher and lower, respectively. Therefore, drought resistance of the majority of genotypes was weak (DRG 7)	C	5–7	13.6
		7	46.04
		7–9	22.67
		9	3.33
Both dry biomass and plant height were highest or higher (highly drought susceptible); thus, drought resistance of genotypes was weak to weakest (DRG 7 to 9)	D	3–5	1.49
		5	0.71
		5–7	6
		7	45.12
		7–9	76.92
		9	96.67

G. N. = number of genotypes; DF = days to flowering; PHL = plant height to leaf; PHP = plant height to panicle; ABP^{-1} = aboveground biomass $plant^{-1}$; GYP^{-1} = grain yield $plant^{-1}$; HI = harvest index; G. = genotypes; SD = standard deviation; SEM = standard error of means. a DRG 7 was skipped in class A to adjust the table.

Table S2 displayed that there were different distribution proportions of genotypes in A and B in terms of regions. Half of the germplasms in those classes distributed in Northmost (Heilongjiang, Jilin, Xinjiang, and Inner Mongolia), one third in North Central (Liaoning, Ningxia, Hebei, Beijing, Shandong, Henan and Tianjin) and exotic regions (Japan and Korea), and one fourth originated from North–South and South China. It was concluded that the majority of the genotypes from China, Korea, and Japan, except for the Northeast and Northwest of China, fall into classes C and D. Table 8 and Table S4 show that the results from the analysis of AHC and DRG for genotypes were highly coherent. Further, out of 25 elite genotypes mentioned in the subheading, 3.5, 13 and 8 belonged to cluster VII and VI (class A), while 4 falls in class B. Concerning the control group, while susceptible control 297-28 was categorized in cluster V, those of resistant controls such as B_1 and Handao 277 were classified in cluster VII.

3.7. Integrated Elite Genotypes Selection

By integrated screening according to the outcomes based on analysis of DRG, AHC and the performance of the six traits, particularly absolute dry biomass (ABP^{-1} and GYP^{-1}) under drought stress, we achieved 42 top genotypes excluding controls from 1118 genotypes (Table S6). All the listed genotypes except Handao 277 and 297-28 exhibited a DRG of 1 to 5. The number of genotypes in DRGs with scores of 1, 1–3, 3, 3–5 and 5 were 1, 3, 6, 17 and 15, respectively. In the elite selection, 23 and 10 screened genotypes were from clusters VII and VI (class A), while there were only five and four genotypes from clusters VIII and IX (class B), respectively. Further, the means of ABP^{-1} and GYP^{-1} under the upland condition for 42 prime genotypes were 13.82 g (9.14 g to 26.47 g) and 5.32 g (3.49 g to 10.08 g) correspondingly that were noticeably higher than those of 7.28 g (1.22 g to 26.47 g) and 2.51 g (0.05 g to 10.08 g) the designated germplasm under study. There were seven improved upland rice lines (D78, LB37-13, NSU77, Handao 385, Handao 306, SF 83, and HF_6-65-119 from CAU), three upland released varieties (Liaojing 27 and Hanfeng 8 from Liaoning, and IRAT109 originated from the Ivory Coast and released in Jiangsu), and three traditional lowland cultivars (Hongmaodao and Weiguo 7 form Liaoning and Xiaohongbandao from Ningxia). However, a maximum of them (29) were released breeding lowland rice varieties of which 24 were from Northmost (Heilongjiang and Jilin) and North Central (Liaoning and Ningxia) China, and the rest of them originated from Beijing, Yunnan, Korea, and Japan.

Furthermore, to visualize the relationship of grain yield to growth duration under drought stress, a scatter plot of those 45 genotypes (controls involved) was drawn (Figure 5) with yield plant^{-1} on x-axis and days to flowering on the y-axis. Two groups among those genotypes were marked for better yield and fewer days to flowering. IRAT109 and Ningzi 786 were elite accessions while Longjing 12, Xaiohongbandao, Ji 85-34, Hongmaodao, Yangjing 22, Jingjing 106, Liajing 27, Songjing 7, Luyu, Songjing 22, Jijing 101 and 2014H020 were attributed with better yield and shorter growth duration. Selected genotypes distinctly expressed better performance from controls particularly from B_1 and 297-28. Susceptible control positioned at right bottom of the graph with more number of DF together with less GYP^{-1}. However, there was a slightly negative relationship between DF and GYP^{-1} which was somewhat according to the correlation among the same traits when it was calculated for the whole germplasm.

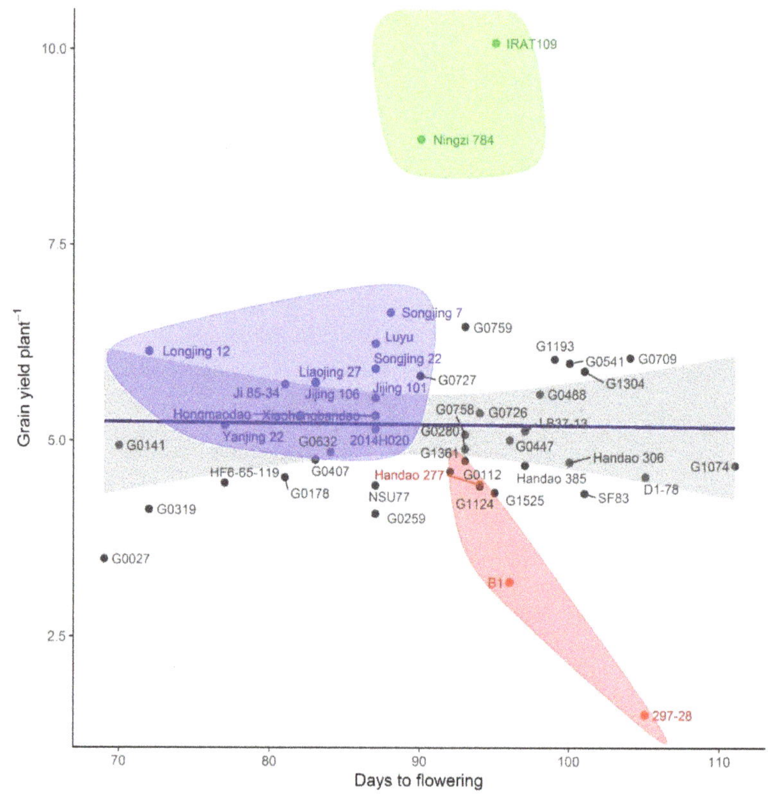

Figure 5. Scatter diagram based on grain yield plant^{-1} (g) versus days to flowering distribution of 45 genotypes including controls (indicated within red circle). Note: two genotypes in the green circle were marked as prime genotypes. The genotypes in the blue circle were characterized by better grain yield along with the shorter growth duration. All encircled genotypes are presented with their names, whereas most of the remaining entries are symbolized by their accession codes.

4. Discussion

4.1. Why Field Identification?

Drought stress is considered the most important constraint in rice production in many rice-growing areas of China [38]. This calls for screening of advanced breeding lines and varieties to provide the farmers with drought-tolerant strains. For this, a single laboratory experiment was questionable and there was a need to conduct multiple stress trials to test accessions against water scarcity [39]. Therefore, 2030 accessions from various regions of China, South Korea, Japan, the Ivory Coast and Brazil were evaluated against drought. The movement behind consideration of large-scale germplasm was that it served as the foundation of a rice breeding program being the source of important traits necessary for improving and developing new breeds of rice varieties [40]. As drought traits were controlled by multiple quantitative trait loci (QTLs) [41], traits such as days to flowering (DF), plant height to leaf and panicle (PHL and PHP), aboveground biomass plant^{-1} (ABP^{-1}), grain yield plant^{-1} (GYP^{-1}) and harvest index (HI) were used to exhibit the mechanism of drought tolerance under field conditions, otherwise it was complex due to variations in plant phenology. The utilization of agro-morphological traits was the most common approach conducted to estimate relationships between genotypes [42]. This approach was already employed to assess diversity on ancestral lines of improved rice varieties in the Philippines [40], the indigenous rice in Yunnan, China [43] and the rice landraces in Nepal [42]. There were watered and water-stressed conditions in the field and studied traits responded differently to both conditions with the influence of all the environmental factors which could not be possible in greenhouse, growth chamber or laboratory experiments. Owing to an experiment in natural drought conditions, this would be beneficial directly for farmers, breeders, and scientists.

4.2. Augmented Randomized Complete Block Design (ARCBD) Application in Rice Phenotyping

Although augmented randomized complete block design (ARCBD) is widely used for the phenotyping of large populations under lowland and upland conditions in maize crops [27–30], in this article we report for the first time on the application of the ARCBD analysis under the same two conditions for rice germplasm phenotyping. Significant differences were observed for genotypes, conditions, years and all their possible interactions (Tables 3 and 4), and obtained results, particularly those for controls were consistent with previous studies in the same station under randomized complete block design (RCBD) [24,44]. However, 297-28, a susceptible control with a long growth duration, never reached maturity in some blocks; therefore, we skipped the days to maturity trait for further analysis. That was the only disadvantage of ARCB as the variation estimate of genotypes solely depends on controls.

4.3. Traits Investigated under Drought Stress

Much of the initial efforts to improve GYP^{-1} under drought focused on the improvement of secondary traits such as DF, PHL, PHP, ABP^{-1}, and HI as studied by Mahalle et al. (2020) [45]. In our study, DF came out with the highest broad sense heritability (H^2) of 94.59 and 93.83 under lowland and upland conditions one-to-one. The trait was selected as a secondary trait because of its high relationship with grain yield [46], convenience in measurement and a highly effective way to improve drought adaptation under terminal stress [47]. Piveta et al. (2020) also measured PH and tillers number to observe drought in rice [48]. Growth duration or days to maturity was another phenologically important trait like DF but it was not used for the final selection of accessions due to limited comparative optimum temperature particularly for rice, and the weather used to become cold earlier in our research station area. We also ignored spike fertility for the same reason. Plant height to leaf and panicle were positively associated (0.21) with ABP^{-1} under *RDS*, making them suitable traits for breeding that was directly associated with grain yield with a significant correlation of 0.8. However, high ABP^{-1} did not always guarantee high grain yield [49]. Nevertheless, to attain GYP^{-1} above 4.5 g under drought stress, it was necessary to accumulate a total ABP^{-1} of at least 9 g. On the other hand, plant height was negatively related to the HI under lowland and upland conditions [50]. However, there was a minimum height below which yield limitation was evident [51]. In a recent study, Lanna et al. (2020) concluded that different secondary traits such as PH and ABP^{-1} should be considered along with grain yield for an effective breeding approach [52]. Under drought, the H^2 for some secondary traits remained high [53], which was also supported by our results by showing H^2 of 93.83, 80.04, 76.94 and 58.57 for DF, PHL, PHP and ABP^{-1}, respectively. This is also endorsed by genetic advance (*GA*) of DF (lowland: 50.29, upland: 56.59), PHL (lowland: 45.27, upland: 31.44), and PHP (lowland: 41.93, upland: 28.25) that is higher than *GA* of ABP^{-1} (lowland: 4.23, upland: 4.08), GYP^{-1} (lowland: 2.64, upland: 1.29) and HI (lowland: 0.15, upland: 0.07). Therefore, using indirect selection based on secondary traits became an attractive strategy in the study.

In our germplasm, some genotypes under measured traits outperformed in the upland condition as compared to the lowland counterparts, particularly DF, by exhibiting shorter duration in drought stress. This could be attributed to the germplasm proportion of North China (947 accessions) which was very sensitive to a slight rise in temperature at the vegetative stage. Secondly, the soil temperature of the water-deficit condition was higher than the watered condition, which also influenced plant growth [54]. Lastly, but not least, the case was credible when investigated genotypes were traditional upland rice because, from many years of field research experience in Shangzhuang, we had already observed taller plant height and higher aboveground biomass and grain yield under drought conditions than normal lowland. A similar phenomenon of higher grain yield under drought stress was expressed by two *indica* accessions in a trial run by IRRI [49].

4.4. Integrated Relative Drought Stress Susceptibility (RDS-I) Derivation and Estimation

Typically, selection should target genotypes with relatively high productivity under lowland and upland conditions. Hence, there was a need to determine the relative drought stress susceptibility (*RDS*; Table 6 and Table S1) for their improved adaptability to contrasting conditions. The *RDS* for GYP^{-1} ranged between −62.47% to 73.11%, discriminating tolerant and susceptible genotypes. Huang et al. (2018) reported the same drought index but for ABP^{-1} with a narrow range of −52.1% to −8.1% as he just assessed six genotypes. Statistical ranking of each trait under *RDS* was based on shared variation, i.e., the higher the variation exhibited by a parameter, the better it would be in the ranking. Instead of that, the practical ranking was based on actual field conditions faced during observation of traits. For instance, statistically, GYP^{-1} ranked first but practically it was placed in the third position due to an earlier decrease in temperature in the period of grain filling stage of rice in Beijing, as practically variation in GYP^{-1} was interfered by an early cool season. Further, several earlier studies reported low selection efficiency for grain yield under drought stress [55–57]. However, ABP^{-1} and PHL were not much affected by that, so ranked first and second, respectively. As discussed earlier, phenology was affected by adaptation; therefore, DF achieved the fifth position and HI was last due to its indirect computation from ABP^{-1} and GYP^{-1}. As the weighted factor (w) was concerned, the shared *RDS* variation for each trait was assigned to w in descending order while keeping the focus of practical ranking. For example, practical ranks of 1, 2 and 3 had values of 37.44, 30.60 and 11.58, respectively, from a shared *RDS* variation (Table 6). That weighed factor of each trait was multiplied directly to the RDS_T to get RDS_{Tw} value as shown in Equation (7). Finally, RDS_{Tw} was summed up to obtain *RDS-I* (8) to rank the genotypes with accuracy. That ranking was true due to adjustment with the actual field conditions.

4.5. Conclusions

For the first time, a large-scale germplasm of 2030 rice (*Oryza sativa* L. ssp. *japonica*) genotypes was subjected to augmented randomized complete block design (ARCBD) under lowland (irrigated) and upland (drought) conditions, which resulted in significant differences for days to flowering (DF), plant height to leaf (PHL), plant height to panicle (PHP), aboveground biomass $plant^{-1}$ (ABP^{-1}), grain yield $plant^{-1}$ (GYP^{-1}) and harvest index (HI). Here, we made an integrated elite selection of 42 genotypes based on integrated relative drought stress susceptibility (*RDS-I*), drought-resistance grade (DRG) and agglomerative hierarchical cluster (AHC) analysis. Findings from these experiments also elucidated that North China was a rich source to find drought-resistant rice germplasms. Further, both ecotypes such as lowland and upland should be researched for genotypes that could withstand the stress. The mean DF of the elite 42 genotypes was 89.38, which was remarkably less than that of the germplasm, which was 100.17. Further, elite genotypes resulted in doubled ABP^{-1} and GYP^{-1} mean performance of 13.54 g and 5.21 g, correspondingly as compared to mean performance of the germplasm, which was 7.99 g and 2.83 g, respectively. Similarly, the mean HI of these 42 genotypes and the investigated germplasm was 0.39 and 0.36, respectively, depicting significant differences. Therefore, these elite genotypes could be used in future breeding programs, and also if a few of them can be grown directly in farmers' fields. Additionally, 12 genotypes namely Longjing 12, Songjing7, Songjing 22, Jijing 101, Jijing 106, Yanjing 22, Ji 85-34, Liaojing 27, Xiaohongband, 2014H020 and Hongmaodao were identified with better yield with less growth duration, so are suitable for cultivation in areas with terminal drought stress. Further narrowing down the selection, IRAT109 and Ningzi 784 are highly recommended for drought resistance with better yield.

Supplementary Materials: The following are available online at https://www.mdpi.com/article/10.3390/agronomy11091740/s1, Figure S1: Quantile-Quantile plots for investigated six traits under lowland (irrigated) and upland (drought) conditions in 2017 and 2018 at Shangzhuang Agricultural Research Station, China Agricultural University, Beijing, China.; Figure S2: Agglomerative hierarchical cluster analysis (AHC) and heat map of agronomic traits of rice genotypes regarding integrated relative drought stress susceptibility.; Table S1: Basic physical property parameters of soil at different depths in Shangzhuang Agricultural Research Station, Beijing.; Table S2. DRG and Clusters results, and *RDS* and *RDS-I* ranking based on RDS_T and RDS_{Tw} performance of the six traits, respectively. Footnote at the end of table., Table S3: All the plant and soil traits measured in 2017 and 2018 for the field experiments under lowland and upland conditions.; Table S4: Drought resistance grades (DRG) along with the genotypes distribution among different ecotypes, regions, and clusters.; Table S5: Genotypes distribution of drought resistance grade (DRG) in different regions of China and other countries.; Table S6: The elite 42 genotypes selected from the designated germplasm of 1118 genotypes, their background information, performance in six traits, RDS ranking, *RDS-I* ranking, drought resistance grade (DRG) and cluster.

Author Contributions: Conceptualization, H.W.; methodology, M.S.A., B.W. and H.W.; software, M.S.A.; validation, M.S.A. and H.W.; formal analysis, M.S.A.; investigation, M.S.A., B.W. and H.W.; resources, M.S.A., B.W. and H.W.; data curation, H.W. and M.S.A.; writing—original draft preparation, H.W. and M.S.A.; writing—review and editing, H.W. and M.S.A.; visualization, M.S.A.; supervision, D.K. and H.W.; project administration, M.S.A., B.W. and H.W.; funding acquisition, H.W. All authors have read and agreed to the published version of the manuscript.

Funding: This work was supported by the National Key Research and Development Program of China (grant numbers 2016YFD0100101-09).

Data Availability Statement: Relevant data are already provided in the article and in Supplementary Materials.

Acknowledgments: We are thankful to the Institute of Crop Sciences, Chinese Academy of Agricultural Sciences, Beijing for providing seeds of 2000 *japonica* rice genotypes. We are also thankful to John Kafwambira for his help with data entry.

Conflicts of Interest: The authors declare no conflict of interest. The funders had no role in the design of the study; in the collection, analyses, or interpretation of data; in the writing of the manuscript; or in the decision to publish the results.

References

1. Estrela, T.; Vargas, E. Drought management plans in the European Union. The case of Spain. *Water Resour. Manag.* **2012**, *26*, 1537–1553. [CrossRef]
2. Liu, J.; Diamond, J. China's environment in a globalizing world. *Nature* **2006**, *435*, 1179–1186. [CrossRef]
3. Rice around the world. In *Rice Almanac*; Global Rice Science Partnership, International Rice Research Institute: Los Banos, Philippines, 2013; pp. 79–260. ISBN 978-9712203008. Available online: http://books.irri.org/9789712203008_content.pdf (accessed on 16 August 2021).
4. Belder, P.; Spiertz, J.H.J.; Bouman, B.A.M.; Lu, G.; Tuong, T.P. Nitrogen economy and water productivity of lowland rice under water-saving irrigation. *Field Crop. Res.* **2005**, *93*, 169–185. [CrossRef]
5. Jockson, M.T. *International Rice Genebank Operations Manual*; International Rice Research Institute: Los Banos, Philippines, 2000.
6. Kumar, A.; Bernier, J.; Verulkar, S.; Lafitte, H.R.; Atlin, G.N. Breeding for drought tolerance: Direct selection for yield, response to selection and use of drought-tolerant donors in upland and lowland-adapted populations. *Field Crop. Res.* **2008**, *107*, 221–231. [CrossRef]
7. Jongdee, B.; Pantuwan, G.; Fukai, S.; Fischer, K. Improving drought tolerance in rainfed lowland rice: An example from Thailand. *Agric. Water Manag.* **2006**, *80*, 225–240. [CrossRef]
8. Lafitte, H.R.; Li, Z.K.; Vijayakumar, C.H.M.; Gao, Y.M.; Shi, Y.; Xu, J.L.; Fu, B.Y.; Yu, S.B.; Ali, A.J.; Domingo, J.; et al. Improvement of rice drought tolerance through backcross breeding: Evaluation of donors and selection in drought nurseries. *Field Crop. Res.* **2006**, *97*, 77–86. [CrossRef]
9. Bernier, J.; Atlin, G.N.; Serraj, R.; Kumar, A.; Spaner, D. Breeding upland rice for drought resistance. *J. Sci. Food Agric.* **2008**, *88*, 927–939. [CrossRef]

10. Luo, L.J. Breeding for water-saving and drought-resistance rice (WDR) in China. *J. Exp. Bot.* **2010**, *61*, 3509–3517. [CrossRef]
11. Vikram, P.; Swamy, B.P.M.; Dixit, S.; Ahmed, H.U.; Teresa Sta Cruz, M.; Singh, A.K.; Kumar, A. *qDTYI*$_{1.1}$, a major QTL for rice grain yield under reproductive-stage drought stress with a consistent effect in multiple elite genetic backgrounds. *BMC Genet.* **2011**, *12*, 89. [CrossRef]
12. Liu, S.; Wang, X.; Wang, H.; Xin, H.; Yang, X.; Yan, J.; Li, J.; Tran, L.-S.P.; Shinozaki, K.; Yamaguchi-Shinozaki, K.; et al. Genome-wide analysis of *ZmDREB* genes and theirassociation with natural variation in drought tolerance at seedling stage of *Zea mays* L. *PLoS Genet.* **2013**, *9*, e1003790. [CrossRef]
13. Huang, Q.; Wang, Y. Overexpression of taNAC2D displays opposite responses to abiotic stresses between seedling and mature stage of transgenic Arabidopsis. *Front. Plant Sci.* **2016**, *7*, 1–13. [CrossRef]
14. Shamsudin, N.A.A.; Swamy, B.P.M.M.; Ratnam, W.; Cruz, M.T.S.; Sandhu, N.; Raman, A.K.; Kumar, A.; Abd, N.; Shamsudin, A.; Swamy, B.P.M.M.; et al. Pyramiding of drought yield QTLs into a high quality Malaysian rice cultivar *MRQ74* improves yield under reproductive stage drought. *Rice* **2016**, *9*, 21. [CrossRef]
15. Oladosu, Y.; Rafii, M.Y.; Samuel, C.; Fatai, A.; Magaji, U.; Kareem, I.; Kamarudin, Z.S.; Muhammad, I.; Kolapo, K. Drought resistance in rice from conventional to molecular breeding: A review. *Int. J. Mol. Sci.* **2019**, *20*, 3519. [CrossRef]
16. Monneveux, P.; Jing, R.; Misra, S.C. Phenotyping for drought adaptation in wheat using physiological traits. *Front. Physiol.* **2012**, *3*, 429. [CrossRef]
17. Passioura, J.B. Phenotyping for drought tolerance in grain crops: When is it useful to breeders? *Funct. Plant Biol.* **2012**, *39*, 851–859. [CrossRef]
18. Lawas, L.M.F.; Shi, W.; Yoshimoto, M.; Hasegawa, T.; Hincha, D.K.; Zuther, E.; Jagadish, S.V.K. Combined drought and heat stress impact during flowering and grain filling in contrasting rice cultivars grown under field conditions. *Field Crop. Res.* **2018**, *229*, 66–77. [CrossRef]
19. Bolaños, J.; Edmeades, G.O. Eight cycles of selection for drought tolerance in lowland tropical maize. II. Responses in reproductive behavior. *Field Crop. Res.* **1993**, *31*, 253–268. [CrossRef]
20. Zhao, Y.; Karypis, G. Evaluation of hierarchical clustering algorithms for document datasets. In Proceedings of the Eleventh International Conference on Information and Knowledge Management, McLean, VA, USA, 4–9 November 2002; Association for Computing Machinery: New York, NY, USA, 2002; pp. 515–524.
21. Kumar, S.S.; Dwivedi, S.K.; Basu, S.; Kumar, G.; Mishra, J.S.; Koley, T.K.; Rao, K.K.; Choudhary, A.K.; Mondal, S.; Kumar, S.S.; et al. Anatomical, agro-morphological and physiological changes in rice under cumulative and stage specific drought conditions prevailed in eastern region of India. *Field Crop. Res.* **2020**, *245*, 107658. [CrossRef]
22. Tardieu, F. Any trait or trait-related allele can confer drought tolerance: Just design the right drought scenario. *J. Exp. Bot.* **2012**, *63*, 25–31. [CrossRef] [PubMed]
23. Ahmad, M.S.; Wu, B.; Wang, H.; Kang, D. Field screening of rice germplasm (*Oryza sativa* L. ssp. japonica) based on days to flowering for drought escape. *Plants* **2020**, *9*, 609. [CrossRef]
24. Zu, X.; Lu, Y.; Wang, Q.; Chu, P.; Miao, W.; Wang, H.; La, H. A new method for evaluating the drought tolerance of upland rice cultivars. *Crop. J.* **2017**, *5*, 488–498. [CrossRef]
25. Xu, J.; Zhang, L.; Li, T.; Wu, H.; Cui, H. Breeding report on the new upland rice variety of Jiaohan No. 1 with high yield and good quality. *J. Henan Agric. Sci.* **2011**, *40*, 64–65.
26. Gong, X.; Liu, X.; Pan, Q.; Mi, G.; Chen, F.; Yuan, L. Combined physiological, transcriptome and genetic analysis reveals a molecular network of nitrogen remobilization in maize. *J. Exp. Bot.* **2020**, *71*, 5061–5073. [CrossRef]
27. Buckler, E.S.; Holland, J.B.; Bradbury, P.J.; Acharya, C.B.; Brown, P.J.; Browne, C.; Ersoz, E.; Flint-Garcia, S.; Garcia, A.; Glaubitz, J.C.; et al. The genetic architecture of maize flowering time. *Science* **2009**, *325*, 714–718. [CrossRef]
28. Kump, K.L.; Bradbury, P.J.; Wisser, R.J.; Buckler, E.S.; Belcher, A.R.; Oropeza-Rosas, M.A.; Zwonitzer, J.C.; Kresovich, S.; McMullen, M.D.; Ware, D.; et al. Genome-wide association study of quantitative resistance to southern leaf blight in the maize nested association mapping population. *Nat. Genet.* **2011**, *43*, 163–168. [CrossRef]
29. Tian, F.; Bradbury, P.J.; Brown, P.J.; Hung, H.; Sun, Q.; Flint-Garcia, S.; Rocheford, T.R.; Mcmullen, M.D.; Holland, J.B.; Buckler, E.S. Genome-wide association study of leaf architecture in the maize nested association mapping population. *Nat. Genet.* **2011**, *43*, 159–162. [CrossRef]
30. Huang, C.; Chen, Q.; Xu, G.; Xu, D.; Tian, J.; Tian, F. Identification and fine mapping of quantitative trait loci for the number of vascular bundle in maize stem. *J. Integr. Plant Biol.* **2016**, *58*, 81–90. [CrossRef]
31. R Core Team. *R: A Language and Environment for Statistical Computing*; R Foundation for Statistical Computing: Vienna, Austria, 2019; Volume 2.
32. De Mendiburu, F. Agricolae: Statistical Procedures for Agricultural Research. 2020.
33. Aravind, J.; Mukesh, S.S.; Wankhede, D.P.; Kaur, V. AugmentedRCBD: Analysis of Augmented Randomized Complete Block Designs. 2020.
34. IRRI. *Standard Evaluation System for Rice*, 5th ed.; International Rice Research Institute: Manila, Philippines, 2013; ISBN 9781612099897.
35. Wickham, H. *ggplot2: Elegant Graphics for Data Analysis*, 2nd ed.; Springer: New York, NY, USA, 2016; ISBN 9783319242750.
36. Kato, Y.; Katsura, K. Rice adaptation to aerobic soils: Physiological considerations and implications for agronomy. *Plant Prod. Sci.* **2014**, *17*, 1–12. [CrossRef]

37. Murtagh, F.; Legendre, P. Ward's hierarchical agglomerative clustering method: Which algorithms implement Ward's criterion? *J. Classif.* **2014**, *31*, 274–295. [CrossRef]
38. Zhang, Q. Strategies for developing green super rice. *Proc. Natl. Acad. Sci. USA* **2007**, *95*, 1663–1668. [CrossRef]
39. Blum, A. Osmotic adjustment is a prime drought stress adaptive engine in support of plant production. *Plant Cell Environ.* **2017**, *40*, 4–10. [CrossRef]
40. Caldo, R.; Sebastian, L.; Hernandez, J. Morphology-based genetic diversity analysis of ancestral lines of Philippine rice cultivars. *Philipp. J. Crop. Sci.* **1996**, *21*, 86–92.
41. Fleury, D.; Jefferies, S.; Kuchel, H.; Langridge, P. Genetic and genomic tools to improve drought tolerance in wheat. *J. Exp. Bot.* **2010**, *61*, 3211–3222. [CrossRef] [PubMed]
42. Bajracharya, J.; Steele, K.A.; Jarvis, D.I.; Sthapit, B.R.; Witcombe, J.R. Rice landrace diversity in Nepal: Variability of agro-morphological traits and SSR markers in landraces from a high-altitude site. *Field Crop. Res.* **2006**, *95*, 327–335. [CrossRef]
43. Yawen, Z.; Shiquan, S.; Zichao, L.; Zhongyi, Y.; Xiangkun, W.; Hongliang, Z.; Guosong, W. Ecogeographic and genetic diversity based on morphological characters of indigenous rice (*Oryza sativa* L.) in Yunnan, China. *Genet. Resour. Crop. Evol.* **2003**, *50*, 567–577. [CrossRef]
44. Huang, M.; Xu, Y.; Wang, H. Field identification of morphological and physiological traits in two special mutants with strong tolerance and high sensitivity to drought stress in upland rice (*Oryza sativa* L.). *J. Integr. Agric.* **2019**, *18*, 970–981. [CrossRef]
45. Mahalle, M.D.; Dey, P.C.; Chetia, S.K.; Baruah, A.R.; Ahmed, T.; Sarma, R.N.; Kaldate, R.C.; Kumar, A.; Singh, S.K.; Modi, M.K. Association mapping for yield traits under drought stress in Autumn rice germplasm collection of Assam. *J. Plant Biochem. Biotechnol.* **2020**, *30*, 26–36. [CrossRef]
46. Bänziger, M.; Edmeades, G.O.; Beck, D.; Bellon, M. Secondary traits. In *Breeding for Drought and Nitrogen Stress Tolerance in Maize: From Theory to Practice*; CIMMYT: Texcoco, Mexico, 2000; pp. 39–46. ISBN 970-648463.
47. Araus, J.L.; Slafer, G.A.; Reynolds, M.P. Plant breeding and drought in C_3 cereals: What should we breed for? *Ann. Bot.* **2002**, *89*, 925–940. [CrossRef] [PubMed]
48. Piveta, L.B.; Roma-Burgos, N.; Noldin, J.A.; Viana, V.E.; de Oliveira, C.; Lamego, F.P.; Avila, L.A. de Molecular and physiological responses of rice and weedy rice to heat and drought stress. *Agriculture* **2020**, *11*, 9. [CrossRef]
49. Torres, R.O.; Mcnally, K.L.; Cruz, C.V.; Serraj, R.; Henry, A. Screening of rice genebank germplasm for yield and selection of new drought tolerance donors. *Field Crop. Res.* **2013**, *147*, 12–22. [CrossRef]
50. Royo, C.; Álvaro, F.; Martos, V.; Ramdani, A.; Isidro, J.; Villegas, D.; García Del Moral, L.F. Genetic changes in durum wheat yield components and associated traits in Italian and Spanish varieties during the 20th century. *Euphytica* **2007**, *155*, 259–270. [CrossRef]
51. Slafer, G.A.; Araus, J.L.J.L.; Royo, C.; García, L.F.; Moral, D.E.L.; Moral, L.F.G. Del Promising eco-physiological traits for genetic improvement of cereal yields in Mediterranean environments. *Ann. Appl. Biol.* **2005**, *146*, 61–70. [CrossRef]
52. Lanna, A.C.; Coelho, G.R.C.; Moreira, A.S.; Terra, T.G.R.; Brondani, C.; Saraiva, G.R.; Lemos, F.D.S.; Guimarães, P.H.R.; Morais, O.P.; Vianello, R.P. Upland rice: Phenotypic diversity for drought tolerance. *Sci. Agric.* **2020**, *78*, 1–14.
53. Bolaños, J.; Edmeades, G.O. The importance of the anthesis-silking interval in breeding for drought tolerance in tropical maize. *Field Crop. Res.* **1996**, *48*, 65–80. [CrossRef]
54. Arai-Sanoh, Y.; Ishimaru, T.; Ohsumi, A.; Kondo, M. Effects of soil temperature on growth and root function in rice. *Plant Prod. Sci.* **2010**, *13*, 235–242. [CrossRef]
55. Rosielle, A.A.; Hamblin, J. Theoretical aspects of selection for yield in stress and non-stress environment. *Crop. Sci.* **1981**, *21*, 943–946. [CrossRef]
56. Blum, A. Plant breeding and yield stability. In *Plant Breeding For Stress Environments*; CRC Press: Boca Raton, FL, USA, 1988; pp. 15–42. ISBN 9781315896618.
57. Edmeades, G.O.; Bolaños, J.; Lafitte, H.R.; Rajaram, S.; Pfeiffer, W.; Fischer, R.A. Traditional approaches to breeding for drought resistance in cereals. In *Drought Resistance in Cereals*; Baker, F.W.G., Ed.; CAB International: Wallingford, UK, 1989; pp. 27–52.

agronomy

MDPI

Article

Phenotypic Characterization of Arabidopsis Ascorbate and Glutathione Deficient Mutants under Abiotic Stresses

Minh Thi Thanh Hoang [1,2,*,†], Mai Thi Anh Doan [1,2,†], Thuong Nguyen [1,2,†], Dong-Phuong Tra [1,2], Thanh Nguyen Chu [1,2], Thi Phuong Thao Dang [2] and Phuong Ngo Diem Quach [1,2]

1 Faculty of Biology and Biotechnology, University of Science, Ho Chi Minh City 700000, Vietnam; dtanhmai@gmail.com (M.T.A.D.); nguyenthuongnt1101@gmail.com (T.N.); tdphuong@hcmus.edu.vn (D.-P.T.); cnthanh14@gmail.com (T.N.C.); qndphuong@hcmus.edu.vn (P.N.D.Q.)

2 Laboratory of Molecular Biotechnology, University of Science, Vietnam National University, Ho Chi Minh City 700000, Vietnam; dtpthao@hcmus.edu.vn

* Correspondence: httminh@hcmus.edu.vn

† Minh Thi Thanh Hoang, Mai Thi Anh Doan and Thuong Nguyen contributed equally to this article.

Abstract: Ascorbic acid (AsA) and glutathione (GSH) are considered important factors to protect plants against abiotic stress. To investigate whether altered endogenous GSH and AsA affect seed germination, plant performance and the abiotic stress tolerance, GSH deficient mutant *cad2-1* and AsA-deficient mutants (*vtc2-4* and *vtc5-2*) were phenotypically characterized for their seed germination, shoot growth, photosynthetic activity and root architecture under abiotic stresses. The *cad2-1*, *vtc2-4* and *vtc5-2* mutants showed a decrease in osmotic and salt stress tolerance, in sensitivity to ABA during seed germination, and in plant performance under severe abiotic stresses. GSH deficiency in the *cad2-1* plants affected plant growth and root development in plants exposed to strong drought, oxidative and heavy metal stress conditions. Plants with lower GSH did not show an increased sensitivity to strong salt stress (100 mM NaCl). In contrast, the mutants with lower AsA enhanced salt stress tolerance in the long-term exposures to strong salt stress since they showed larger leaf areas, longer primary roots and more lateral root numbers. Limitations on AsA or GSH synthesis had no effect on photosynthesis in plants exposed to long-term strong salt or drought stresses, whereas they effected on photosynthesis of mutants exposed to $CdCl_2$. Taken together, the current study suggests that AsA and GSH are important for seed germination, root architecture, shoot growth and plant performance in response to different abiotic stresses, and their functions are dependent on the stress-inducing agents and the stress levels.

Keywords: abiotic stress tolerance; ascorbate (AsA); *cad2-1*; glutathione (GSH); leaf area; photosynthesis; root architecture; seed germination; *vtc2-4*; *vtc5-2*

Citation: Hoang, M.T.T.; Doan, M.T.A.; Nguyen, T.; Tra, D.-P.; Chu, T.N.; Dang, T.P.T.; Quach, P.N.D. Phenotypic Characterization of Arabidopsis Ascorbate and Glutathione Deficient Mutants under Abiotic Stresses. *Agronomy* **2021**, *11*, 764. https://doi.org/10.3390/agronomy11040764

Academic Editor: Santiago Signorelli

Received: 1 March 2021
Accepted: 12 April 2021
Published: 14 April 2021

1. Introduction

Abiotic stresses, such as salinity, drought, temperature extremes and heavy metals toxicity, are major factors in limiting plant growth and decreasing crop productivity. The exposure of plants to unfavorable environmental conditions increases the production of reactive oxygen species (ROS) such as, superoxide ($O^{2\bullet-}$), hydrogen peroxide (H_2O_2), and hydroxyl radical ($OH^\bullet$) [1]. The production of excessive ROS in plant cells leads to oxidative cellular damage, which ultimately affects the plant growth and productivity [2]. To protect themselves from adverse conditions, plants have evolved a number of cellular defense mechanisms, including the employment of antioxidants such as ascorbate (AsA), glutathione (GSH) and tocopherols, as well as ROS-detoxifying enzymes such as superoxide dismutases, peroxidases and catalases [3]. Among these stress-related molecules, the two soluble antioxidants, AsA and GSH, are central components of the AsA-GSH cycle, which regulates the cellular redox homeostasis, and are involved in plant tolerance against abiotic and biotic stresses [4].

AsA plays important roles in stress tolerance, cellular redox regulation and redox signaling [4,5]. An increase in AsA content contributes to abiotic stress tolerance in Arabidopsis and tobacco [6,7]. Previous studies have shown that overexpressing genes of the AsA biosynthetic pathway lead to an increase in AsA content, and enhances abiotic stress tolerance [8–10]. Plants with lower AsA content displayed reduced tolerance to salt, heat and high light stress [11–13]. Furthermore, AsA also involves in many biological processes, including cell wall biosynthesis, elongation, cell division, iron uptake, hormone biosynthesis, anthocyanin accumulation, and the xanthophyll cycle [12,14–17]. AsA is synthesized on the plant mitochondrial inner membrane, and then distributed throughout different cellular compartments/organelles, such as apoplast, vacuoles, mitochondria, cytosol and chloroplasts. In higher plants, L-galactose pathway, a major ascorbic acid biosynthesis pathway, was characterized and confirmed by genetic analysis in Arabidopsis [18]. The central reaction in this pathway is the conversion of GDP-L-galactose to L-Galactose 1-P by the enzyme GDP-L-galactose phosphorylase encoded by the *Vitamin C-defective 2* (*VTC2*) and *VTC5* genes [19]. Two At4g26850 (*VTC2*) T-DNA insertion mutants (SALK_146824 or SAIL_ 769_H05 named *vtc2-4*) showed severe reduction of AsA biosynthesis, producing only 20–30% of the wild-type (WT) ascorbate level while two At5g55120 (*VTC5*) T-DNA insertion mutants (*vtc5-1* and *vtc5-2*) displayed a slight reduction of AsA level with 80% of the WT ascorbate level [5,15,19]. The *vtc2-4* (SAIL_ 769_H05) and *vtc5-1*(SALK_000989) mutants showed no phenotypic growth difference to WT on standard conditions [5,19]. Although the physiological function of AsA in abiotic stress tolerance has been discovered, the mechanism of AsA involvement in the control of plant growth and development under abiotic stress conditions is largely unknown. Therefore, the present study was designed to explore further the effects of altered endogenous AA on germination, shoot growth, root development and photosynthesis in response to abiotic stresses by phenotyping of Arabidopsis AsA-deficient mutants with moderate (*vtc5-2*) or very low (*vtc2-4*) AsA content, compared to WT.

GSH is a low molecular weight thiol tripeptide (γ-glutamyl-cysteinyl-glycine) with multiple functions in plants. It is involved in cell differentiation, cell growth/division, cell death, detoxification and expression of stress responsive genes [1]. GSH is an important component of the plant antioxidant system, which is involved in abiotic stress signaling and tolerance [20]. An increased GSH level is commonly observed in plants under stress, and exogenously applied GSH can improve abiotic stress tolerances in plants [21]. The metabolism of glutathione has been extensively characterized in plants and other organisms. Glutathione is synthesized in a two-step process catalyzed by the gammaglutamyl cysteine synthase (GSH1) and the glutathione synthase (GSH2) [22]. GSH acts as a crucial regulator of normal plant metabolism since the loss-of-function of GSH biosynthesis genes causes an embryonic lethal phenotype [23]. The cadmium-sensitive 2 mutant (*cad2-1*) carries a 6 bp deletion within an exon of GSH1, has about 20–30% of the WT GSH content and exhibits the hypersensitivity to cadmium [24]. However, the *cad2-1* displayed no difference to the WT throughout vegetative development [25]. Another mutation at AtGSH1 is *pad2-1* mutant, which contains only 22% of the WT level of glutathione, showing enhanced sensitivity to pathogens [26]. The *cad2-1* and *pad2-1* showed the negative impacts on leaf area when it was exposed long-term on high salt and sorbitol treatment [22]. In contrast, the *pad2-1* mutant plants had a lower survival rate compared to WT plants after a two-week drought treatment [27].

Although many studies on mutants and /or transgenic plants with altered levels of AsA or GSH levels proved the roles of AsA and GSH in abiotic stress responses, the functions of AsA and GSH in the control of germination, plant growth, photosynthesis and root development under abiotic stress conditions require further investigations. Furthermore, there are several contradictory results on the phenotypes of glutathione deficient mutants in response to abiotic stress conditions [20]. The apparent contradictory findings among these studies may have resulted from variations in stress conditions and scoring system [20]. Moreover, several previous studies that used a *vtc2-1* mutant line containing an independently cryptic mutation, which affected the growth and physiological responses in different conditions, should be re-evaluated [5]. In addition, although AsA and GSH are both main important antioxidants for AsA-GSH cycle, the GSH- and AsA-related biosynthetic pathways in the response to abiotic stress conditions used to be investigated separately. Therefore, to further elucidate the physiological effects of altered endogenous AsA and GSH levels in the response of Arabidopsis to abiotic stress conditions, this study was designed for: (1) characterization of the seed germination and leaf area phenotypes in different AsA-deficient mutants (*vtc2-4* and *vtc5-2*), and the GSH-deficient mutant *cad2-1* under abiotic stresses; (2) measurement of the photosynthetic activity of AsA and GSH-deficient mutants; and (3) determination of the roles of AsA and GSH biosynthesis in the root development under control and abiotic stresses (salt, drought, oxidative stresses and $CdCl_2$ toxicity). The findings demonstrate that AsA and GSH play various roles in plant abiotic stress tolerance, but their functions are dependent on stress-inducing agents and stress levels.

2. Results

2.1. AtGSH1 and AtGSH2 Can Complement Yeast Δ*gsh1* and Δ*gsh2* Mutants

In the glutathione-biosynthetic-deficient mutants, the yeast mutants lacking the *GSH1* or *GSH2* genes involved in two-step glutathione synthesis (Δ*gsh1* and Δ*gsh2*) were reported to be sensitive to oxidative stress and $CdCl_2$ [28,29]. To understand the function of GSH1 and GSH2 in response to abiotic stress conditions, yeast Δ*gsh1* and Δ*gsh2* strains were grown on different abiotic stress conditions was observed. The current study showed that yeast GSH-deficient mutants also displayed sensitivity to hyper-osmotic stresses (Figure 1). Therefore, this experiment was performed to determine whether the heterologous expression of AtGSH1 on Δ*gsh1* and the expression of AtGSH2 on Δ*gsh2* can rescue the growth defect phenotype of glutathione-deficient mutants on high concentration of $CdCl_2$, NaCl or sorbitol. Both AtGSH1-expressing Δ*gsh1* yeast cells and AtGSH2-expressing Δ*gsh2* yeast cells were grown to the same level as WT cells on medium containing 100 μM $CdCl_2$ or hypertonic medium containing 1 M NaCl or 1.5 M Sorbitol (Figure 1). These results indicated that the expression of AtGSH1 on Δ*gsh1* and AtGSH2 on Δ*gsh2* restored the cadmium and hyper-osmotic tolerance of GSH-deficient mutants.

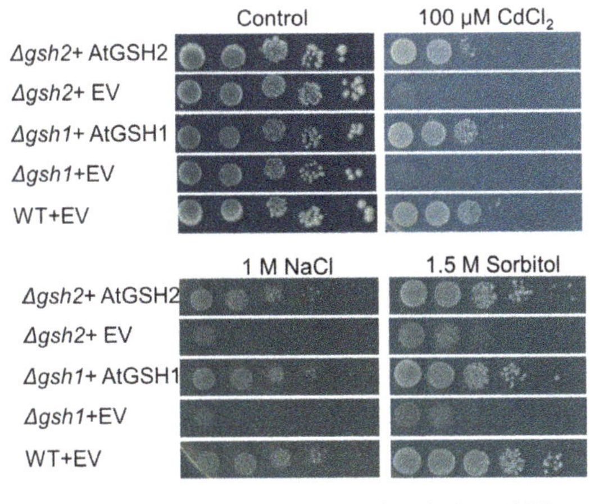

Figure 1. Complementation of Δ*gsh1* and Δ*gsh2* by AtGSH1 and AtGSH2, respectively under heavy metal and hypertonic conditions. An empty vector pDR195 (EV), a vector expressing AtGSH1 (AtGSH1), a vector expressing AtGSH2 (AtGSH2) were transformed into yeast WT cells or yeast mutant cells Δ*gsh1* or Δ*gsh2*. 10-fold serial dilutions of the transformed yeast cells were dropped onto different media. Photographs were taken after 3–4 days incubated at 30 °C. Similar results were observed in three independent experiments.

2.2. The Lower GSH and AsA Concentration in Arabidopsis Showed a Reduced Salt, Osmotic Stress Tolerance and an Increased Insensitivity to ABA during Germination

AtGSH1 and AtGSH2 enhanced salt and osmotic stress tolerance in yeast. This observation raised a question about the functions of AtGSH1 in response to abiotic stress conditions in different growth stages of plants such as germination and vegetative growth. To determine the roles of GSH during germination in response to different environmental stresses, germination tests under salt stress and osmotic stress were performed. First, seed germination in response to salt stress was investigated in *cad2-1* and WT by sowing their seeds on $\frac{1}{2}$ MS medium supplemented with different concentrations of NaCl (0, 50, 100, 150, 170 and 200 mM). As shown in Figure 2A, the WT and *cad2-1* mutant seed germination rates were similar under standard condition and up to 100 mM NaCl treatment. When being exposed to 150 mM NaCl and higher NaCl concentrations, the germination rates of *cad2-1* seeds were significantly lower than that of WT seeds on the second day after sown on plates (Figure 2A). These results suggest that *cad2-1* mutant was more sensitive to salt stress than WT. Second, the current study investigated the *cad2-1* mutant seed germination in response to osmotic stress. The WT seeds and the *cad2-1* seeds were sown on $\frac{1}{2}$ MS medium supplement with 0, 200, 300, 400, 500 and 600 mM sorbitol. The WT seeds showed a higher germination rate than the *cad2-1* mutant in the presence of sorbitol. In 300 mM sorbitol, the *cad2-1* seeds showed a 21–22% reduction in germination rate than WT seeds after 6 days (Figure 2B). These results indicated that the *cad2-1* mutant was more sensitive to osmotic stress than WT during seed germination stage. Taken together, the lower glutathione concentration in Arabidopsis increased the sensitivity to salt and osmotic stress at germination stage.

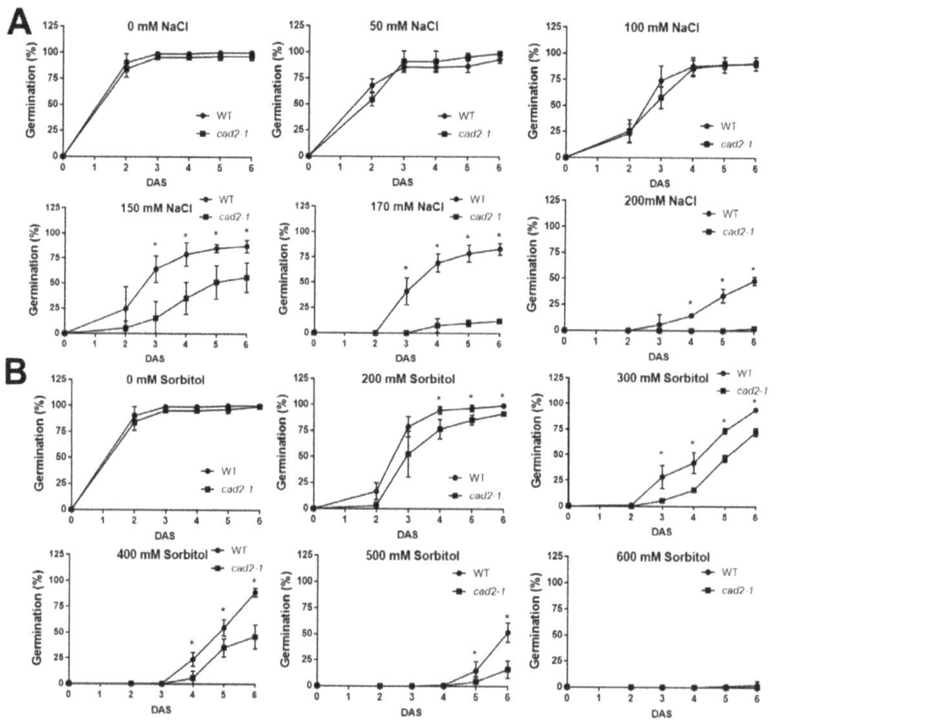

Figure 2. The *Arabidopsis thaliana* GSH-deficient mutant *cad2-1* is more sensitive to salt and osmotic stress conditions during germination. (**A**). Germination percentages of WT and *cad2-1* seeds under salt stress. (**B**). Germination percentages of WT and *cad2-1* seeds under osmotic stress. Germination percentages were counted at the indicated times. Data are shown as the means ± SD of three independent experiments using 50–100 seeds of each genotype. Experiments were repeated at least twice with similar results. Asterisks indicate significant differences (Student's *t*-test; * $p < 0.05$).

AsA is not only an important component of human nutrition but an antioxidant and H_2O_2-scavenger that defends plants against abiotic stress [9]. The maintenance of AsA level is required for oxidative stress tolerance in Arabidopsis [9]. Germination with various NaCl or sorbitol concentrations on WT seeds showed that 150 mM NaCl or 300 mM sorbitol decreased germination rates by about 65% or 72%, respectively after 72 h in above experiments. Therefore, to determine whether the lower AsA concentration in Arabidopsis alters NaCl or sorbitol sensitivity during germination, AsA-deficient mutants were germinated in the presence of 150 mM NaCl or 300 mM Sorbitol, and the germination rates were evaluated after 72 h. In control treatment (no stressor), the germination rates of *vtc2-4* and *vtc5-2* mutant seeds were similar to that of WT seeds. The germination rate of *vtc2-4* seeds was much lower than *vtc5-2* and WT seeds in the presence of NaCl or sorbitol. After 72 h of NaCl treatment, the germination rates of *vtc2-4* and *vtc5-2* mutants were 5.56% and 47.78%, respectively, while that of WT was 58.89% (Figure 3). Similarly, the germination rates of *vtc2-4* and *vtc5-2* mutants after 72 h under sorbitol treatment were 16.67% and 52.22%, respectively, while that of WT was 83.33% (Figure 3). Together, the current study showed that the deficiencies in AsA synthesis decreased the salt and osmotic tolerance during seed germination stage.

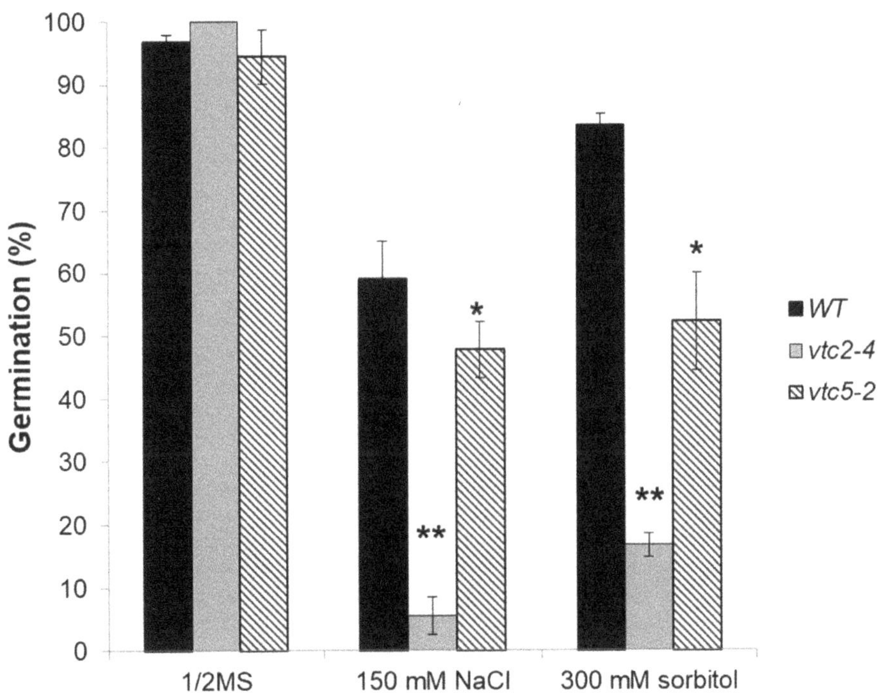

Figure 3. Seeds of *Arabidopsis thaliana* AsA-deficient mutants are more sensitive to salt and osmotic stress conditions during seed germination. Seeds of WT, *vtc2-4* and *vtc5-2* were assayed for germination on control (black), or in response to with 150 mM NaCl (gray) or 300 mM sorbitol (stripped). Germination percentages were recorded after 3 days sowing. Data are shown as the means ± SD of three independent experiments using 50 seeds of each genotype. Experiments were repeated at least twice with similar results. Asterisks indicate significant differences (Student's *t*-test; * $p < 0.05$ and ** $p < 0.01$).

Abscisic acid (ABA) plays important roles in abiotic stress response and regulation of seed germination [30]. Therefore, this study investigated whether the seed germination was affected by exogenous ABA in the GSH-deficient mutant *cad2-1* and AsA-deficient *vtc2-4* and *vtc5-2* mutants. The WT, *cad2-1*, *vtc2-4* and *vtc5-2* seeds were sown on $\frac{1}{2}$ MS medium supplement with 0, 2.5, 5, 10 and 15 μM ABA. In contrast to the effect of sorbitol or salt, *cad2-1*, *vtc2-4* and *vtc5-2* were less sensitive to the inhibitory effect of ABA on seed germination compared to WT. All of the mutants had higher germination rate than WT, when the seeds were treated with different concentrations of ABA (Figure 4 and Supplementary Table S1). The current study indicates that GSH-deficient mutant and AsA-deficient mutants displayed the reduction of salt and osmotic tolerance during germination and an increased insensitivity to ABA, the key phytohormone in maintaining the seed dormancy stage.

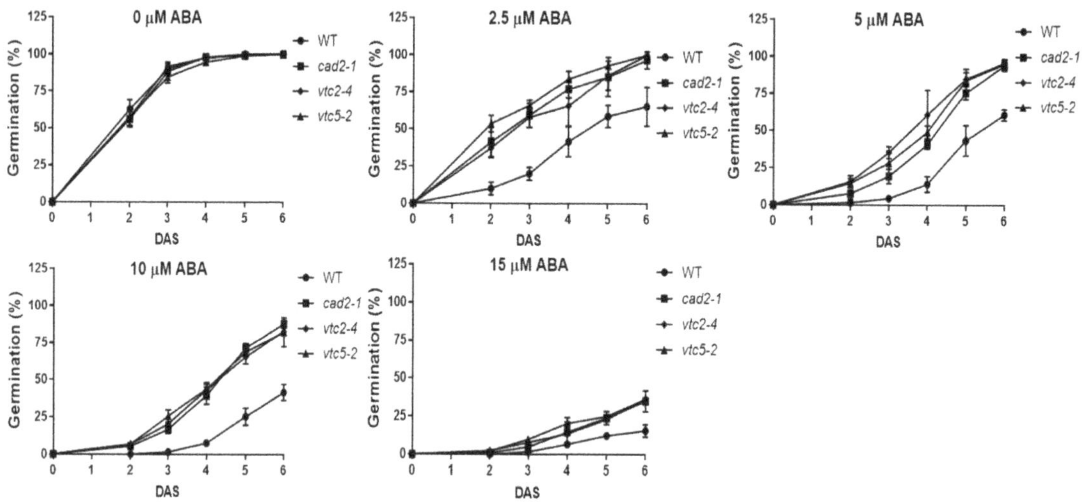

Figure 4. Seeds of *Arabidopsis thaliana cad2-1*, *vtc2-4* and *vtc5-2* are more resistant to ABA-mediated inhibition of seed germination. Germination rates (%) were recorded at the indicated time-points. Data are shown as the means ± SD of three independent experiments using 50–100 seeds of each genotype. The experiments were repeated at least twice with similar results.

2.3. *Leaf Areas of AsA and GSH-Deficient Mutants Were Differently Affected by Long-Term Exposures to Various Abiotic Stresses*

To investigate the importance of AsA and GSH in leaf growth under different abiotic conditions, the AsA-deficient *vtc2-4* and *vtc5-2* mutants and the GSH-deficient *cad2-1* mutant were exposed to various long-term abiotic stresses and leaf areas of plants were measured and analyzed. Shoot fresh weight was also analyzed but showed the same trends as leaf area (Supplementary Figure S1). Under control condition ($\frac{1}{2}$ MS), there was no significant difference among WT, *cad2-1*, *vtc2-4* and *vtc5-2* (Figure 5). After 18–21 days exposed to salt stress (100 mM NaCl), leaf areas of the *vtc2-4* and *vtc5-2* mutants were 91% and 82% larger than that of WT plants, respectively, while the results showed no difference between WT and *cad2-1* plants. Under osmotic stress (225 mM sorbitol), leaf area of the *cad2-1* was reduced by 16% compared to WT, while leaf area of *vtc2-4* and *vtc5-2* showed no significant difference. Leaf areas of *cad2-1*, *vtc2-4* and *vtc5-2* were reduced by 77%, 18% and 31%, respectively, compared to WT under oxidative stress condition (1 mM H_2O_2) (Figure 5). When exposed to $CdCl_2$, *cad2-1* leaf area decreased significantly, up to 81% compared to WT. GSH is involved in cell cycle regulation in leaves of cadmium-exposed plants [31]. The stronger leaf growth inhibition observed in *cad2-1* mutants compared to WT plants upon prolonged Cd exposure underlines the importance of GSH in plant defense against Cd. Interestingly, leaf areas of *vtc2-4* and *vtc5-2* were larger than WT when they were exposed to $CdCl_2$, 64% and 97%, respectively. Together with the results observed under oxidative stress, these results confirmed the critical functions of ASA and GSH: both are important antioxidants in redox balance mechanism, while GSH alone is a key player in heavy metal ($CdCl_2$) detoxification. These data indicated that an impaired GSH synthesis adversely affected leaf growth in plants exposed long-term osmotic, oxidative and heavy metal stresses but not salt stress while AsA deficiency negatively affected leaf growth in plants exposed to salt and heavy metal stresses but not oxidative stress.

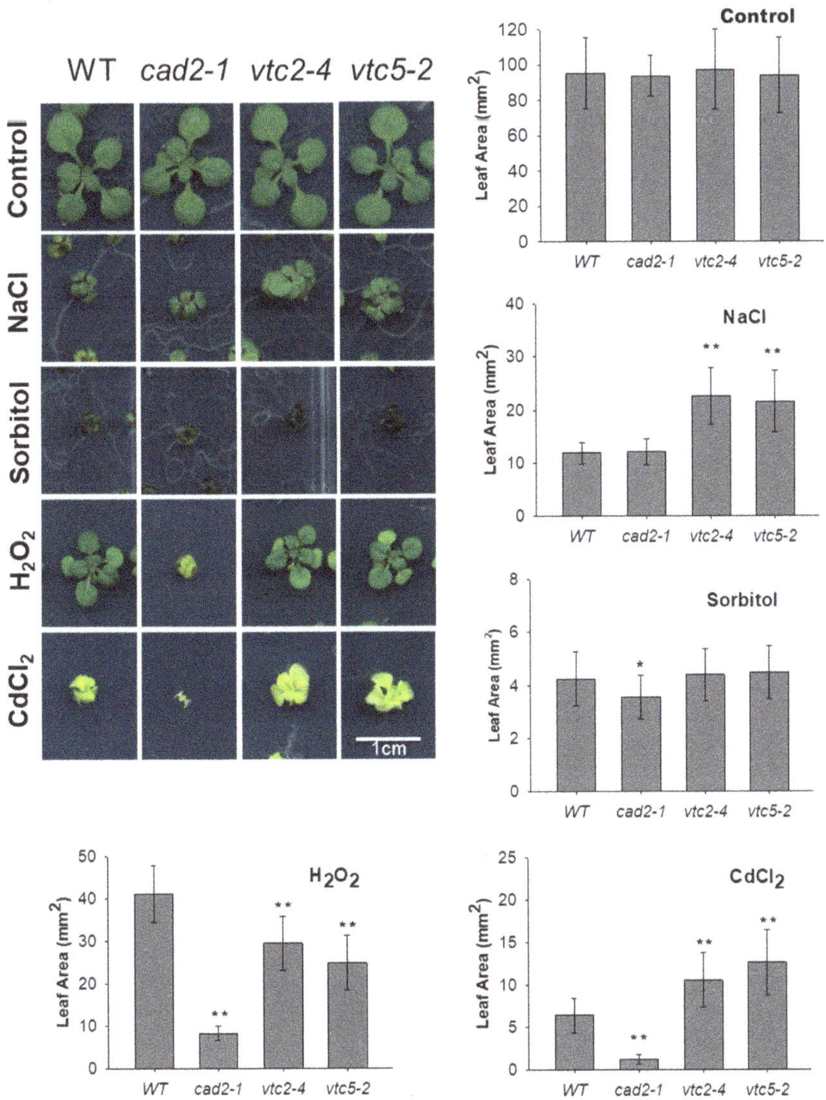

Figure 5. Rosette leaf area of the *cad2-1*, *vtc2-4*, *vtc5-2* mutants and WT under normal conditions or different strong stress conditions. The *cad2-1*, *vtc2-4* and *vtc5-2* mutant as well as the WT were exposed to different strong abiotic stress conditions: salt stress (100 mM NaCl), osmotic stress (225 mM sorbitol), oxidative stress (1 mM H_2O_2), and heavy metal stress (40 μM $CdCl_2$) for 18–21 days. Bar graphs represent mean of treatments, with error bars indicate Standard Deviation (N = 25–35). Experiments were performed twice with similar results. Asterisks indicated the statistical significance of differences between WT and mutants by Student's *t*-test: * $p < 0.05$ and ** $p < 0.01$. White bar in the photograph corresponds to 1 cm.

2.4. GSH-Deficient Mutants Had a Significantly Lower Photosynthesis than the WT under Oxidative and Heavy Metal Stresses

Photosynthesis is one of the major determinants of plant growth and yield formation [32]. To investigate the link between the reduction of AsA and GSH level and photosynthesis under abiotic stress, maximum PSII efficiency (F_v/F_m) of rosette leaves of the WT, *cad2-1*, *vtc2-4* and *vtc5-2* grown for 18-21 days in different abiotic conditions was measured (Figure 6). Arabidopsis WT and mutants plants grown on control conditions (no stressor) showed F_v/F_m values around 0.77. Maximum PSII efficiencies of WT and mutants were not significantly different in control conditions versus salt stress and osmotic stress. However, in oxidative stress, *cad2-1* Fv/Fm was reduced by 26% compared to WT, while Fv/Fm of *vtc2-4* and *vtc5-2* was not dramatically affected, indicating the importance of GSH in oxidative stress tolerance. In heavy metal stress, the results showed that Fv/Fm of *cad2-1* and *vtc2-4* was respectively 58% and 67% compared to WT, while Fv/Fm of *vtc5-2* showed no difference. While Fv/Fm of *vtc2-4* in heavy metal stress was affected, the PSII efficiency of *vtc5-2* was similar to WT. In conclusion, the *cad2-1*, *vtc2-4* and *vtc5-2* mutants responded similarly in terms of photosynthetic activity in strong stress level of NaCl and sorbitol, whereas the photosynthetic activity of the *cad2-1* was affected under $CdCl_2$ and H_2O_2, and photosynthetic activity of the *vtc2-4* was significantly decreased under $CdCl_2$.

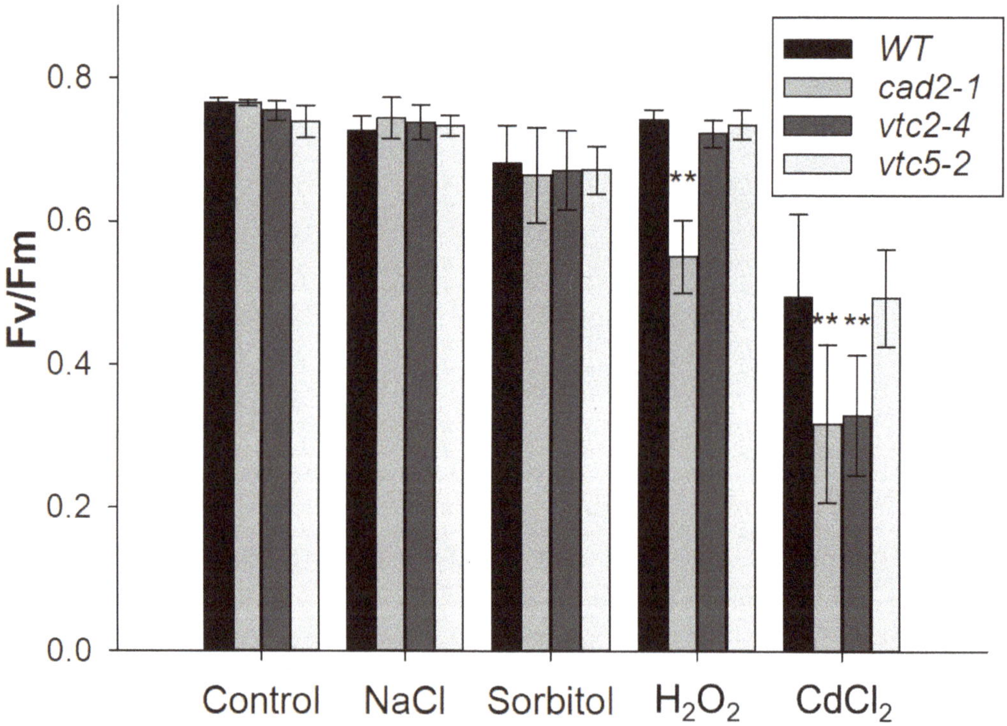

Figure 6. Photosynthetic activity of Arabidopsis WT, *cad2-1*, *vtc2-4* and *vtc5-2* mutants grown under control and different abiotic stress conditions. Maximum PSII efficiency (F_v/F_m) measured on rosette leaves of the WT, *cad2-1*, *vtc2-4* and *vtc5-2* grown on different strong abiotic stress conditions: salt stress (100 mM NaCl), osmotic stress (225 mM sorbitol), oxidative stress (1 mM H_2O_2), and heavy metal stress (40 µM $CdCl_2$) for 18–21 days. FluorCam FC 800-O (Photon Systems Instruments) was used to estimate the maximal photochemical efficiency of PSII [$F_v/F_m = (F_m - F_o)/F_m$] which revealed PSII activity. Data are shown as the means ± SD (n = 10–25). Experiments were performed twice with similar results. Asterisks indicate values significantly different from those of the WT (Student's t-test, ** $p < 0.01$).

2.5. GSH and AsA Deficiency Differentially Altered Root Architecture of Plants Exposed to Long-Term Abiotic Stresses

In order to determine the effects of abiotic stresses on Arabidopsis root architecture of GSH-deficient mutants and AsA-deficient mutants, Arabidopsis seeds were germinated and grown on control or strong stress media containing 100 mM NaCl or 225 mM sorbitol or 1.0 mM H_2O_2 or 40 μM $CdCl_2$. Primary root length and lateral root (LR) number were identified from the images of the plantlets at 13–14 day (Figures 7 and 8 and Supplemental Figure S2).

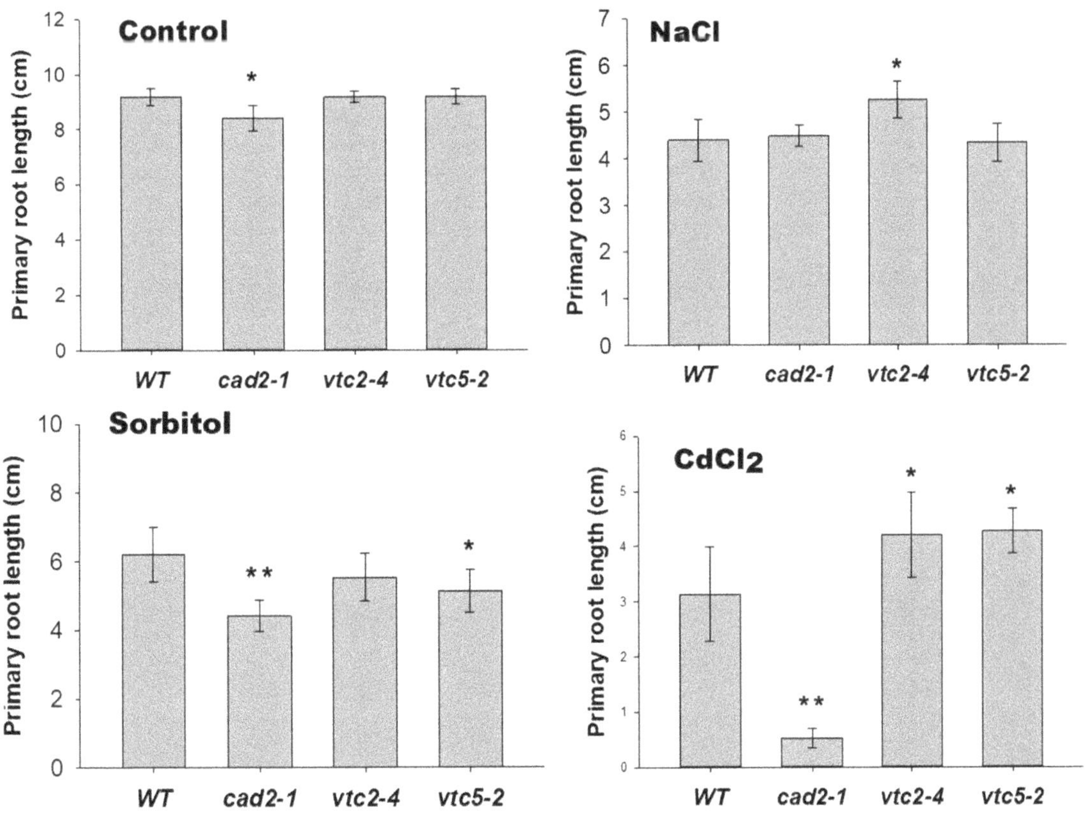

Figure 7. Primary root length of the *cad2-1*, *vtc2-4*, *vtc5-2* mutants and WT under normal conditions or different strong stress conditions. The *cad2-1*, *vtc2-4* and *vtc5-2* mutants as well as the WT seeds were exposed to different strong abiotic stress conditions: salt stress (100 mM NaCl), osmotic stress (225 mM sorbitol), oxidative stress (1 mM H_2O_2), and heavy metal stress (40 μM $CdCl_2$) for 18–21 days. Data are shown as the means ± SD (n = 5–10). Experiments were performed twice with similar results. Asterisks indicated the statistical significance of differences between WT and mutants by Student's t-test: * $p < 0.05$ and ** $p < 0.01$.

Under control conditions, the primary root lengths of the mutants were almost indistinguishable from the WT, while the *cad2-1* and *vtc2-4* mutants had lower numbers of LRs compared to the WT. Primary root elongation and lateral root production were significantly decreased in all stressed conditions compared to the unstressed conditions. Compared to the WT, the *cad2-1* mutants displayed no difference in primary root growth and lateral root production in medium with 100 mM NaCl. However, under these conditions, *vtc2-4* mutant showed an enhanced root system with longer primary roots and more lateral roots, which may help this mutant to enhance their salt stress tolerance. Thus, the *vtc2-4* mutant was less sensitive to salt stress than the WT in primary root growth, lateral root production and leaf growth. On $\frac{1}{2}$ MS medium with 225 mM sorbitol, the lengths of *cad2-1* and *vtc5-2* primary roots were significantly decreased, whereas the length of *vtc2-4* primary roots was not different to the WT. Compared to the WT, the numbers of LRs of the *vtc2-4* and *vtc5-2* mutants were significantly increased, whereas the number of the *cad2-1* was not changed; when plants were grown on medium with 225 mM sorbitol. When grown on $\frac{1}{2}$ MS medium containing 40 µM $CdCl_2$, the primary roots of *cad2-1* was significantly shorter, while primary roots of *vtc2-4* and *vtc5-2* were longer than those of WT. The current study observed no significant difference in LRs production in *vtc2-5* and WT under $CdCl_2$ treatment. However, LRs number was markedly decreased in the *cad2-1* mutant, while it was significantly increased in the *vtc2-4* mutant in comparison to the WT. Primary root elongation of all genotypes was inhibited by supplementing 1 mM H_2O_2 (Supplementary Figure S2). These results suggested that GSH deficiency in plants altered the root architecture under unstressed conditions and heavy metal stress while AsA deficiency in plants altered the root architecture under strong salt, osmotic and heavy metal stresses. Taken together, the AsA and GSH-deficient mutants exposed to long-term strong abiotic stresses displayed pleiotropic effects on root architecture.

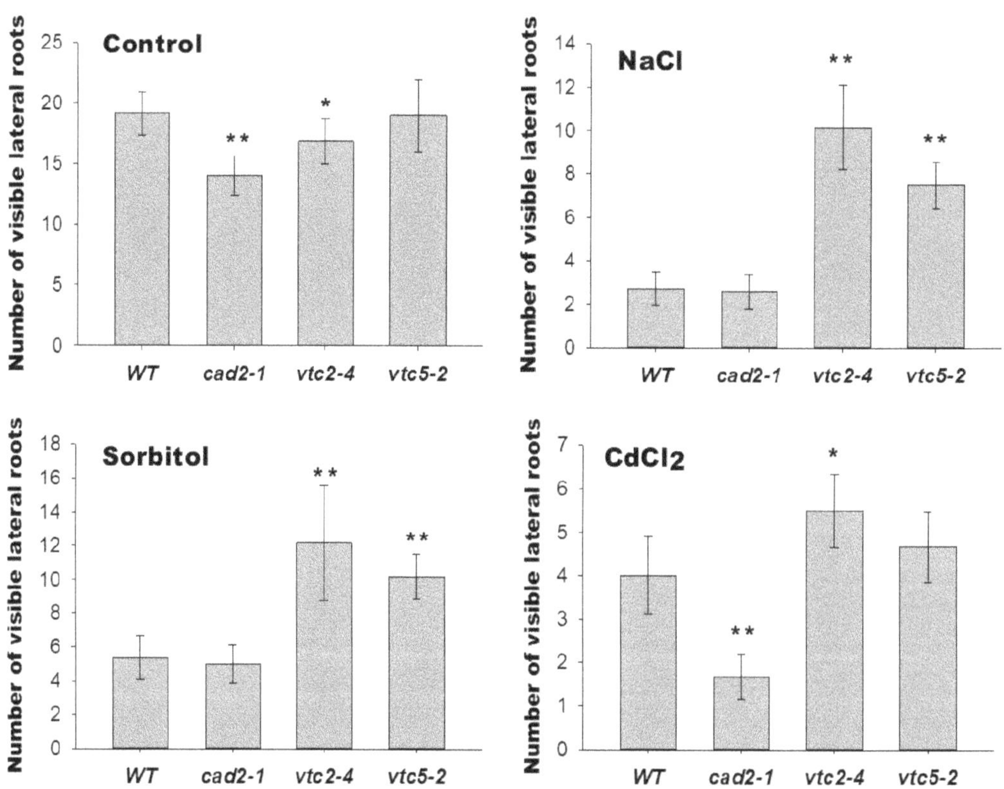

Figure 8. Lateral root number of the *cad2-1*, *vtc2-4*, *vtc5-2* mutants and WT under normal conditions or different strong stress conditions. The *cad2-1*, *vtc2-4* and *vtc5-2* mutants as well as the WT seeds were exposed to different strong abiotic stress conditions: salt stress (100 mM NaCl), osmotic stress (225 mM sorbitol), oxidative stress (1 mM H_2O_2), and heavy metal stress (40 µM $CdCl_2$) for 18–21 days. Data are shown as the means ± SD (n = 5–10). Experiments were performed twice with similar results. Asterisks indicated the statistical significance of differences between WT and mutants by Student's t-test: * $p < 0.05$ and ** $p < 0.01$.

2.6. Plants with Lower AsA and GSH Levels Showed a Shorter Primary Root under Severe Abiotic Stresses

To further characterize and evaluate the responses of *cad2-1*, *vtc2-4* and *vtc5-2* after germination and seedling stages to different abiotic stresses under severe stress conditions, a transference system was applied. Arabidopsis seeds were sown and germinated on plates with $\frac{1}{2}$ MS media, then six-day-old seedlings of WT and mutants were transferred to the media supplemented with 150 mM NaCl, 375 mM sorbitol, 1.5 mM H_2O_2 or 100 μM $CdCl_2$, and primary root length was measured after five days. Under normal growth conditions, the primary root growth of the mutants was almost indistinguishable from the WT (Figure 9). When seedlings were transferred to severe stress conditions, primary root growth of *cad2-1*, *vtc2-4* and *vtc5-2* mutants was significantly more inhibited than the WT (Figure 9). These data suggest that AsA and GSH contents in plants are important for the primary root growth under severe stress conditions.

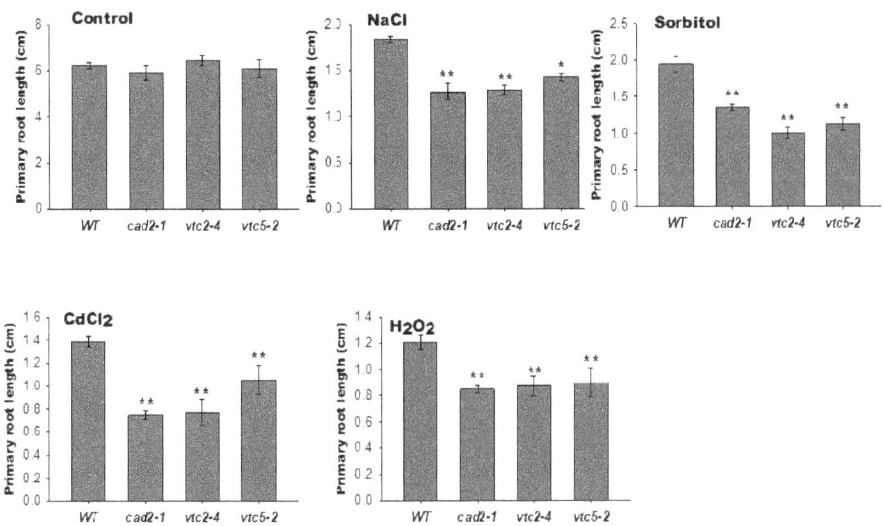

Figure 9. Primary root length of the *cad2-1*, *vtc2-4*, *vtc5-2* mutants and WT under normal conditions or different severe stress conditions. The *cad2-1*, *vtc2-4* and *vtc5-2* mutants as well as the WT seeds were grown on $\frac{1}{2}$ MS medium for 6 days. Six-day old seedlings were transferred to different severe abiotic stress conditions: salt stress (150 mM NaCl), osmotic stress (375 mM sorbitol), oxidative stress (1.5 mM H_2O_2), and heavy metal stress (100 μM $CdCl_2$) for 5 days and then photographed. Data are shown as the means ± SD (n = 5–7). Experiments were performed twice with similar results. Asterisks indicated the statistical significance of differences between WT and mutants by Student's *t*-test: * $p < 0.05$ and ** $p < 0.01$.

3. Discussion

This study demonstrates the importance of AsA and GSH during seed germination, root elongation in severe salt and osmotic stress conditions. However, the GSH contents did not affect the plant tolerance in strong salt stress conditions (100 mM NaCl) for long-term exposures. Deficiency of AsA enhanced salt stress tolerance in plants exposed to long-term strong salt stress as determined by leaf area, root elongation and lateral root development. GSH-deficient mutant plants increased their sensitivity to 40 μM $CdCl_2$ and 1.0 mM H_2O_2 for long-term exposures, suggesting the importance of GSH contents on detoxification of high $CdCl_2$ and H_2O_2. The AsA-deficient *vtc2-4* and *vtc5-2* mutants, however, decreased sensitivity toward $CdCl_2$ in term of leaf area. This study suggests that AsA and GSH may have various functions in abiotic stress tolerance and detoxification of ROS in plants, but their functions are dependent on stress-inducing agents and stress levels.

Antioxidant enzymes, such as catalases, superoxide dismutases and peroxidases, are the main ROS scavengers in plants [3]. Moreover, plants also possess antioxidant molecules, GSH and AsA, which effectively scavenge ROS directly and indirectly through enzymatic reactions [4]. In addition, AsA plays a role in Fe uptake through Fe^{3+} reduction and is an important cofactor of enzymes involved in auxin degradation and synthesis of plant hormones (ethylene, abscisic acid, gibberellins), as well as anthocyanins and glucosinolates [17]. Moreover, GSH is also important in the synthesis of phytochelatins and detoxification of heavy metals, GSH interacts with hormones, and its redox state triggers signal transduction [21]. Here, this study characterized the GSH-deficient mutant (*cad2-1*) and AsA-deficient mutants (*vtc2-4*, *vtc5-2*) during germination, leaf growth, photosynthetic activity and root development under different abiotic stresses and stress levels. Interestingly, the GSH-deficient mutant and AsA-deficient mutants displayed altered sensitivity to salt, osmotic, oxidative stresses and heavy metal toxicity, which was depend on the stress levels. Therefore, this study suggested that GSH and AsA may play different roles in the tolerance to salinity, drought and Cd toxicity. These stresses can indirectly produce ROS, which leads to oxidative stress [33]. However, apart from the deleterious effects of oxidative damage, there is evidence that ROS are critical to and being continually produced during all phases of seed development, from desiccation to germination. Therefore, ROS may cover important roles in seed germination [34]. In fact, it has been shown that an optimal range of H_2O_2 is crucial for the dormancy release and a disturbance in ROS homeostasis decreases the seed germination [35,36]. Therefore, the balance between ROS production and scavenging should be strictly controlled during seed germination. Interestingly, *cad2-1*, *vtc2-4* and *vtc5-2* are not affected under no-stress conditions, whereas those mutants show a strongly reduction in germination rate under salinity and osmotic stress. Moreover, the most likely explanation for the differences between both *vtc* mutants is that *vtc2-4* mutant only contains 20–30%, while *vtc5-2* still contains 80% of the WT AsA level [5,19]. These results suggest that GSH and AsA synthesis mutants are still able to partially regulate seed ROS levels in the absence of external stress. Therefore, GSH and AsA may be important ROS scavengers implicated in dormancy release and seed germination under salt and osmotic stress.

In addition to ROS, another important player in seed germination is ABA. Germination begins with the release of dormancy, which is controlled by ABA. ROS accumulation acts as a positive signal for dormancy release by altering the synthesis and signaling of ABA. It has been demonstrated that H_2O_2 accumulation in germinating seeds is associated with ABA degradation, likely through the activation of ABA catabolic enzyme (ABA-8′-hydroxylase) or by the direct oxidation of ABA as well as antagonize ABA signaling [35,37,38]. In contrast, multiple studies in germinated seeds have shown that the direct inhibition of ABA blocks ROS production in seeds [39]. $OH^{\bullet}$ also promotes dormancy release by contributing to the cell wall loosening required for germination [40]. Dry seeds accumulate GSH and a very low amount of AsA and the AsA is synthetized de novo upon dormancy release and during germination [34]. Germinating seeds of AsA-deficient and GSH-deficient mutants very likely produce lower levels of AsA and GSH, respectively, compared to WT. Therefore, the disruption of GSH and AsA synthesis could reduce ABA-sensitivity during seed germination by the presence of high ROS levels, which may also alter the ABA homeostasis and signaling. However, the detailed molecular mechanism still remains unclear. Hence, further studies need to verify whether ROS and ABA levels are affected in GSH and AsA synthesis mutant seeds.

Regarding to seedling development, contradictory results related to GSH deficient mutants have been reported. While several studies showed that *cad2-1* mutant is significantly smaller than WT [22,41], recent works have reported that no distinct growth phenotype was observed between *cad2-1* and WT [20,25]. This study showed shorter primary roots and a decrease in lateral root numbers, but the leaf area showed no phenotypic difference between *cad2-1* mutant and WT seedlings. Likewise, previous experiments showed that *cad2-1* mutant is not affected by salt and osmotic stress [20,22]. However, in the present study, the *cad2-1* exhibits strong sensitivity to osmotic stress. These pronounced differences emphasized the critical role of growth conditions. Indeed, GSH levels may vary significantly between experiments due to several non-controlled experimental differences [42]. Nevertheless, taking into account the importance of GSH in the root cell division [43] and auxin signaling [41,44], we speculate that the decrease of GSH causes an inhibition on root development by diminishing the cell proliferation and the auxin-dependent root growth. In contrast, under high ROS production provoked by osmotic stress, the inhibition of root elongation and enhancement of lateral roots might be related to the regulation of GSH concentration in the root apical meristem (RAM) and pericycle, respectively, which modulates the auxin signaling differently [44]. In the case of cadmium toxicity, GSH plays a key role as a chelator due to the high affinity of Cd for its thiol group, and also as a precursor for phytochelatins synthesis [31]. Therefore, a decreased capacity to chelate Cd ions due to a decrease in GSH levels and, by extent, low phytochelatins may contribute to the strong Cd sensitivity of *cad2-1* mutant.

Contrarily to *cad2-1*, *vtc2-4* and *vtc5-2* mutants were tolerant to 100 mM NaCl and 40 μM $CdCl_2$, but unaffected by sorbitol. AsA can potentially be oxidized and acts as a cofactor of 1-aminocyclopropane-1-carboxylic acid oxidase (ACO), the enzyme that catalyzes the last step of ethylene biosynthesis [45]. In addition, being a cofactor of dioxygenases, AsA could also be involved in the ABA biosynthesis by modulating the activity of the 9-cis-epoxycarotenoid dioxygenase (NCED3) responsible for the oxidative cleavage of neoxanthin to xanthoxin [17]. Recently, it was proposed that under salt stress, both ABA and ethylene production are induced, which could control ROS levels by regulating AsA biosynthesis through *VTC2* gene expression [46]. The *vtc2-4* and *vtc5-2* mutants could accumulate less ethylene and/or ABA, which could cause changes in plant response to abiotic stresses, leading to a decrease in growth inhibition at early stages. Intriguingly, the *vtc2-1* mutant accumulates more ABA and shows small size in non-stressed conditions [47]. However, it has been reported its small size is due to another mutation rather than VTC2 itself [5], suggesting that further analyses performed in different *vtc2* mutant alleles are required to determine how ethylene and ABA levels are affected in these mutant alleles.

Another hypothesis is based on the results observed in *vtc1-1* mutant, corresponding to a mutation on the gene *VTC1* encoded for GDP–d-Mannose pyrophosphorylase (VTC1), another AsA synthesis enzyme located upstream of VTC2 and VTC5 in the biosynthesis pathway [48]. Interestingly, this mutant accumulates higher amount of GSH than WT [49,50]. In addition, AsA deficiency was previously suggested to provide a primed state that decreases pathogen infection and abiotic stresses [51–53]. In addition to the elevated GSH levels, phytochelatin levels in *vtc1-1* were approximately twice in *vtc1-1* roots compared to WT plants upon Cd exposure [52]. However, mannose metabolism may be changed in this *vtc1-1* mutant, which might affect its physiological roles and responses to stress [54]. As one of the major defense mechanisms in Cd-exposed plants is chelation and sequestration by thiols [31], *vtc2-4* and *vtc5-2*, which are present also lower levels of AsA concentration, could also have higher GSH amount that increase the capacity to chelate Cd, contributing to a less Cd-sensitive phenotype.

Surprisingly, although *cad2-1*, *vtc2-4* and *vtc5-2* mutants are affected by oxidative stress, only *cad2-1* displays a stronger reduction in leaf area and lower Fv/Fm value. AsA and GSH are important players in the protection against ROS produced during photosynthesis [55]. Therefore, since AsA synthesis mutants may accumulate more GSH, these results also suggest that GSH is more active in the detoxification of ROS in chloroplasts, under strong oxidative stress, than AsA.

Photosynthesis is fundamental for plant growth. Photosynthetic activity of AsA-deficient mutants and GSH-deficient mutant was similar to WT at normal condition and under salt and osmotic stress. Under oxidative stress, Fv/Fm value was significantly reduced in *cad2-1* mutant while it was not changed in *vtc2-4* and *vtc5-2* mutants. Fv/Fm values of *cad2-1* mutant and *vtc2-4* mutant were decreased under Cd toxicity. Although AsA works as a reductant of violaxanthin de-epoxidase (VDE) during photo-oxidative stress and a protective role in photoinhibition in heat-stressed leaves [13]. The AsA and GSH accumulation in mutants is possible to change their level in subcellular compartment such as chloroplasts, which might affect their photosynthetic activity [56].

Although the root growth of AsA-deficient mutants and GSH-deficient mutant responded differently to abiotic stress conditions at strong level (100 mM NaCl, 225 mM sorbitol, 1 mM H_2O_2 or 40 μM $CdCl_2$), all mutants showed a significant decrease of primary root growth response to severe abiotic stress conditions (150 mM NaCl, 375 mM sorbitol, 1.5 mM H_2O_2 or 100 μM $CdCl_2$). These results suggested that 20 to 80% AsA or 20–30% GSH was sufficient to maintain normal plant physiological activity under control or under osmotic and salt stress at strong level. However, the AsA and GSH contents or ratio of reduced form to oxidized form might change in AsA-deficient mutants and GSH-deficient mutant, compared to WT under severe stress conditions, which strongly affected the root growth. Interestingly, the enhanced root system observed in AsA-deficient mutants when they were exposed to strong salt stress (100 mM NaCl). Hence, these mutants may be beneficial and significant importance for studies on salt tolerance of economically important crops in the future.

This study showed that AsA and GSH could have different functions in detoxification of ROS in plants. This may include different intracellular compartmentation of both molecules and the ratio of reduced form to oxidized form under different abiotic stress conditions. Moreover, the AsA-GSH pathway is one of the main defense systems to protect the plants from multiple abiotic stresses [57]. Both AsA and GSH are strong antioxidants and the modulation of their redox state may work as a sensitive machinery to assess the stress levels and fine-tune molecular responses. Thus, the alteration in the synthesis of one of these molecules will strongly affect the stability of AsA-GSH pathway, leading to changes in the tolerance to abiotic stresses. Hence, the effects of altered GSH-AsA homeostasis in plant seed germination and seedling development need to be examined in further details, which will investigate the concentration and redox state of GSH and AsA in single or double mutants defective in AsA-biosynthesis gene and GSH-biosynthesis gene in response to the single and combinatorial stresses.

4. Conclusions

This study suggested AsA and GSH may have different functions in plant abiotic stress tolerance and ROS detoxification. Both AsA and GSH are important for seed germination under salt and osmotic stresses. Limitations of GSH and AsA synthesis reduce ABA-sensitivity during seed germination. Furthermore, both AsA and GSH are crucial factors for the primary root growth under severe stresses. GSH alone is a key player in heavy metal detoxification. Interestingly, deficiency of AsA favored larger leaf areas, enhancer root systems in plants exposed to long-term strong salt stress. Taken together, the findings shed new light on the functions of AsA in salt stress tolerance of plants, showing that deficiency of AsA might enhance plant salt stress tolerance. Therefore, further investigations of altered GSH-AsA homeostasis in plants need to be addressed in detail.

5. Materials and Methods

5.1. Plants Materials and Growth Conditions

A. thaliana ecotype Columbia (Col-0), *cad2-1* [24], *vtc2-4* (SALK_146824) and *vtc5-2* (SALK_135468) [15] were kindly provided by Stephane Mari (BPMP Montpellier, France). Seeds were surface sterilized by soaking in 70% ethanol for 1 min and 2.5% NaClO for 5 min, then washed four times with sterilized water. Sterilized seeds were sown on half-strength Murashige and Skoog (MS) medium containing 1.0% sucrose and 0.8% agar (adjusted to pH 5.8 with MES-KOH) ($\frac{1}{2}$ MS medium) [58]. Seeds were stratified for two days at 4 °C and then transferred to a growth chamber at 22 °C with light intensity of 120 μE m^{-2} s^{-1} and a 16 h light/8 h dark cycle [59]. Each experiment was repeated at least twice times.

5.2. Seed Germination Test

For germination assay, sterilized seeds of WT and *cad2-1* were sown on half strength MS medium containing 0.8% agar supplemented with or without NaCl (0, 50, 100, 150, 170, 200 mM), sorbitol (0, 200, 300, 400, 500, 600 mM) [60]. Sterilized seeds of WT, *vtc2-4* and *vtc5-1* were sown on the $\frac{1}{2}$ MS media supplemented with 150 mM NaCl or with 300 mM sorbitol. Sterilized seeds of WT, *cad2-1*, *vtc2-4* and *vtc5-1* were sown on $\frac{1}{2}$ MS media supplemented with different concentrations of ABA (0, 2.5, 5, 10 or 15 μM) [58]. The plates were kept at 4 °C for two days and then placed horizontally at growth chamber at 22 °C under a 16 h light and 8 h dark photoperiod. Germination rates were recorded at 2, 3, 4, 5, and 6 day, based on the radicle tip emergence.

5.3. Abiotic Stress Treatments

5.3.1. Strong Long-Term Stress

Sterilized seeds were sown on plates in $\frac{1}{2}$ MS medium (control), or media containing 100 mM NaCl or 225 mM sorbitol, 1 mM H_2O_2, or 40 μM $CdCl_2$. Seedlings were grown for 18–20 days under these conditions [22,59]. The plates were scanned, shoot fresh weight was determined and the other analysis performed as described below.

5.3.2. Severe Short-Term Stress

Seeds were sterilized and sown on MS $\frac{1}{2}$ media. Uniform 6-day-old seedlings were transferred into $\frac{1}{2}$ MS media or $\frac{1}{2}$ MS media supplemented with 150 mM NaCl or 375 mM sorbitol or 1.5 mM H_2O_2 or 100 μM $CdCl_2$ and grown for a further five days in vertically placed petri plate [22,61].

5.4. Phenotypic Analysis

5.4.1. Leaf Area Measurement

For rosette leaf area measurements, the seeds were sown on $\frac{1}{2}$ MS agar plates containing the indicated stress treatment as described above and placed horizontally. Plants were grown at 22 °C with 16 h light/8 h dark photoperiod for 18–21 days. Total leaf area was measured and determined using the ImageJ software (https://imagej.nih.gov/ij/, accessed and downloaded on 19 May 2019) [22,59].

5.4.2. Primary Root Length and Number of Lateral Root

For long-term exposure to strong abiotic stress conditions, the seeds were sown on $\frac{1}{2}$ MS agar plates containing the indicated stress treatment as described above and placed vertically for 18–22 days. Primary root length and number of lateral roots were analyzed using the ImageJ software (https://imagej.nih.gov/ij/, accessed and downloaded on 19 May 2019) from the images of the plants at day 14 [22].

For short-term exposure to severe abiotic stress conditions, the seeds were germinated and grown on $\frac{1}{2}$ MS for six days, and six-day-old seedlings were transferred to different severe abiotic stress conditions: salt stress (150 mM NaCl), osmotic stress (375 mM sorbitol), oxidative stress (1.5 mM H_2O_2), and heavy metal stress (100 μM $CdCl_2$) for 5 days and then photographed. Primary root length was analyzed using the ImageJ software (https://imagej.nih.gov/ij/, accessed and downloaded on 19 May 2019) [61].

5.4.3. Photosynthesis Fluorescence Measurements

Maximum PSII quantum yield [$F_v/F_m = (F_m - F_o)/F_m$] of rosette leaves was determined using FluorCam FC 800-O (Photon Systems Instruments) and photosynthetic activity was measured at F_v/F_m. 18–21 days old plants were dark-adapted for 30 min before measurements and all fluorescence measurements were performed in vivo at room temperature. A saturating light of 1000 μmol photons m^{-2} s^{-1} was applied to the measure the maximum fluorescence. The data were analyzed using the Fluorcam 7 software (https://fluorcams.psi.cz/, accessed and downloaded on 08 March 2019) [62].

5.5. Plasmid, Yeast Strains and Yeast Growth

The open reading frames (ORF) of AtGSH1 and AtGSH2 were amplified from Arabidopsis cDNA using the primers 5′ATGGCGCTCTTGTCTCAAGC3′ and 5′TTAGTACAGC AGCTCTTCGAAC3′ and primers 5′ATGGGCAGTGGCTGCTCTTC3′ and TCAAATCA-GATATATGCTGTCC, respectively. The amplified products were A-tailed and cloned into pGEM-T Easy vector (Promega), and sub-cloned into the yeast expression vector pDR195 at the NotI sites, resulting in the pDR195-AtGSH1 and pDR195-AtGSH2 constructs [63].

Yeast cells were grown at 30 °C in yeast extract peptone dextrose YPD medium (2% glucose, 2% tryptone, and 1% yeast extract) or synthetic defined (SD) medium (2% glucose, 0.7% yeast nitrogen base with ammonium sulfate, pH 5.5). WT BY4741 (*MATa; his3, leu2, met15, ura3*), *gsh1Δ* (*MATa; his3, leu2, met15, ura3*, gsh1*::kanMX4*), *gsh2 Δ (MATa; his3, leu2, met15, ura3*, gsh12*::kanMX4)* were kindly provided by Dr. Léon Dirick (BPMP Montpellier). Yeast cells were transformed with pDR195 or pDR195-AtGSH1 construct or pDR195-AtGSH2, using the LiAc ssDNA/PEG method [64]. Drop assay was performed as previously described with minor modifications [62,65]. Transformed yeast cells were cultured in liquid SD medium containing amino acids without uracil (URA) overnight at 30 °C. Yeast cells were harvested by centrifugation at 2500 rpm for 4 min and washed twice with sterile water. The cells were resuspended and adjusted to $OD_{600\ nm} = 1$, and 8 μl of 10-fold serial dilutions were spotted onto 2% (*w*/*v*) agar plates containing SD-URA medium alone or supplemented 100 μM $CdCl_2$, 1 M NaCl or 1 M sorbitol. The plates were incubated at 30 °C for 3–4 days before being photographed.

5.6. Statistical Analysis

The data were statistically analyzed using Excel version 2010 and GraphPad Prism 7 program (https://www.graphpad.com/scientific-software/prism/, accessed and downloaded on 14 July 2020). Statistically significant differences were performed by Student's *t*-test (* $p < 0.05$; ** $p < 0.01$) or by ANOVA using the Tukey's Honestly Significant Difference (HSD) test [66].

Supplementary Materials: The following are available online at https://www.mdpi.com/article/10.3390/agronomy11040764/s1, Figure S1: The effects of long-term exposures to different abiotic stress treatments on fresh weight of *cad2-1*, *vtc2-4* and *vtc5-2* mutants, Figure S2: The effects of long-term exposures to different abiotic stress treatments on root architecture of *cad2-1*, *vtc2-4* and *vtc5-2* mutants, Table S1: Seeds of *Arabidopsis thaliana cad2-1*, *vtc2-4* and *vtc5-2* are more resistant to ABA-mediated inhibition of seed germination.

Author Contributions: Conceptualization and Supervision, M.T.T.H.; formal analysis, methodology, software and validation, data curation, investigation, writing—original draft preparation M.T.T.H., M.T.A.D. and T.N.; writing—review and editing, M.T.T.H.; visualization, D.-P.T.; data curation, T.N.C.; project administration, and funding acquisition, P.N.D.Q. and T.P.T.D. All authors have read and agreed to the published version of the manuscript.

Funding: This research was funded by Vietnam National University-Ho Chi Minh City (VNU-HCM) under Grant Number NV2018-18-03 to Thi Phuong Thao Dang.

Institutional Review Board Statement: Not applicable.

Informed Consent Statement: Not applicable.

Data Availability Statement: Data sharing not applicable.

Acknowledgments: We would like to thank Stephane Mari (BPMP, Montpellier, France) for providing *Arabidopsis thaliana* Col-0 and line mutants including *vtc2-4* and *vtc5-2*, Léon Dirick (BPMP, Montpellier, France) for providing yeast mutant strains. We also thank Santiago Alejandro (Martin Luther University, Halle-Wittenberg, Germany), Tuan Minh Tran (Nanyang Technological University, Singapore), Thanh-Hao Nguyen (University of Science, VNU, Ho Chi Minh City, Vietnam) and Ngoc Hoang Bao Bui (University Health Network, Toronto, ON, Canada) for critical reading, editing, valuable suggestions and comments on improving the quality of the article.

Conflicts of Interest: The authors declare no conflict of interest.

References

1. Das, K.; Roychoudhury, A. Reactive oxygen species (ROS) and response of antioxidants as ROS-scavengers during environmental stress in plants. *Front. Environ. Sci.* **2014**, *2*, 53. [CrossRef]
2. Wang, W.; Vinocur, B.; Altman, A. Plant responses to drought, salinity and extreme temperatures: Towards genetic engineering for stress tolerance. *Planta* **2003**, *218*, 1–14. [CrossRef] [PubMed]
3. Foyer, C.H.; Noctor, G. Oxidant and antioxidant signalling in plants: A re-evaluation of the concept of oxidative stress in a physiological context. *Plant Cell Environ.* **2005**, *28*, 1056–1071. [CrossRef]
4. Foyer, C.H.; Noctor, G. Ascorbate and glutathione: The heart of the redox hub. *Plant Physiol.* **2011**, *155*, 2–18. [CrossRef]
5. Lim, B.; Smirnoff, N.; Cobbett, C.S.; Golz, J.F. Ascorbate-deficient VTC2 mutants in arabidopsis do not exhibit decreased growth. *Front. Plant Sci.* **2016**, *7*, 1025. [CrossRef]
6. Wang, Z.; Xiao, Y.; Chen, W.; Tang, K.; Zhang, L. Increased vitamin C content accompanied by an enhanced recycling pathway confers oxidative stress tolerance in Arabidopsis. *J. Integr. Plant Biol.* **2010**, *52*, 400–409. [CrossRef]
7. Yin, L.; Wang, S.; Eltayeb, A.E.; Uddin, M.I.; Yamamoto, Y.; Tsuji, W.; Takeuchi, Y.; Tanaka, K. Overexpression of dehydroascorbate reductase, but not monodehydroascorbate reductase, confers tolerance to Aluminum stress in transgenic Tobacco. *Planta* **2010**, *231*, 609–621. [CrossRef] [PubMed]
8. Liu, W.; An, H.M.; Yang, M. Overexpression of Rosa roxburghii L-galactono-1,4-lactone dehydrogenase in tobacco plant enhances ascorbate accumulation and abiotic stress tolerance. *Acta Physiol. Plant.* **2013**, *35*, 1617–1624. [CrossRef]
9. Bu, Y.; Sun, B.; Zhou, A.; Zhang, X.; Takano, T.; Liu, S. Overexpression of AtOxR gene improves abiotic stresses tolerance and vitamin C content in Arabidopsis thaliana. *BMC Biotechnol.* **2016**, *16*, 69. [CrossRef] [PubMed]
10. Acosta-Gamboa, L.M.; Suxing, L.; Jarrod, W.C.; Zachary, C.C.; Raquel, T.; Walter, P.S.; Jessica, P.Y.C.; Argelia, L. Characterization of the response to abiotic stresses of high ascorbate Arabidopsis lines using phenomic approaches. *Plant Physiol. Biochem.* **2020**, *151*, 500–515. [CrossRef]
11. Zhang, Z.; Wang, J.; Zhang, R.; Huang, R. The ethylene response factor AtERF98 enhances tolerance to salt through the transcriptional activation of ascorbic acid synthesis in Arabidopsis. *Plant J.* **2012**, *71*, 273–287. [CrossRef] [PubMed]
12. Tóth, S.Z.; Puthur, J.T.; Nagy, V.; Garab, G. Experimental evidence for ascorbate-dependent electron transport in leaves with inactive oxygen-evolving complexes. *Plant Physiol.* **2009**, *149*, 1568–1578. [CrossRef] [PubMed]
13. Tóth, S.Z.; Nagy, V.; Puthur, J.T.; Kovács, L.; Garab, G. The physiological role of ascorbate as photosystem II electron donor: Protection against photoinactivation in heat-stressed leaves. *Plant Physiol.* **2011**, *156*, 382–392. [CrossRef] [PubMed]
14. Sultana, N.; Florance, H.V.; Johns, A.; Smirnoff, N. Ascorbate deficiency influences the leaf cell wall glycoproteome in Arabidopsis thaliana. *Plant Cell Environ.* **2015**, *38*, 375–384. [CrossRef] [PubMed]
15. Grillet, L.; Ouerdane, L.; Flis, P.; Hoang, M.T.T.; Isaure, M.P.; Lobinski, R.; Curie, C.; Mari, S. Ascorbate efflux as a new strategy for iron reduction and transport in plants. *J. Biol. Chem.* **2014**, *289*, 2515–2525. [CrossRef]
16. Müller-Moulé, P.; Conklin, P.L.; Niyogi, K.K. Ascorbate deficiency can limit violaxanthin de-epoxidase activity in vivo. *Plant Physiol.* **2002**, *128*, 970–977. [CrossRef]
17. Smirnoff, N. Ascorbic acid metabolism and functions: A comparison of plants and mammals. *Free Radic. Biol. Med.* **2018**, *122*, 116–129. [CrossRef]
18. Smirnoff, N. Vitamin C: The Metabolism and Functions of Ascorbic Acid in Plants. *Adv. Bot. Res.* **2011**, *59*, 107–177.
19. Dowdle, J.; Ishikawa, T.; Gatzek, S.; Rolinski, S.; Smirnoff, N. Two genes in Arabidopsis thaliana encoding GDP-L-galactose phosphorylase are required for ascorbate biosynthesis and seedling viability. *Plant J.* **2007**, *52*, 673–689. [CrossRef]
20. Bangash, S.A.K.; Müller-Schüssele, S.J.; Solbach, D.; Jansen, M.; Fiorani, F.; Schwarzländer, M.; Kopriva, S.; Meyer, A.J. Low-glutathione mutants are impaired in growth but do not show an increased sensitivity to moderate water deficit. *PLoS ONE* **2019**, *14*, e0220589. [CrossRef]
21. Hasanuzzaman, M.; Nahar, K.; Anee, T.I.; Fujita, M. Glutathione in plants: Biosynthesis and physiological role in environmental stress tolerance. *Physiol. Mol. Biol. Plants* **2017**, *23*, 249–268. [CrossRef]
22. Schnaubelt, D.; Schulz, P.; Hannah, M.A.; Yocgo, R.E.; Foyer, C.H. A phenomics approach to the analysis of the influence of glutathione on leaf area and abiotic stress tolerance in Arabidopsis thaliana. *Front. Plant Sci.* **2013**, *4*, 1–9. [CrossRef]
23. Cairns, N.G.; Pasternak, M.; Wachter, A.; Cobbett, C.S.; Meyer, A.J. Maturation of Arabidopsis seeds is dependent on glutathione biosynthesis within the embryo. *Plant Physiol.* **2006**, *141*, 446–455. [CrossRef]
24. Cobbett, C.S.; May, M.J.; Howden, R.; Rolls, B. The glutathione-deficient, cadmium-sensitive mutant, cad2-1, of Arabidopsis thaliana is deficient in γ-glutamylcysteine synthetase. *Plant J.* **1998**, *16*, 73–78. [CrossRef]
25. Bashandy, T.; Guilleminot, J.; Vernoux, T.; Caparros-Ruiz, D.; Ljung, K.; Meyer, Y.; Reichheld, J.P. Interplay between the NADP-linked thioredoxin and glutathione systems in Arabidopsis auxin signaling. *Plant Cell* **2010**, *22*, 376–391. [CrossRef] [PubMed]
26. Dubreuil-Maurizi, C.; Vitecek, J.; Marty, L.; Branciard, L.; Frettinger, P.; Wendehenne, D.; Meyer, A.J.; Mauch, F.; Poinssot, B. Glutathione deficiency of the Arabidopsis mutant pad2-1 affects oxidative stress-related events, defense gene expression, and the hypersensitive response. *Plant Physiol.* **2011**, *157*, 2000–2012. [CrossRef]

27. Cheng, M.C.; Ko, K.; Chang, W.L.; Kuo, W.C.; Chen, G.H.; Lin, T.P. Increased glutathione contributes to stress tolerance and global translational changes in Arabidopsis. *Plant J.* **2015**, *83*, 926–939. [CrossRef] [PubMed]
28. Ullmann, P.; Gondet, L.; Potier, S.; Bach, T.J. Cloning of Arabidopsis thaliana glutathione synthetase (GSH2) by functional complementation of a yeast gsh2 mutant. *Eur. J. Biochem.* **1996**, *669*, 662–669. [CrossRef]
29. Grant, C.M.; MacIver, F.H.; Dawes, I.W. Glutathione synthetase is dispensable for growth under both normal and oxidative stress conditions in the yeast Saccharomyces cerevisiae due to an accumulation of the dipeptide γ-glutamylcysteine. *Mol. Biol. Cell* **1997**, *8*, 1699–1707. [CrossRef] [PubMed]
30. Tuan, P.A.; Kumar, R.; Rehal, P.K.; Toora, P.K.; Ayele, B.T. Molecular mechanisms underlying abscisic acid/gibberellin balance in the control of seed dormancy and germination in cereals. *Front. Plant Sci.* **2018**, *9*, 1–14. [CrossRef]
31. Hendrix, S.; Jozefczak, M.; Wójcik, M.; Deckers, J.; Vangronsveld, J.; Cuypers, A. Glutathione: A key player in metal chelation, nutrient homeostasis, cell cycle regulation and the DNA damage response in cadmium-exposed Arabidopsis thaliana. *Plant Physiol. Biochem.* **2020**, *154*, 498–507. [CrossRef]
32. Simkin, A.J.; López-Calcagno, P.E.; Raines, C.A. Feeding the world: Improving photosynthetic efficiency for sustainable crop production. *J. Exp. Bot.* **2019**, *70*, 1119–1140. [CrossRef] [PubMed]
33. Mittler, R. ROS Are Good. *Trends Plant Sci.* **2017**, *22*, 11–19. [CrossRef] [PubMed]
34. Chen, C.; Letnik, I.; Hacham, Y.; Dobrev, P.; Ben-Daniel, B.H.; Vanková, R.; Amir, R.; Miller, G. ASCORBATE PEROXIDASE6 protects arabidopsis desiccating and germinating seeds from stress and mediates cross talk between reactive oxygen species, Abscisic acid, And auxin. *Plant Physiol.* **2014**, *166*, 370–383. [CrossRef] [PubMed]
35. Liu, Y.; Ye, N.; Liu, R.; Chen, M.; Zhang, J. H_2O_2 mediates the regulation of ABA catabolism and GA biosynthesis in Arabidopsis seed dormancy and germination. *J. Exp. Bot.* **2010**, *61*, 2979–2990. [CrossRef] [PubMed]
36. Pagano, A.; de Sousa Araújo, S.; Macovei, A.; Dondi, D.; Lazzaroni, S.; Balestrazzi, A. Metabolic and gene expression hallmarks of seed germination uncovered by sodium butyrate in Medicago truncatula. *Plant Cell Environ.* **2019**, *42*, 259–269. [CrossRef]
37. Bahin, E.; Bailly, C.; Sotta, B.; Kranner, I.; Corbineau, F.; Leymarie, J. Crosstalk between reactive oxygen species and hormonal signalling pathways regulates grain dormancy in barley. *Plant Cell Environ.* **2011**, *34*, 980–993. [CrossRef]
38. Zhou, R.; Cutler, A.J.; Ambrose, S.J.; Galka, M.M.; Nelson, K.M.; Squires, T.M.; Loewen, M.K.; Jadhav, A.S.; Ross, A.R.S.; Taylor, D.C.; et al. A New Abscisic Acid Catabolic Pathway. *Plant Physiol.* **2004**, *134*, 361–369. [CrossRef]
39. Bailly, C. The signalling role of ROS in the regulation of seed germination and dormancy. *Biochem. J.* **2019**, *476*, 3019–3032. [CrossRef]
40. Muller, K.; Linkies, A.; Vreeburg, R.A.M.; Fry, S.C.; Krieger-Liszkay, A.; Leubner-Metzger, G. In vivo cell wall loosening by hydroxyl radicals during cress seed germination and elongation growth. *Plant Physiol.* **2009**, *150*, 1855–1865. [CrossRef]
41. Marquez-Garcia, B.; Njo, M.; Beeckman, T.; Goormachtig, S.; Foyer, C.H. A new role for glutathione in the regulation of root architecture linked to strigolactones. *Plant Cell Environ.* **2014**, *37*, 488–498. [CrossRef]
42. Parisy, V.; Poinssot, B.; Owsianowski, L.; Buchala, A.; Glazebrook, J.; Mauch, F. Identification of PAD2 as a γ-glutamylcysteine synthetase highlights the importance of glutathione in disease resistance of Arabidopsis. *Plant J.* **2007**, *49*, 159–172. [CrossRef]
43. Vernoux, T.; Wilson, R.C.; Seeley, K.A.; Reichheld, J.P.; Muroy, S.; Brown, S.; Maughan, S.C.; Cobbett, C.S.; Van Montagu, M.; Inzé, D.; et al. The ROOT MERISTEMLESS1/CADMIUM SENSITIVE2 gene defines a glutathione-dependent pathway involved in initiation and maintenance of cell division during postembryonic root development. *Plant Cell* **2000**, *12*, 97–109. [CrossRef]
44. Pasternak, T.; Palme, K.; Paponov, I.A. Glutathione enhances auxin sensitivity in arabidopsis roots. *Biomolecules* **2020**, *10*, 1550. [CrossRef]
45. Brisson, L.; El Bakkali-Taheri, N.; Giorgi, M.; Fadel, A.; Kaizer, J.; Réglier, M.; Tron, T.; Ajandouz, E.H.; Simaan, A.J. 1-Aminocyclopropane-1-carboxylic acid oxidase: Insight into cofactor binding from experimental and theoretical studies. *J. Biol. Inorg. Chem.* **2012**, *17*, 939–949. [CrossRef] [PubMed]
46. Yu, Y.; Wang, J.; Li, S.; Kakan, X.; Zhou, Y.; Miao, Y.; Wang, F.; Qin, H.; Huang, R. Ascorbic acid integrates the antagonistic modulation of ethylene and abscisic acid in the accumulation of reactive oxygen species. *Plant Physiol.* **2019**, *179*, 1861–1875. [CrossRef]
47. Kerchev, P.I.; Pellny, T.K.; Vivancos, P.D.; Kiddle, G.; Hedden, P.; Driscoll, S.; Vanacker, H.; Verrier, P.; Hancock, R.D.; Foyer, C.H. The transcription factor ABi4 Is required for the ascorbic acid-dependent regulation of growth and regulation of jasmonate-dependent defense signaling pathways in arabidopsis. *Plant Cell* **2011**, *23*, 3319–3334. [CrossRef]
48. Giovannoni, J.J. Completing a pathway to plant vitamin C synthesis. *Proc. Natl. Acad. Sci. USA* **2007**, *104*, 9109–9110. [CrossRef] [PubMed]
49. Pavet, V.; Olmos, E.; Kiddle, G.; Mowla, S.; Kumar, S.; Antoniw, J.; Alvarez, M.E.; Foyer, C.H. Ascorbic acid deficiency activates cell death and disease resistance responses in Arabidopsis. *Plant Physiol.* **2005**, *139*, 1291–1303. [CrossRef] [PubMed]
50. Veljovic-Jovanovic, S.D.; Pignocchi, C.; Noctor, G.; Foyer, C.H. Low ascorbic acid in the vtc-1 mutant of arabidopsis is associated with decreased growth and intracellular redistribution of the antioxidant system. *Plant Physiol.* **2001**, *127*, 426–435. [CrossRef]
51. Huang, C.; He, W.; Guo, J.; Chang, X.; Su, P.; Zhang, L. Increased sensitivity to salt stress in an ascorbate-deficient Arabidopsis mutant. *J. Exp. Bot.* **2005**, *56*, 3041–3049. [CrossRef]
52. Jozefczak, M.; Bohler, S.; Schat, H.; Horemans, N.; Guisez, Y.; Remans, T.; Vangronsveld, J.; Cuypers, A. Both the concentration and redox state of glutathione and ascorbate influence the sensitivity of arabidopsis to cadmium. *Ann. Bot.* **2015**, *116*, 601–612. [CrossRef] [PubMed]

53. Mukherjee, M.; Larrimore, K.E.; Ahmed, N.J.; Bedick, T.S.; Barghouthi, N.; Traw, M.B.; Barth, C. Ascorbic acid deficiency in arabidopsis induces constitutive priming that is dependent on hydrogen peroxide, salicylic acid, and the NPR1 gene. *Mol. Plant-Microbe Interact.* **2010**, *23*, 340–351. [CrossRef] [PubMed]
54. Lukowitz, W.; Nickle, T.C.; Meinke, D.W.; Last, R.L.; Conklin, P.L.; Somerville, C.R. Arabidopsis cyt1 mutants are deficient in a mannose-1-phosphate guanylyltransferase and point to a requirement of N-linked glycosylation for cellulose biosynthesis. *Proc. Natl. Acad. Sci. USA* **2001**, *98*, 2262–2267. [CrossRef] [PubMed]
55. Foyer, C.H.; Shigeoka, S. Understanding oxidative stress and antioxidant functions to enhance photosynthesis. *Plant Physiol.* **2011**, *155*, 93–100. [CrossRef] [PubMed]
56. Zechmann, B. Compartment-Specific Importance of Ascorbate During Environmental Stress in Plants. *Antioxid. Redox Signal* **2018**, *29*, 1488–1501. [CrossRef]
57. Mittler, R.; Vanderauwera, S.; Gollery, M.; Van Breusegem, F. Reactive oxygen gene network of plants. *Trends Plant Sci.* **2004**, *9*, 490–498. [CrossRef]
58. Wu, J.; Zhang, N.; Liu, Z.; Liu, S.; Liu, C.; Lin, J.; Yang, H.; Li, S.; Yukawa, Y. The AtGSTU7 gene influences glutathione-dependent seed germination under ABA and osmotic stress in Arabidopsis. *Biochem. Biophys. Res. Commun.* **2020**, *528*, 538–544. [CrossRef]
59. Claeys, H.; Van Landeghem, S.; Dubois, M.; Maleux, K.; Inzé, D. What Is Stress? Dose-response effects in commonly used in vitro stress assays. *Plant Physiol.* **2014**, *165*, 519–527. [CrossRef]
60. Yan, A.; Wu, M.; Yan, L.; Hu, R.; Ali, I.; Gan, Y. AtEXP2 is involved in seed germination and abiotic stress response in Arabidopsis. *PLoS ONE* **2014**, *9*, e85208. [CrossRef] [PubMed]
61. Robles, P.; Navarro-Cartagena, S.; Ferrández-Ayela, A.; Núñez-Delegido, E.; Quesada, V. The characterization of arabidopsis mTERF6 mutants reveals a new role for mterf6 in tolerance to abiotic stress. *Int. J. Mol. Sci.* **2018**, *19*, 2388. [CrossRef] [PubMed]
62. Alejandro, S.; Cailliatte, R.; Alcon, C.; Dirick, L.; Domergue, F.; Correia, D.; Castaings, L.; Briat, J.F.; Mari, S.; Curie, C. Intracellular distribution of manganese by the trans-golgi network transporter NRAMP2 is critical for photosynthesis and cellular Redox homeostasis. *Plant Cell* **2017**, *29*, 3068–3084. [CrossRef] [PubMed]
63. Schaaf, G.; Honsbein, A.; Meda, A.R.; Kirchner, S.; Wipf, D.; Von Wirén, N. AtIREG2 encodes a tonoplast transport protein involved in iron-dependent nickel detoxification in Arabidopsis thaliana roots. *J. Biol. Chem.* **2006**, *281*, 25532–25540. [CrossRef] [PubMed]
64. Gietz, R.D.; Schiestl, R.H. High-efficiency yeast transformation using the LiAc/SS carrier DNA/PEG method. *Nat. Protoc.* **2007**, *2*, 31–34. [CrossRef] [PubMed]
65. Fasano, R.; Gonzalez, N.; Tosco, A.; Dal Piaz, F.; Docimo, T.; Serrano, R.; Grillo, S.; Leone, A.; Inzé, D. Role of arabidopsis UV RESISTANCE LOCUS 8 in plant growth reduction under osmotic stress and low levels of UV-B. *Mol. Plant* **2014**, *7*, 773–791. [CrossRef]
66. Chu, T.N.; Van Bui, L.; Hoang, M.T.T. Pseudomonas PS01 isolated from maize rhizosphere alters root system architecture and promotes plant growth. *Microorganisms* **2020**, *8*, 471. [CrossRef] [PubMed]

agronomy

MDPI

Article

Diversity in Root Architecture of Durum Wheat at Stem Elongation under Drought Stress

Ieva Urbanavičiūtė, Luca Bonfiglioli and Mario A. Pagnotta *

Department of Agricultural and Forest Sciences, Tuscia University, Via S. C. de Lellis, snc, 01100 Viterbo, Italy; ieva.urbanaciviute@unitus.it or ievaurbanaviciute@yahoo.com (I.U.); luca.bonfiglioli@unitus.it or luca.bonf91@gmail.com (L.B.)

* Correspondence: pagnotta@unitus.it

Abstract: Durum wheat is a major crop in the Mediterranean basin, where water deficit is the most important factor affecting its production. Under drought conditions, the root system has a crucial role in crop productivity as a water and nutrition supplier. The aim of the study was to analyze root system diversity in six contrasting durum wheat accessions, including two hydric stress-tolerant genotypes, and to evaluate root traits using the high-throughput phenotyping scanner Win-RHIZO in order to determine the main traits to be used in breeding programs. Six durum wheat accessions were subjected to two drought events under greenhouse conditions from the seedlings stage (BBCH12) for 49 days. Root phenotyping data were validated with results from plants grown in the rainfed field. This study highlighted a great variability among the analyzed genotypes in terms of development, distribution, and architecture of the root system under difficult environments, underlining a good resilience to climate change. Interestingly, the two hydric stress-tolerant genotypes, Cham1 and J. Khetifa, showed different root system ideotypes and rooting patterns under drought conditions. The late flowering landrace J. Khetifa (as also genotypes; Pelsodur and Vulci) showed a steep and long root system ideotype that led to the maintaining of the highest root biomass, length, and volume under drought conditions, while the early flowering genotype Cham1 (as also genotype; Sebatel) was distinguished by a wider root system ideotype, and by increasing the root volume in the topsoil as a strategy to tolerate drought. Moreover, a significant positive correlation was obtained between the root angle of plants grown under greenhouse conditions and plants from the field. Our results demonstrated that screening plant roots in early stages grown under greenhouse conditions using high-throughput phenotyping systems can speed up the selection for crop improvement and future drought stress breeding programs.

Keywords: *Triticum durum*; *Triticum turgidum*; abiotic stress; phenotyping; root architecture; Win-RHIZO

Citation: Urbanavičiūtė, I.; Bonfiglioli, L.; Pagnotta, M.A. Diversity in Root Architecture of Durum Wheat at Stem Elongation under Drought Stress. *Agronomy* **2022**, *12*, 1329. https://doi.org/10.3390/agronomy12061329

Academic Editor: Santiago Signorelli

Received: 22 April 2022
Accepted: 28 May 2022
Published: 30 May 2022

1. Introduction

As a result of climate changes, drought is probably the most severe and unpredictable abiotic stress, significantly affecting crop production due to decreased water and fertilizer availability [1,2]. Drought negatively affects the grain yield and crop production, with a variation reaching 82%, as plants experience both water deficit and the ability for nitrogen uptake and assimilation, which reduce plant vigour [3,4]. Moreover, drought reduces grain yield depending on its intensity, duration, and timing [5,6]. In the Mediterranean basin, water deficit is the most important factor affecting grain yield. Durum wheat (*Triticum turgidum* L. subsp. *durum* (Desf) Husn.) is one of the major cultivated crops according to the importance and cultivated area [7,8]. Furthermore, the Mediterranean basin also has high volatility in rainfall distribution, so water scarcity, with a negative effect on productivity, can also occur in the early growth stages. Several studies revealed that drought affects grain yield more during the vegetative stage (tillering and stem elongation) than in reproductive stages, and can cause about 72% of yield loss [9–11]. Since spikelet initiation starts at the seedling stage to tillering, and floret initiation begins from tillering to stem elongation, both these stages determine spike and spikelet numbers per plant, and in turn grain yield [12]. Therefore, studies on tillering and stem elongation stages should get more attention, as they have a significant impact on grain yield; moreover, selection in early stages can save labour and time.

Root system architecture (RSA) plays a crucial role in crop productivity, especially in drought conditions as a water and nutrition supplier [13,14]. Previous studies have shown that early vigour is important for crop development in dry areas [15,16], and that RSA is directly involved in the resilience of wheat in drought-prone environments [17,18]. The root system cannot be considered only as a whole, as the roots could be divided into seminal, lateral, crown and primary roots, and different root typology could react differently to different external stimuli across different genotypes [19]. However, roots are not easily studied; they are not accessible by non-destructive analyses as opposed to aboveground plant organs, and different methods to study root systems have not always given consistent results. Difficulties in studying root systems increase with plant growth, since a wider root system has a higher percentage of damage during the measurement. Researchers try to avoid these problems by (i) analysing plants at an earlier stage (seedlings) and (ii) using artificial systems or an easy-to-clean substrate. Moreover, it was reported that RSA traits detected on the seedling stage can be used to predict the RSA of adult plants in the field-grown [20] or in an artificial system [21] in order to evaluate the crop adaptation under water stress conditions, but these reports were validated only for what concerned the roots angle and seminal root apparatus.

The studies of root system characteristics have recently received more attention since several high-throughputs, multifunctional root phenotyping platforms have been developed, and the studies of root system characteristics have recently received more attention. Consequently, some root system ideotypes and their growth patterns under different environments have been identified [22]. For example, the root angle is considered a very important feature to select wheat accessions for drought tolerance, since it has high heritability, and is able to give some indications about root ideotypes' capacity for soil water extraction under a drought regime [23,24]. In early stages, wheat genotypes with a narrow root angle have grown deeper compared with varieties that have wider root angles [21,25]. Such findings suggest that wheat with longer roots and with a narrower root angle will be more drought tolerant, as it can reach water from deeper soil layers. In addition, it has been observed that drought-tolerant genotypes have a great number of nodal and seminal roots concentrated in the crown region and located near the surface, while in susceptible genotypes, roots are located far from the top of the soil [26]. Root phenotyping using the high-throughput technique could help to determine root system architectures, traits or ideotypes and rooting patterns, which can be used to select durum wheat for conditions similar to the Mediterranean basin.

The present study was conducted to better understand the genotypic diversity of root architecture in six contrasting durum wheat accessions by evaluating root traits using a high-throughput scanner Win-RHIZO. High-throughput analysis can accelerate the determination of root traits for crop improvement under drought conditions and enable the selection of tolerant and susceptible durum wheat genotypes for breeding programs at the early stages.

2. Materials and Methods

2.1. Plant Materials

The plant materials consisted of six durum wheat accessions with contrasting morphological and stress resistance characteristics, including two hydric stress-tolerant genotypes Cham1 and Jennah Khetifa (J. Khetifa). J. Khetifa is a landrace grown in the dry areas of Algeria and Tunisia that shows specific adaptation to the North African continental dry land; it is tall and resistant to abiotic and biotic stresses [27,28]. The semi-dwarf variety Cham1, selected at ICARDA (the International Center for Agricultural Research in the Dry

Areas) and released for commercial production in several countries of the Mediterranean basin, is characterized by both salt and drought tolerance and yield stability [29,30]. Several studies report that both of these genotypes are hydric stress-tolerant [31–34], and only one mentioned abiotic stress effect on roots [27].

The other four genotypes were chosen according to the flowering time, as Mv-Pelsodur (from now on just Pelsodur) and Vulci were chosen for late, and Azeghar 2-1 (56) (from now on just Azeghar), and Sebatel2 (45) (from now on just Sebatel) were chosen for early [35].

2.2. Greenhouse Experimental Design and Conditions

The six genotypes were sown, with one seed per pot placed with the embryo facing down, in the greenhouse at the Tuscia University experimental farm (Viterbo, Italy), on 20 January 2021. The air temperature ranged from 22 to 28 °C during the day and from 14 to 17 °C during the night. The pots (17 cm diameter, 16 cm high) were filled with 2.5 L of sand. Pots were irrigated three times per week to keep them at 50% soil field capacity (FC), and 80 mL/per pot were filled with a water nutrition solution composed of nitric acid (0.286 mL/L), calcium nitrate (0.432 g/L), potassium nitrate (0.436 g/L), dihydrogen phosphate potassium (0.13 g/L), potassium sulphate (0.04 g/L), magnesium nitrate (0.244 g/L) and Mikron (0.23 g/L) (Prof. Giuseppe Colla, personal communication). The experiment lasted 49 days, and drought treatment was applied two times by discontinuing irrigation; treated plants were re-watered to prevent death. The first drought treatment started when all plants reached early seedling stages (two leaves unfolded—BBCH12) and continued for 14 days. The first drought treatment plants were re-watered (80 mL/per pot) three times per week for two weeks. The second drought treatment, which continued for seven days, started after re-watering (Figure 1).

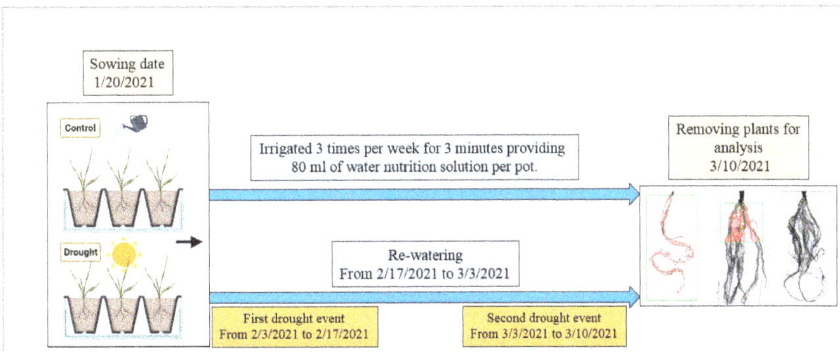

Figure 1. Experimental model.

Drought conditions were managed by discontinuing watering, and the field capacity (FC, %) of the soil in the pots was determined during the drought period by weighing the pots regularly according to Grewal et al. [36] (Figure 2).

2.3. Phenotyping

The stress effects on plant growth and development in durum wheat genotypes were evaluated by several morphological characters above and below ground. All plants from both treatments were collected 49 days after sowing; roots were carefully washed to remove the sand using a soft spray watering head. The morphological traits of the roots were recorded both for the whole root system (W) and separately in the first 5 cm (5) below ground (Figure 3A,B), such as root length (RLW, RL5 sum of all roots lengths, cm), surface area (SAW, SA5 total root surface area, cm^2), root volume (RVW, RV5 total root volume, cm^3), number of tips (TIW, TI5), forks (FRW, FR5), and crossings (CRW, CR5) using Win-RHIZO Pro software v2009 (version 4.0b; Regent Instruments, Montreal, QC, Canada). The root angle (RA°) was measured using the software *ImageJ* when the angle between the two extreme sides of the roots with the center set in the middle of the crown, as shown in Figure 3C. After measurements, the plants were separated into roots and shoots, dried in an oven (80 °C for 12 h), and weighed for shoot and root dry weight (SDW, RDW). In addition, the number of leaves (NL), and number of tillers (NT) were assessed. Moreover, some ratios were calculated, such as the root/shoot (RS) ratio, to determine which above-ground or below-ground part of the plant is dominant. The ratio of total root length and number of tips (RL/TI) was also evaluated, since it could show root system pattern, i.e., the capacity to generate a longer root system or to increase the number of new roots under stress conditions. To determine root system distribution at different soil layers, the ratio of some root traits between the whole system and topsoil was evaluated, such as RL5/RLW, SA5/SAW, and RV5/RVW. Root length density (RLD) was calculated using the following equation: RLW/Soil volume (cm/cm^{-3}). The recorded raw data are public available (after an access request) at the Zenodo web site with DOI: 10.5281/zenodo.5883299.

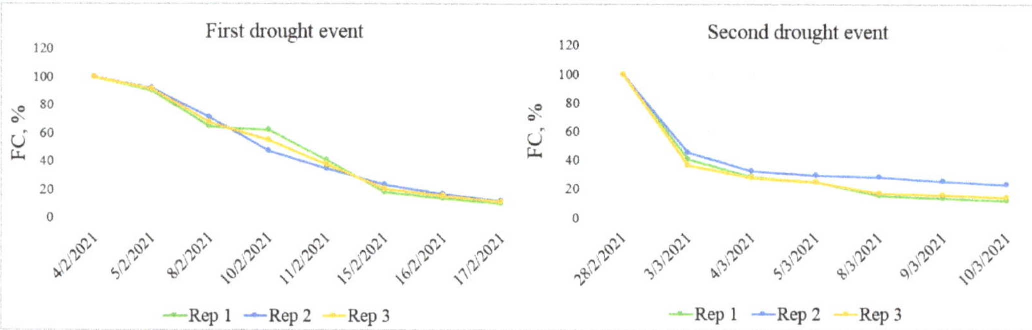

Figure 2. Pots' water content in terms of percentage of field capacity (FC, %) for first and second drought treatments.

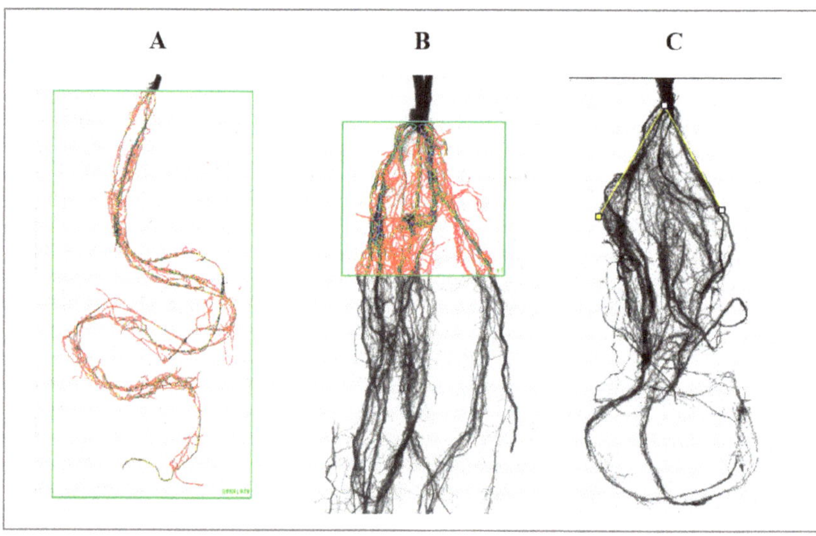

Figure 3. The measurements of morphological traits for whole roots (**A**), topsoil (**B**) using Win-RHIZO system, and root angle (**C**) using *ImageJ* (°).

2.4. Field Experiment

To validate the results obtained in greenhouse conditions, the same genotypes were sown in rain fed field conditions without any specific treatment (i.e., natural agronomic conditions). A field experiment was performed at Tuscia University experimental farm (Viterbo, Italy). The experimental design was a randomized complete block design with three replicates. Plot size was 1.5 × 5.2 m (7.8 m^2) and sowed with 234.7 g seeds per plot on 11 November 2021. The soil was sandy loam, the pH was 6.9 and organic matter was 14.8 g kg^{-1}.

Plants were collected on 17 March 2022, when all genotypes reached the end of tillering and the beginning of stem elongation stages. Plant samples from the field were collected very carefully so as not to damage the root system of the topsoil, at 5 cm deep. Roots were carefully washed to remove soil and analyzed for root angle (°) using *ImageJ* software. The morphological traits of the roots of plants from field such as root length (RLF), surface area (SAF), root volume (RVF), number of tips (TIF), forks (FRF), and crossings (CRF) were recorded using Win-RHIZO Pro software v2009 (version 4.0b; Regent Instruments, Montreal, QC, Canada).

2.5. Statistical Analysis

All of the statistical analyses were performed using R Studio (Version R-4.1.0). Two-way analysis of variance (ANOVA) was conducted at a significance level of 5% using the aov() function, while, one-way ANOVA was used to test the variance component of each trait under each treatment, with genotype as a factor. Fischer's least significant difference (LSD) test was used for mean comparisons. The correlation matrix between all traits was constructed and Pearson correlation coefficients were calculated using the corrplot function R package [37]. Principal component analysis was performed using the prcomp() function and then biplot was generated with the ggbiplot function R package [38].

3. Results

3.1. Drought Effect on Shoot Growth and Development in Greenhouse Experiment

Recurrent short events of drought at the vegetative stage, from two fully expanded leaves to stem elongation, had adverse effects and growth reductions both below and above-ground in all genotypes. Compared with the control condition, water deficit significantly reduced shoot dry weight (SDW) in the six investigated durum wheat genotypes (Table 1). The ANOVA revealed the presence of highly significant differences among genotypes and treatments for shoot traits, such as number of leaves (NL) and number of tillers (NT) per plant, while there were no statistical differences among genotypes for shoot dry weight (in control condition) (Table 1). Under control conditions, the studied genotypes were divided into two groups in accord with the number of leaves produced: J. Khetifa, Pelsodur, and Vulci, with a higher number of leaves (on average 32 leaves), and Cham1, Azeghar, and Sebatel with a lower number of leaves (on average 19 leaves) (Table 1). Furthermore, J. Khetifa and Pelsodur also produced a higher number of tillers (on average 10 tillers) in comparison with the other genotypes (on average four tillers). Not surprisingly, in general, NL and NT are reduced by the drought stress (Table 1). Even so, these decreases under drought conditions have different degrees in the different genotypes. Vulci and Cham1 have a stronger drought effect on NL (−47.2% and −62.3% respectively) than on NT (−23.5% and −42.9% respectively), while Azeghar loses more tillers than leaves (−66.7% and −55.3% respectively) (Table 1).

Interestingly, J. Khetifa and Pelsodur have similar degrees of reduction in both the number of leaves and number of tillers, while the main effect of treatment was not significant for the Sebatel genotype. Results confirm that the development dynamics of the number of tillers and number of leaves highly depend not only on environment conditions but also by genotype inner capacity. This is also highlighted by the significant G × T interaction (Table 1). Moreover, regarding shoot biomass (SDW), Sebatel, J. Khetifa, Pelsodur and Vulci maintained significantly higher SDW than Cham1 and Azeghar under drought conditions. That means that these genotypes are more tolerant to drought according to their successful development of shoots under stress conditions.

Table 1. Drought effect on shoot growth and development in greenhouse experiment. D/C—drought vs. control.

Genotype	Shoot Dry Weight (g)			Number of Leaves			Number of Tillers		
	Control	Drought	D/C, %	Control	Drought	D/C, %	Control	Drought	D/C, %
Azeghar	2.1 ± 0.1a	0.51 ± 0.1c	−75.7 ***	16 ± 3d	7b	−55.3 *	3d	1c	−66.7 *
Cham1	2.1 ± 0.1a	0.50 ± 0.1bc	−76.3 ***	23 ± 2c	99 ± 1b	−62.3 ***	5 ± 1bc	3 ± 1b	−42.9 *
J. Khetifa	2.6 ± 0.2a	0.63 ± 0.2a	−75.9 ***	37 ± 1a	16 ± 1a	−56.4 ***	9a	4 ± 1a	−51.9 ***
Pelsodur	1.6 ± 0.1b	0.58 ± 0.1a	−63.7 *	31 ± 1ab	15 ± 2a	−51.1 ***	10 ± 1a	5 ± 1a	−50.0 ***
Sebatel	2.1 ± 0.1a	0.66 ± 0.1a	−68.4 **	17 ± 4d	10 ± 1b	−43.1ns	4 ± 1cd	2 ± 1b	−36.4ns
Vulci	2.0 ± 0.1a	0.55 ± 0.1ab	−72.1 **	30 ± 6b	16 ± 3a	−47.2 ***	6 ± 1b	4 ± 1a	−23.5ns
Genotype		ns			***			***	
Treatments		***			***			***	
G × T		ns			**			***	

Values are means ± standard deviations (n = 3). Means with same letter in each column are not significantly different between genotypes ($p < 0.05$) (LSD test). ns—Not significant; *, **, and *** indicate significance at $p < 0.05$, $p < 0.01$, and $p < 0.001$ levels, respectively.

3.2. Drought Effect on Whole Root System in Greenhouse Experiment

Under control conditions, no significant differences between genotypes for some root traits, such as total root length (RLW), root surface area (SAW), root tips (TIW), and root length density (RLD), were found. However, the effect of recurrent short events of drought was significant and varied between genotypes (Table 2).

Table 2. Drought effect on the whole root growth and plants development.

Genotype	RVW (cm^3)		RLW (cm)		TIW		RDW (g)	
	Control	Drought	Control	Drought	Control	Drought	Control	Drought
Azeghar	5.3 ± 1.1b	0.97 ± 0.3c	989 ± 84	569 ± 186b	1989 ± 217	1314 ± 329b	0.53 ± 0.1c	0.17 ± 0.1c
Cham1	7.0 ± 0.3b	1.5 ± 0.5bc	1056 ± 182	816 ± 125ab	2009 ± 312	2299 ± 524a	0.83 ± 0.1a	0.26 ± 0.1ab
J. Khetifa	9.7 ± 1.9a	2.2 ± 0.2ab	973 ± 124	887 ± 107a	2218 ± 309	1970 ± 168ab	0.86 ± 0.1a	0.30 ± 0.1a
Pelsodur	6.4 ± 0.9b	2.7 ± 0.4a	835 ± 187	912 ± 190a	1775 ± 556	1812 ± 438ab	0.68 ± 0.1b	0.36 ± 0.1a
Sebatel	6.1 ± 1.6b	1.5 ± 0.1abc	921 ± 231	806 ± 168ab	1704 ± 385	1799 ± 489ab	0.62 ± 0.1b	0.25 ± 0.1ab
Vulci	9.4 ± 0.5a	2.4 ± 1.5ab	992 ± 93	768 ± 36ab	2024 ± 68	1742 ± 316ab	0.72 ± 0.1b	0.29 ± 0.1a
Genotype	***		ns		ns		*	
Treatments	***		**		ns		***	
G × T	**		ns		ns		ns	

Genotype	Root/Shoot Ratio		SAW (cm^2)		Root Angle (°)		RLD ($cm\ cm^{-3}$)	
	Control	Drought	Control	Drought	Control	Drought	Control	Drought
Azeghar	0.25 ± 0.1c	0.33 ± 0.1c	256 ± 33	83 ± 26c	125 ± 4a	108 ± 5b	0.39 ± 0.03	0.23 ± 0.07b
Cham1	0.42 ± 0.2ab	0.51 ± 0.0b	304 ± 28	123 ± 24bc	115 ± 8ab	121 ± 1a	0.42 ± 0.07	0.33 ± 0.05ab
J. Khetifa	0.33 ± 0.1abc	0.48 ± 0.1b	345 ± 56	156 ± 15ab	102 ± 2c	105 ± 4b	0.39 ± 0.05	0.35 ± 0.04a
Pelsodur	0.46 ± 0.1a	0.63 ± 0.1a	281 ± 72	175 ± 15a	94 ± 2c	98 ± 3c	0.33 ± 0.07	0.36 ± 0.08a
Sebatel	0.3 ± 0.02bc	0.38 ± 0.0c	265 ± 68	124 ± 14bc	114 ± 9b	121 ± 1a	0.37 ± 0.09	0.32 ± 0.07ab
Vulci	0.43 ± 0.02ab	0.67 ± 0.1a	315 ± 36	147 ± 50ab	114 ± 6b	106 ± 1b	0.31 ± 0.04	0.31 ± 0.01ab
Genotype	***		*		***		ns	
Treatments	***		***		ns		**	
G × T	ns		ns		***		ns	

Values are means ± standard deviations (n = 3). Means with same letter in each column are not significantly different between genotypes ($p < 0.05$) (LSD test). RVW—total root volume; RLW—sum of all root lengths; TIW—total number of tips; RDW—root dry weight; SAW—total root surface area; RLD—root length density. ns—Not significant; *, **, and *** indicate significance at $p < 0.05$, $p < 0.01$, and $p < 0.001$ levels, respectively.

The RLW of Azeghar shortened by about 40% in drought compared with control (Table 2), while, Pelsodur, which had the lowest RLW under control conditions, grew slightly during the drought and had the longest roots. RLW decreased by about 23% in Vulci and Cham1, while in J. Khetifa it was most stable, having the smallest difference between drought and control (Table 2). The root surface area (SAW) appeared to be a quite important characteristic, as it has a crucial role for water and nutrition uptake efficiency. Interestingly, no significant differences were detected between genotypes for SAW under control conditions (Table 2). Contrarily, drought affected the root surface area in the different genotypes differently. Pelsodur maintained the highest value of SAW, while Azeghar the lowest. Significant differences clustered genotypes into those with higher

(Pelsodur, Vulci, and J. Khetifa) and those with lower (Azeghar, Cham1, and Sebatel) root surface area under drought, where two putative hydric stress-tolerant genotypes J. Khetifa and Cham 1 were separated. In addition, it was observed that genotypes were allocated into the same groups according to phenological development under control conditions. Cham1 was assigned to the early flowering group together with Sebatel and Azeghar. Tolerant landrace J. Khetifa showed a late flowering habit and was grouped with late Pelsodur and Vulci. No significant differences were detected between genotypes for RLD (Table 2). However, root length density was affected by drought. The trends indicated that J. Khetifa and Pelsodur were affected the least, and maintained the highest RLD. Meanwhile, the RLD of Azeghar was above 40 percent, and it had the lowest RLD in drought conditions.

Highly significant differences among genotypes and treatments were obtained for total root volume (RVW). Under control conditions, J. Khetifa and Vulci had significantly larger RVW compared to other genotypes (Table 2). However, under drought conditions, Pelsodur maintained the highest RVW, while Azeghar experienced great losses and had the lowest RVW. Genotypes were divided by root volume values under drought treatment into three groups; (i) Pelsodur, Vulci and J. Khetifa, with a root volume over 2 cm^3; (ii) Cham1 and Sebatel, with RVW less than 2 cm^3; while (iii) Azeghar was less than 1 cm^3 (Table 2). Drought effects also varied between genotypes in terms of number of tips (TIW). Although Sebatel and Pelsodur had the lowest number of tips under control conditions, under drought TIW remained the same. Azeghar had highest TIW losses (around 33.9%), followed by Vulci (−14.4%) and J. Khetifa (−9%), while only Cham1 increased the number of TIW (+14%) under drought conditions. According to the ratio of root length by the number of tips (RLW/TIW), J. Khetifa and Pelsodur had the smallest drought effect on both, growing new roots and maintaining the root length (Figure 4). However, Cham1 and Sebatel increased the number of new roots under drought conditions, since the RLW/TIW ratio decreased from 33% to 17% (Figure 4).

The short drought events had a significant negative effect on root biomass (RDW), even if the effect was different among genotypes. Under control conditions, J. Khetifa and Cham1 produced the highest root biomass, but Pelsodur, Vulci, and J. Khetifa had the highest root mass under drought. Azeghar had the lowest root biomass under both conditions. Genotypes could be distinguished by different root system angles, where Azeghar had the widest root angle, J. Khetifa, and Pelsodur the steepest, and Vulci, Sebatel, and Cham1 were of medium root angle. However, there were highly significant differences between genotypes and their response, as indicated by the significant interaction (G × T) (Table 2). The RA of Azeghar and Vulci narrowed, passing from control to drought, while in all other genotypes it flattened (Table 2). According to RDW and root system angle, six investigated genotypes showed five root ideotypes under control (C) conditions. These are: (iC) widest angle and lowest root biomass (i.e., Azeghar); (iiC) moderate root angle moderate biomass (i.e., Sebatel and Vulci); (iiiC) moderate root angle and high biomass (i.e., Cham1); (ivC) narrow root angle moderate biomass (i.e., Pelsodur); and (vC) narrow root angle high biomass (i.e., J. Khetifa). However, under drought (D) conditions, four root ideotypes were highlighted according to the same traits, root angle and biomass. These are: (iD) widest root angle and retained moderate root biomass (i.e., Sebatel and Cham1); (iiD) moderate root angle and low root biomass (i.e., Azeghar); (iiiD) moderate root angle and high root biomass (i.e., J. Khetifa, and Vulci); and (ivD), narrowest root angle and high root biomass (i.e., Pelsodur). In terms of overall plant development, as shown by the proportion of shoots and roots, three groups were highlighted under drought stress. (i) Pelsodur with Vulci had the highest root shoot ratio, (ii) Sebatel with Azeghar had the lowest, and (iii) two tolerant genotypes, J. Khetifa with Cham1, fell in between.

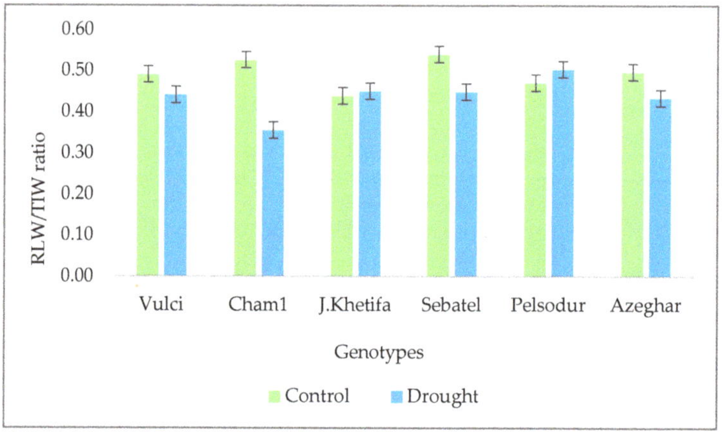

Figure 4. The ratio of sum of all root lengths and total number of tips (RLW/TIW) under greenhouse conditions.

3.3. Drought Effect on Root Parameters at Topsoil in Greenhouse Experiment

At a depth of the first 5 cm, the total root length (RL5) had no significant differences between genotypes under control conditions (Table 3), just like the total length of the whole root (Table 2).

Table 3. Root characteristics at a depth of the first 5 cm below ground in greenhouse experiment.

Genotype	RL5 Root Length (cm)		SA5 Root Surface Area (cm^2)		TI5 Tips		RV5 Root Volume (cm^3)	
	Control	Drought	Control	Drought	Control	Drought	Control	Drought
Azeghar	144 ± 35	96 ± 12c	40 ± 12b	17. ± 1.5b	341 ± 60	181 ± 43b	0.9 ± 0.3	0.24 ± 0.03b
Cham1	149 ± 33	124 ± 7bc	46 ± 9ab	21 ± 3b	311 ± 82	375 ± 38a	1.14 ± 0.2	0.28 ± 0.07b
J. Khetifa.	102 ± 27	181 ± 28ab	57 ± 15ab	34 ± 3a	231 ± 79	451 ± 60a	2.88 ± 0.9	0.52 ± 0.07ab
Pelsodur	133 ± 4	217 ± 49a	51 ± 3ab	36 ± 4a	283 ± 63	396 ± 89a	1.58 ± 0.2	0.49 ± 0.01ab
Sebatel	120 ± 42	202 ± 58a	43 ± 15b	36 ± 7a	246 ± 97	435 ± 137a	1.26 ± 0.4	0.52 ± 0.05ab
Vulci	138 ± 71	140 ± 16bc	63 ± 3a	33 ± 13a	261 ± 97	326 ± 57a	3.04 ± 2.3	0.65 ± 0.44a
Genotype	ns		**		ns		ns	
Treatments	***		***		**		***	
G × T	ns		ns		ns		ns	
Genotype	**(RL5/RLW) * 100**		**(SA5/SAW) * 100**		**RL5/TI5**		**(RV5/RVW) * 100**	
	Control	Drought	Control	Drought	Control	Drought	Control	Drought
Azeghar	14.7	17.8bc	15.6ab	21.4bc	0.4	0.5	16.6	26.5
Cham1	14.2	15.4c	15.1b	17.2c	0.5	0.3	16.3	19.4
J. Khetifa	10.6	20.5ab	16.5ab	22.0bc	0.4	0.4	28.3	23.7
Pelsodur	16.5	23.7a	18.9ab	20.7bc	0.5	0.6	21.7	18.1
Sebatel	12.8	24.8a	16.2ab	29.1a	0.5	0.5	20.6	34.3
Vulci	13.6	18.2bc	20.1a	22.5b	0.5	0.4	37.2	28.3
Genotype	ns		**		ns		ns	
Treatments	***		***		ns		ns	
G × T	ns		*		ns		ns	

Values are means ± standard deviations (n = 3). Means with same letter in each column are not significantly different between genotypes ($p < 0.05$) (LSD test). RL5—sum of root lengths in the first 5 cm below the ground; SA5—root surface area in the first 5 cm below the ground; TI5—number of tips in the first 5 cm below the ground; RV5—root volume in the first 5 cm below the ground. Description is added i.e. "ns—Not significant; *, **, and *** indicate significance at $p < 0.05$, $p < 0.01$, and $p < 0.001$ levels, respectively".

In order to determine the peculiarities of rooting in different soil layers, the ratio between the root traits was also recorded on the whole root and the ones recorded on the topsoil layer (first 5 cm) were calculated and expressed as a percentage. Under drought, the root reduction in the different layers varied among genotypes, while at 5 cm topsoil the total root length significantly increased for J. Khetifa, Pelsodur, and Sebatel. The largest decrease of RL5 was found in Azeghar (about 30%), then in Cham1 (about 17%). Vulci remained almost the same RL5 under drought as under control conditions. According to the RL5/RLW ratio, in the first 5 cm below ground, the root length ranged from 10 to 16 percent of the whole root length under control conditions, while under drought it increased (from 15 to 25%); indicating a higher root concentration in the first soil layer.

Nevertheless, the effect of drought on root length distribution in different layers varied among genotypes, even if not statistically. According to the RL5/RLW ratio, J. Khetifa, Sebatel, and Pelsodur showed an increase of root length at topsoil under drought conditions, which almost doubled, while Cham1 maintained almost the same root distribution between layers under both conditions. Although the effect of drought on the roots length at different depths varied, all genotypes under drought conditions have increased concentration of the roots at topsoil (Table 3). The number of TI5 at a depth of first 5 cm (TI5, Table 3) had no significant differences between genotypes under control conditions, but the response to drought showed Azeghar to have the lowest tip number. Under stress conditions, Azeghar lost about 50% of TI5, while other genotypes increased it, especially J. Khetifa and Sebatel, where TI5 doubled. Regarding the ratio between root length and number of TI5 at the first 5 cm below ground (RL5/TI5), some genotypes maintained the same pattern under drought as the ratio of the whole plant (RLW/TIW) (Figure 4). J. Khetifa had the smallest drought effect on developing new roots and maintaining the length in both the upper and deeper layers of the soil (Table 3). Sebatel, in contrast, showed smaller changes at the topsoil level. Interestingly, Azeghar applied different strategies at different layers; when the whole plant RLW/TIW ratio decreased, at the topsoil level this ratio (RL5/TI5) increased. According to ratio SAW/SA5, at a depth of five centimeters under control conditions, the surface area (SA5) accounted for 15 to 20 percent of the whole root SAW, where Vulci had the highest SA5 and Azeghar with Sebatel had the lowest SA5. Under drought conditions the ratio SAW/SA5 increased (about 17 to 30%) and varied between genotypes (Table 3). Pelsodur, Sebatel, J. Khetifa, and Vulci maintained significantly higher SA5 compared with Azeghar and Cham1. The smallest changes in SA5 between control and drought conditions were observed for Sebatel.

Comparing the distribution of SAW under drought conditions among soil layers, the smallest changes in the SA5/SAW ratio were found for Pelsodur, Cham1, and Vulci. Furthermore, drought root volume at a depth of the first 5 cm (RV5) significantly decreased. Also, the distribution of root volume among layers varied between genotypes. Sebatel had the most concentrated root volume at 5 cm depth under drought conditions (Table 3). Moreover, Azeghar and Cham1 also increased RV5 concentration at topsoil under drought, while J. Khetifa, Vulci, and Pelsodur maintained a higher volume of the whole root systems.

3.4. The Comparison of Root System between Greenhouse and Field

The total root volume (RVF) and the number of crossings (CRF) of plants collected from field experiments, at a depth of first 5 cm (Figure 5) had no significant differences between genotypes (Table 4). However, the two tolerant genotypes, Cham1 and J. Khetifa, had significantly higher root length (RLF) and root surface area (SAF). J. Khetifa also had a significantly higher number of root tips (TIF) and forks (FRF).

Figure 5. Pictures of six durum wheat genotypes grown in field.

Table 4. Root characteristics of plants collected from field at a depth of the first 5 cm below ground.

	Azeghar	Cham1	J. Khetifa	Pelsodur	Sebatel	Vulci	ANOVA
RVF, cm^3	0.2 ± 0.1	0.3 ± 0.1	0.2 ± 0.1	0.2 ± 0.1	0.2 ± 0.1	0.3 ± 0.0	ns
RLF, cm	59 ± 4cd	89 ± 6a	84 ± 15ab	49 ± 6de	38 ± 3e	69 ± 11bc	***
SAF, cm^2	11.8 ± 2.1bc	19.1 ± 3.4a	14.5 ± 1.3b	11.3 ± 2.7bc	8.2 ± 2.9c	15.0 ± 2.2ab	**
TIF	224 ± 48bc	266 ± 44b	390 ± 65a	218 ± 83bc	142 ± 30c	261 ± 20b	**
FRF	223 ± 5bc	345 ± 34ab	449 ± 88a	248 ± 84bc	114 ± 50c	446 ± 155a	**
CRF	22 ± 8	28 ± 7	50 ± 29	15 ± 3	8 ± 2	44 ± 29	ns
RAF, °	112 ± 6a	102 ± 5ab	93 ± 6bc	84 ± 7cd	105 ± 5ab	76 ± 13d	***

Values are means ± standard deviations ($n = 3$). Means with same letter in the line are not significantly different between genotypes ($p < 0.05$) (LSD test). RVF—root volume in the first 5 cm below the ground of plants from field; RLF—sum of all root lengths in the first 5 cm below the ground of plants from field of plants from field; SAF—root surface area in the first 5 cm below the ground of plants from field of plants from field; TIF—number of tips in the first 5 cm below the ground of plants from field of plants from field; FRF—number of forks in the first 5 cm below the ground of plants from field of plants from field; CRF—number of crossings in the first 5 cm below the ground of plants from field of plants from field; RAF—root angle of plants grown in field. ns—Not significant; **, and *** indicate significance at $p < 0.01$ and $p < 0.001$ levels, respectively.

The comparison between genotype root angles, measured using *ImageJ* software in the greenhouse under control condition and in the field, showed the same trend, where Azeghar had the widest root angle, J. Khetifa and Pelsodur the steepest, and Sebatel and Cham1 were between wide and narrow (Tables 2 and 4).

Although all genotypes showed lower values of root angle in the field than in the greenhouse, Vulci had the greatest differences compared with control conditions in the greenhouse, where the root angle was wider. This could be explained by the fact that the plants were grown by a single plant per pot in a greenhouse, without any competition, unlike in the field, where the plants were sown in dense conditions.

3.5. Correlation among Traits

The correlation matrix among traits under the control condition (Figure 6) shows that the number of leaves (NL) and number of tillers (NT) had a highly significant positive correlation between themselves and the root volume at topsoil (RV5). Moreover, a significant positive correlation was found for traits such as root shoot ratio (RS), root dry weight (RDW), root volume (RVW), and root surface area (SAW; SA5).

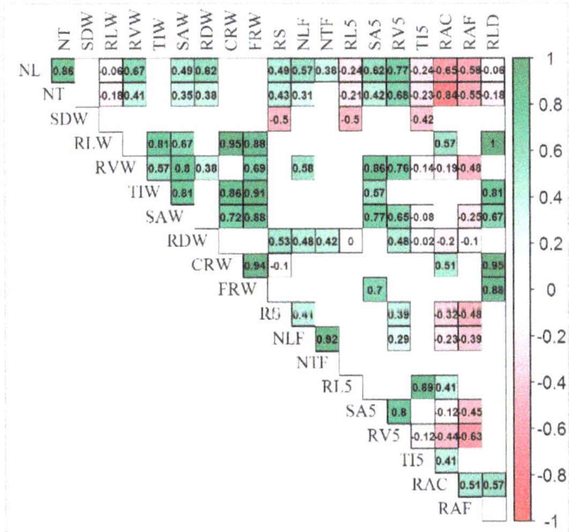

Figure 6. Positive significant correlation (in green) and negative significant correlation (in red) among all traits under control conditions. NL—number of leaves; NT—number of tillers; SDW—shoot dry weight; RVW—total root volume; RLW—sum of all root lengths; TIW—total number of tips; RDW—root dry weight; SAW—total root surface area, RLD—root length density, CRW—total number of crossings; FRW—total number of forks; RS—root shoot ratio; RL5—sum of root lengths in the first 5 cm below the ground; SA5—root surface area in the first 5 cm below the ground; TI5—number of tips in the first 5 cm below the ground; RV5—root volume in the first 5 cm below the ground; RAC—root angle of plants grown in greenhouse as control; RAD—root angle of plants grown in greenhouse under drought conditions; RAF—root angle of plants grown in the field.

Furthermore, both NL and NT had highly significant negative correlation with root angle under both greenhouse (RAC) and field conditions (RAF). Root length (RLW) under control conditions had a highly significant positive correlation with root angle (RA), number of leaves (NL), and number of root elements such as tips (TIW), forks (FRW), and crosses (CRW). Root volume (RVW) under control conditions had a highly significant positive correlation with RV5, SAW, SA5, and FRW.

Interesting results were found in the relation between traits under drought conditions and root system angle under all conditions: control (RAC), drought (RAD), and field (RAF) (Figure 7). For example, between almost all traits, statistically positive significant correlations were obtained, except shoot biomass (SDW), and root angle under all conditions, which had a significant negative correlation. However, root angles from all conditions (RAC, RAD, and RAF), had a significant positive correlation between each other and showed a similar trend in the relation with other traits such as a significant negative correlation with almost all traits under drought conditions (Figure 7). Furthermore, a significant negative correlation (-0.57) between root angle under control (RAC) and root length under drought stress (RLW) was obtained, which means that genotypes with a narrower root angle in control conditions retained longer roots in water deficit environments. Moreover, a significant negative correlation was obtained between the root angle of plants under all conditions (control, drought, and field) and plants root biomass under drought conditions (RDW), which means that genotypes with a narrower root angle can grow higher root biomass. Highly statistically ($p < 0.001$) significant positive correlations were detected between the main traits of the whole root system and topsoil, such as RL and RL5 (0.8), RV and RV5 (0.8), SA and SA5 (0.8), TI and TI5 (0.7).

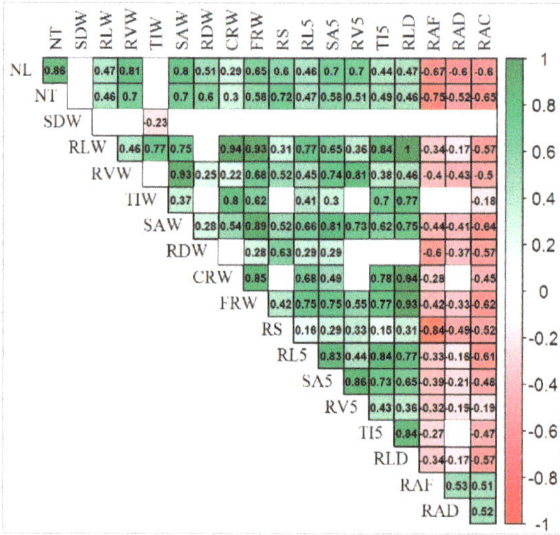

Figure 7. Positive significant correlation (in green) and negative significant correlation (in red) among all traits under drought conditions and root angle under different conditions. NL—number of leaves; NT—number of tillers; SDW—shoot dry weight; RVW—total root volume; RLW—sum of all root lengths; TIW—total number of tips; RDW—root dry weight; SAW—total root surface area, RLD—root length density, CRW—total number of crossings; FRW—total number of forks; RS—root shoot ratio; RL5—sum of root lengths in the first 5 cm below the ground; SA5—root surface area in the first 5 cm below the ground; TI5—number of tips in the first 5 cm below the ground; RV5—root volume in the first 5 cm below the ground; RAC—root angle of plants grown in greenhouse as control; RAD—root angle of plants grown in greenhouse under drought conditions; RAF—root angle of plants grown in the field.

3.6. PCA in Response to Drought

The three top PCA account for 93.9% of the total variation of shoot root traits under drought stress and root angle from field plants (RAF) (Table S1). PC1 accounted for 66.0% of the total variation and was strongly influenced by several traits (i.e., RLW, RVW, SAW, SA5, NL, NT, and RDW). PC2 accounted for 16.8% of the total variation and was mainly associated with RS, -RL5, -TI5, -RAD, AND -RAF, while PC3 accounted for 11.1% and was strongly associated with TIW, -SDW, and -RV5. The first two principal components, explaining 82.8% of the total variation, divided genotypes into two groups (I) and (II) (Figure 8).

Pelsodur and Vulci were assigned to cluster (I) with the drought tolerant landrace J. Khetifa, due to their positive association with traits on the positive side of PC1, while the Sebatel and Azeghar with tolerant Cham1 under drought conditions are on the negative side of PC1, associated with root angle under drought (RAD) and field (RAF), and assigned to cluster (II). Interestingly, these two groups overlapped with genotype shoot development either early or late flowering. Moreover, the same groups were highlighted regarding RV changes at different soil layers; early flowering growth genotypes had the lowest RV, but maintained higher root volume at 5 cm depth compared with a group of late genotypes. However, cluster (II) was divided into two subgroups, where Azeghar (IIA) was separated from the Cham1 and Sebatel cluster (IIB) due to the largest losses in the aboveground part of the plant under drought conditions.

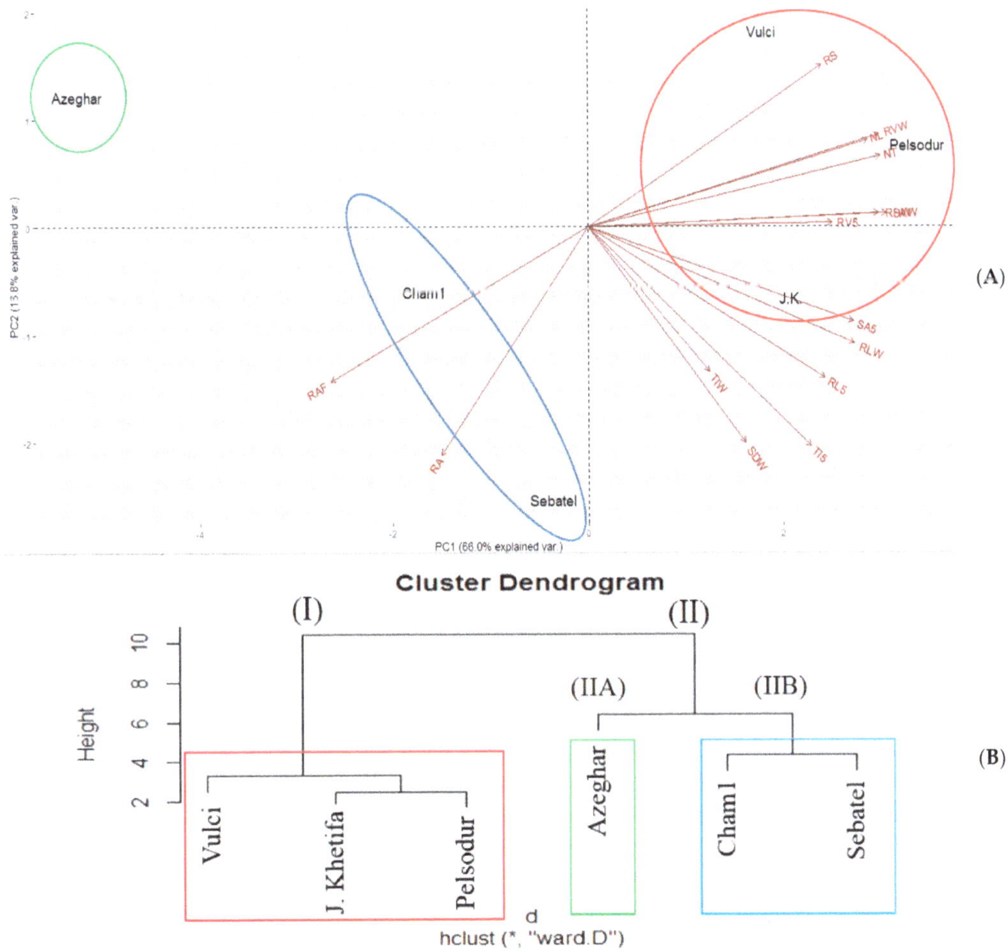

Figure 8. Principal component analysis (**A**) and hierarchical cluster dendrogram applying Ward's method (**B**) of the six investigated durum wheat genotypes under drought conditions. NL—number of leaves; NT—number of tillers; SDW—shoot dry weight; RVW—total root volume; RLW—sum of all root lengths; TIW—total number of tips; RDW—root dry weight; SAW—total root surface area; RS—root shoot ratio; RL5—sum of root lengths in the first 5 cm below the ground; SA5—root surface area in the first 5 cm below the ground; TI5—number of tips in the first 5 cm below the ground; RV5—root volume in the first 5 cm below the ground; RA—root angle of plants grown in greenhouse under drought conditions; RAF—root angle of plants grown in the field. * indicates significance at $p < 0.05$.

The first three components of PCA considering all traits under both drought and control conditions accounted for 89.9% of total variation: 57.7% to PC1, and 17.0% to PC2, and 14.4% to PC3 (Table S2). Interestingly, all the genotypes under drought stress are grouped on the positive side of PC1 with more or less the same values (ranging from about 2 to 4). While the same genotypes under control conditions are on the negative side of PC1 with different values, AzegarC is close to zero, while J. Khetifa (JKC) is almost at −5 of PC1. Considering that PC1 is mainly determined by –RVW, -RV5, -RDW, -SAW, SA5, -NL, and -SDW, these traits are different among the genotypes under control (ending with C in Figure 9) and drought (ending with D in Figure 9) conditions, and in particularly are negatively associated (decreased) in the case of drought. On the other hand, PC2, which is mainly determined by RLW, RS, RL5, TIW, and TI5, is very wide (ranging from 2 to −4) for the genotypes under drought while it is more or less constant (ranging from about 1 to −1) for the same genotypes under controlled conditions.

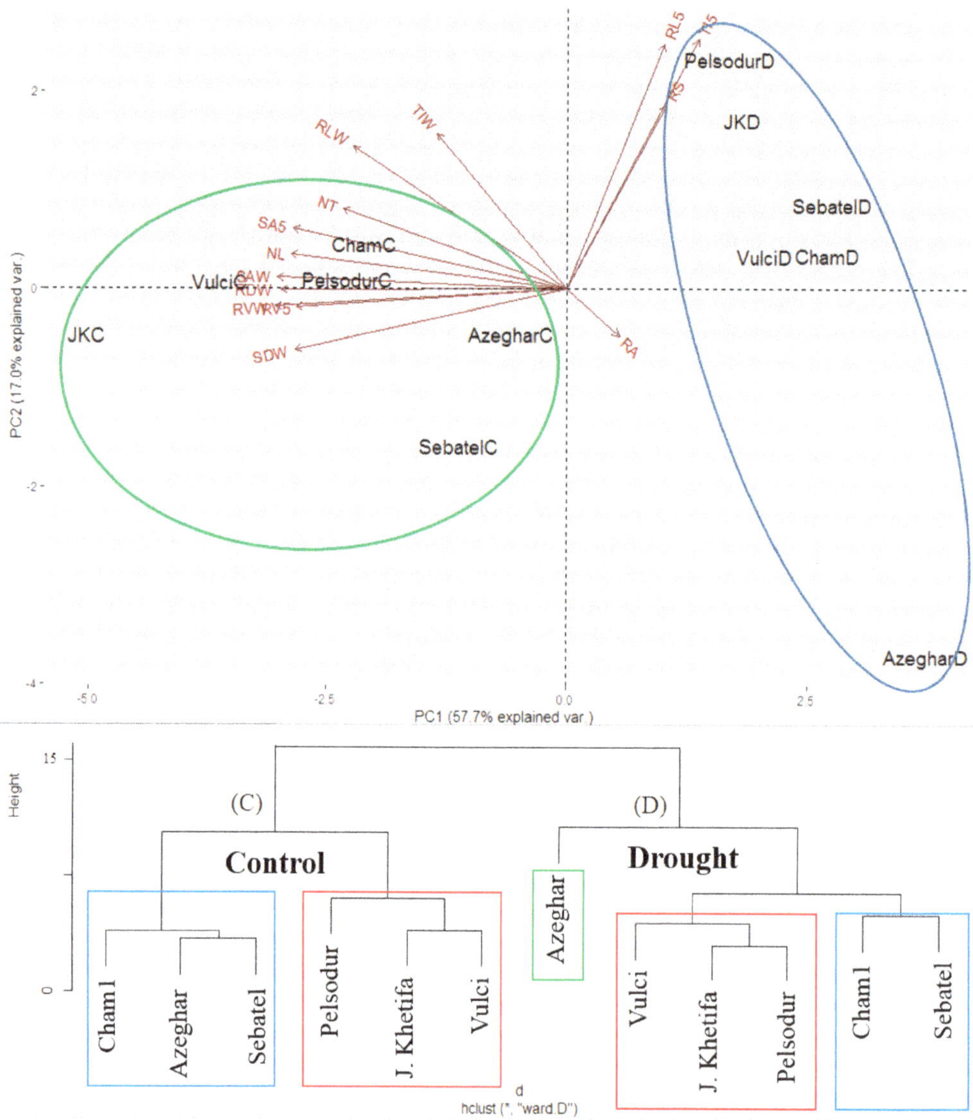

Figure 9. Principal component analysis and cluster dendrogram of 6 investigated durum wheat genotypes under control (C) and drought (D) conditions. NL—number of leaves; NT—number of tillers; SDW—shoot dry weight; RVW—total root volume; RLW—sum of all root lengths; TIW—total number of tips; RDW—root dry weight; SAW—total root surface area; RS—root shoot ratio; RL5—sum of root lengths in the first 5 cm below the ground; SA5—root surface area in the first 5 cm below the ground; TI5—number of tips in the first 5 cm below the ground; RV5—root volume in the first 5 cm below the ground; RA—root angle of plants grown in greenhouse under drought conditions; RAF—root angle of plants grown in the field. * indicates significance at $p < 0.05$.

The tolerant landrace J. Khetifa was assigned together with Pelsodur and Vulci to the same cluster under both conditions. However, under control conditions J. Khetifa was closer to the Vulci, while under drought conditions it was closer to the late flowering time Pelsodur. Early flowering genotypes, such as Cham1, Sebatel, and Azeghar formed one cluster under control conditions, while the drought separate Azeghar, which is probably a susceptible genotype to hydric stress.

4. Discussion

According to Reynolds and Langridge [39], several crucial steps are required to improve crop tolerance for abiotic stress; among them, trait-based breeding and a "crop ideotype design" are the first to be considered. Breeding programs have focused largely on above ground traits, leaving the root system importance overlooked [40]. Some of the previous studies reported that RSA traits on the seedling stage can be used to predict the RSA of adult plants in field-grown [20] or in an artificial system [21] to evaluate the crop adaptation under water stress conditions; unfortunately, these were validated only for what concerned the root angle and seminal root apparatus. Conversely, in the present study, the analysis was performed on a more advanced stage than seedling and on several root traits since different root types could perform differently under different conditions [23], as also highlighted by the present results. For what concerns the substrate, in the present study we used sand that allows roots to grow around the 360° rather than rhizotrons (*GrowScreen-Rhizo 1*) [41], growth pouch, or clear pot [21] which fundamentally allow a root development in only two dimensions. Moreover, Gregory et al. [42] recorded a significant soil by genotype interaction, when wheat root length among accessions grown on agar plates showed positive differences of about 40%, while in a sandy loam soil the differences were negative and at about 30%. However, in our study, we obtained a positive correlation between the root angle of genotypes grown under greenhouse and field conditions, which are good traits for selection for drought tolerance [23,24]. In addition, the highly statistically significant correlations between the main traits of the whole root system and topsoil indicate that it would be possible to extrapolate the root phenotyping of the whole root system by analyzing only the topsoil root system.

After simulations of several root system traits and ideotypes for drought conditions, it was proposed that for cereals a "deep, steep, and cheap" root system is the most proper for water scarcity environments [43,44]. A deep root system is the most important characteristic for the drought tolerant crop ideotype, as it allows access to residual water resources and N from deeper soil layers [45]. Moreover, recent studies observed an association between steep root angle and the depth of RSA, when narrower angles allow for an increase in deep rooting [14,46,47]. The same relationship was found in our study, as genotypes from the same cluster with the narrowest root angle in control conditions, such as J. Khetifa and Pelsodur, retained longer roots in water deficit environments and vice versa, while Azeghar with the widest angle in control conditions had the shortest RLW under drought conditions. Furthermore, a significant negative correlation between root angle under control (RAC) conditions and root length under drought stress (RLD) support the methods proposing to analyze the root angle at a seedling stage and select the genotypes adapted to drought on the basis of that data. These root system characteristics are often associated with the germplasm origin and the relative climate conditions. In general, a narrow root angle and deep root system are more common for rain-fed crops, which depend on stored water in the soil. However, the analysis of durum wheat landraces from the Mediterranean basin showed large variability in root system architecture [48–50]. Cultivars originating from the western Mediterranean showed a narrow root angle, the same as genotype J. Khetifa, which originated from Tunisia [34,48]. In agreement with Ober et al. [51], we demonstrated that a narrow root angle was associated with higher root biomass; for example, J. Khetifa and Pelsodur had a narrow root angle under control conditions (RAC) and maintained higher root biomass (RDW) under water scarcity.

Manschadi et al. [14] demonstrated how the yield increases of 55 kg/ha for each millimeter of water extracted from the soil after anthesis (i.e., in grain filling stage); hence a deep root system, but also wide root system, are the most desirable in environments with terminal drought [40]. In addition, genotypes from cluster (II) Cham1 and Sebatel showed different root ideotypes than genotypes from cluster (I). They had a wider root angle under control conditions, and under drought stress maintained moderate root biomass, length, surface area, and volume. However, regarding shoot biomass (SDW), Sebatel maintained significantly higher SDW than Cham1 under drought conditions. Interestingly, most of the differences between the Sebatel and Cham1 genotypes were found in topsoil root traits. Sebatel maintained significantly longer root length, volume, and surface area at the topsoil compared to Cham1, which helps to develop a significantly higher shoot biomass. Similar results were obtained in other studies, where the late flowering genotypes (in our case J. Khetifa, Pelsodur and Vulci) had a more uniform distribution of the root system compared to the early flowering genotypes (in our case Cham1 and Sebatel), which kept a higher root volume at 5 cm depth under drought [52].

The deep root allocation for durum wheat was determined to be a good strategy for drought avoidance in Mediterranean-type environments, which facilitates access to subsoil water resources [43]. In agreement with the "deep, steep, and cheap" root system ideotype, the genotypes Pelsodur, J. Khetifa, and Vulci can be considered drought tolerant; confirming the suitability for drought-tolerant breeding, as it was reported that this root system had significant functional relation to water absorption and drought avoidance in rice [53]. Also, the genotypes Sebatel and Cham1 with compact topsoil rooting patterns under drought condition could be good candidates for breeding, as it was demonstrated that higher root mass and root length density in subsoil layers contribute to the grain yield of winter wheat under drought conditions [54]. Finally, Azeghar with the lowest root biomass, surface area, length, and volume under drought stress and the widest root angle under control conditions shows susceptible root ideotype for hydric stress.

5. Conclusions and Perspectives

This study highlighted huge variability among a small number of genotypes in terms of the development, distribution, and architecture of the root system in order to tolerate difficult environments and to increase resilience to climate change. Even two hydric stress-tolerant genotypes showed different root system ideotypes and rooting patterns under drought. Moreover, it was demonstrated that screening plants' roots in the early stage grown under control conditions using a high-throughput scanner can expedite the selection of novel traits for crop improvement in plant breeding.

Supplementary Materials: The following are available online at https://www.mdpi.com/article/10.3390/agronomy12061329/s1, Table S1: Principal Components Analysis of traits detected under drought stress, Table S2: Principal Components Analysis of traits detected under both drought and control conditions.

Author Contributions: Conceptualization, I.U., L.B. and M.A.P.; methodology, M.A.P.; software, L.B.; validation, I.U., L.B. and M.A.P.; formal analysis, L.B.; investigation, I.U. and L.B.; resources, M.A.P.; data curation, I.U. and L.B.; writing—original draft preparation, I.U.; writing—review and editing, M.A.P.; visualization, I.U.; supervision, M.A.P.; project administration, M.A.P.; funding acquisition, M.A.P. All authors have read and agreed to the published version of the manuscript.

Funding: This research was funded by the European Union's Horizon 2020 research and innovation programme under grant agreement No 771367 (ECOBREED project).

Data Availability Statement: The recorded raw data are public available (after an access request) at the Zenodo web site with DOI: 10.5281/zenodo.5883299.

Conflicts of Interest: The authors declare no conflict of interest.

Abbreviations

CR5	number of crossings in the first 5 cm below the ground
CRF	number of crossings in the first 5 cm below the ground of plants from field
CRW	total number of crossings
FR5	number of forks in the first 5 cm below the ground
FRF	number of forks in the first 5 cm below the ground of plants from field
FRW	total number of forks
NL	number of leave
NT	number of tillers
RAC	root angle of plants grown in greenhouse as control
RAD	root angle of plants grown in greenhouse under drought condition
RAF	root angle of plants grown in the field
RDW	root dry weight
RL5	sum of roots lengths in the first 5 cm below the ground
RLD	root length density
RLF	sum of roots lengths in the first 5 cm below the ground of plants from field
RLW	sum of all root lengths
RS	root shoot ratio
RV5	root volume in the first 5 cm below the ground
RVF	root volume in the first 5 cm below the ground of plants from field
RVW	total root volume
SA5	root surface area in the first 5 cm below the ground
SAF	root surface area in the first 5 cm below the ground of plants from field
SAW	total root surface area
SDW	shoot dry weight
TI5	numbers of tips in the first 5 cm below the ground
TIF	number of tips in the first 5 cm below the ground of plants from field
TIW	total number of tips

References

1. Nam, W.H.; Hayes, M.J.; Svoboda, M.D.; Tadesse, T.; Wilhite, D.A. Drought hazard assessment in the context of climate change for South Korea. *Agric. Water Manag.* **2015**, *160*, 106–117. [CrossRef]
2. Mann, M.E.; Gleick, P.H. Climate change and California drought in the 21st century. *Proc. Natl. Acad. Sci. USA* **2015**, *112*, 3858–3859. [CrossRef] [PubMed]
3. Lawlor, D.W. Stress metabolism: Its implication in breeding programmes. In *Drought Tolerance in Winter Cereals*; Srivastava, J.P., Porceddu, E., Acevedo, E., Varma, S., Eds.; John Wiley & Sons, Inc.: Hoboken, NJ, USA, 1987.
4. Baird, J.; Gentry, J.; Lawrence, D.; Erbacher, A.; Aisthorpe, D.; Bell, L.; Anderson, B.; Brook, G.; Verrell, A.; Dunn, M.; et al. Nitrogen in the Farming System: Drought Implications on the Availability of Nitrogen. How Much Is There and Where Is It within the Profile? Does Cropping Sequence Influence Fallow Mineralisation Activity? Available online: https://grdc.com.au/resources-and-publications/grdc-update-papers/tab-content/grdc-update-papers/2020/03/nitrogen-in-the-farming-system-drought-implications-on-the-availability-of-nitrogen.-how-much-is-there-and-where-is-it-within-the-profile-does-cropping-sequence-influence-fallow-mineralisation-activity (accessed on 21 April 2022).
5. Sarto, M.V.M.; Sarto, J.R.W.; Rampim, L.; Rosset, J.S.; Bassegio, D.; da Costa, P.F.; Inagaki, A.M. Wheat phenology and yield under drought: A review. *Aust. J. Crop Sci.* **2017**, *11*, 941–946. [CrossRef]
6. Daryanto, S.; Wang, L.; Jacinthe, P.A. Global synthesis of drought effects on maize and wheat production. *PLoS ONE* **2016**, *11*, e0156362. [CrossRef]
7. Carvalho, P.; Azam-Ali, S.; Foulkes, M.J. Quantifying relationships between rooting traits and water uptake under drought in Mediterranean barley and durum wheat. *J. Integr. Plant Biol.* **2014**, *56*, 455–469. [CrossRef]
8. Rezzouk, F.Z.; Gracia-Romero, A.; Kefauver, S.C.; Nieto-Taladriz, M.T.; Serret, M.D.; Araus, J.L. Durum wheat ideotypes in Mediterranean environments differing in water and temperature conditions. *Agric. Water Manag.* **2022**, *259*, 107257. [CrossRef]
9. Saeidi, M.; Ardalani, S.; Jalali-Honarmand, S.; Ghobadi, M.E.; Abdoli, M. Evaluation of drought stress at vegetative growth stage on the grain yield formation and some physiological traits as well as fluorescence parameters of different bread wheat cultivars. *Acta Biol. Szeged.* **2015**, *59*, 35–44.
10. Blum, A.; Ramaiah, S.; Kanemasu, E.T.; Paulsen, G.M. Wheat recovery from drought stress at the tillering stage of development. *F Crop. Res.* **1990**, *24*, 67–85. [CrossRef]
11. Ding, J.; Huang, Z.; Zhu, M.; Li, C.; Zhu, X.; Guo, W. Does cyclic water stress damage wheat yield more than a single stress? *PLoS ONE* **2018**, *13*, e0195535. [CrossRef] [PubMed]

12. Khadka, K.; Earl, H.J.; Raizada, M.N.; Navabi, A. A Physio-Morphological Trait-Based Approach for Breeding Drought Tolerant Wheat. *Front. Plant Sci.* **2020**, *11*, 715. [CrossRef] [PubMed]
13. Passioura, J.B. Water Transport in and to Roots. *Annu. Rev. Plant Physiol. Plant Mol. Biol.* **1988**, *39*, 245–265. [CrossRef]
14. Manschadi, A.M.; Christopher, J.; deVoil, P.; Hammer, G.L.; Manschadi, A.M.; Christopher, J.; deVoil, P.; Hammer, G.L. The role of root architectural traits in adaptation of wheat to water-limited environments. *Funct. Plant Biol.* **2006**, *33*, 823–837. [CrossRef] [PubMed]
15. Lopez-Castaneda, C.; Richards, R.A.; Farquhar, G.D. Variation in Early Vigor between Wheat and Barley. *Crop Sci.* **1995**, *35*, 472–479. [CrossRef]
16. Reynolds, M.; Tuberosa, R. Translational research impacting on crop productivity in drought-prone environments. *Curr. Opin. Plant Biol.* **2008**, *11*, 171–179. [CrossRef]
17. Sanguineti, M.C.; Li, S.; MacCaferri, M.; Corneti, S.; Rotondo, F.; Chiari, T.; Tuberosa, R. Genetic dissection of seminal root architecture in elite durum wheat germplasm. *Ann. Appl. Biol.* **2007**, *151*, 291–305. [CrossRef]
18. Manschadi, A.M.; Hammer, G.L.; Christopher, J.T.; DeVoil, P. Genotypic variation in seedling root architectural traits and implications for drought adaptation in wheat (*Triticum aestivum* L.). *Plant Soil* **2008**, *303*, 115–129. [CrossRef]
19. Hochholdinger, F.; Tuberosa, R. Genetic and genomic dissection of maize root development and architecture. *Curr. Opin. Plant Biol.* **2009**, *12*, 172–177. [CrossRef]
20. Maccaferri, M.; El-Feki, W.; Nazemi, G.; Salvi, S.; Canè, M.A.; Colalongo, M.C.; Stefanelli, S.; Tuberosa, R. Prioritizing quantitative trait loci for root system architecture in tetraploid wheat. *J. Exp. Bot.* **2016**, *67*, 1161–1178. [CrossRef]
21. Richard, C.A.I.; Hickey, L.T.; Fletcher, S.; Jennings, R.; Chenu, K.; Christopher, J.T. High-throughput phenotyping of seminal root traits in wheat. *Plant Methods* **2015**, *11*, 13. [CrossRef]
22. Clark, R.T.; Famoso, A.N.; Zhao, K.; Shaff, J.E.; Craft, E.J.; Bustamante, C.D.; Mccouch, S.R.; Aneshansley, D.J.; Kochian, L.V. High-throughput two-dimensional root system phenotyping platform facilitates genetic analysis of root growth and development. *Plant Cell Environ.* **2013**, *36*, 454–466. [CrossRef]
23. El Hassouni, K.; Alahmad, S.; Belkadi, B.; Filali-Maltouf, A.; Hickey, L.T.; Bassi, F.M. Root system architecture and its association with yield under different water regimes in Durum wheat. *Crop Sci.* **2018**, *58*, 2331–2346. [CrossRef]
24. Chen, X.; Li, Y.; He, R.; Ding, Q. Phenotyping field-state wheat root system architecture for root foraging traits in response to environment × management interactions. *Sci. Rep.* **2018**, *8*, 2642. [CrossRef] [PubMed]
25. Wasson, A.P.; Richards, R.A.; Chatrath, R.; Misra, S.C.; Prasad, S.V.S.; Rebetzke, G.J.; Kirkegaard, J.A.; Christopher, J.; Watt, M. Traits and selection strategies to improve root systems and water uptake in water-limited wheat crops. *J. Exp. Bot.* **2012**, *63*, 3485–3498. [CrossRef] [PubMed]
26. Gesimba, R.M.; Njoka, E.; Kinyua, M. Root Characteristics of Drought Tolerant Bread Wheat (*Triticum aestivum*) Genotypes at Seedling Stage. *Asian J. Plant Sci.* **2004**, *4*, 512–515. [CrossRef]
27. Mondini, L.; Nachit, M.; Porceddu, E.; Pagnotta, M.A. Identification of SNP mutations in DREB1, HKT1, and WRKY1 genes involved in drought and salt stress tolerance in durum wheat (*Triticum turgidum* L. var *durum*). *Omics A J. Integr. Biol.* **2012**, *16*, 178–187. [CrossRef] [PubMed]
28. Nachit, M.M.; Elouafi, I.; Pagnotta, M.A.; El Saleh, A.; Iacono, E.; Labhilili, M.; Asbati, A.; Azrak, M.; Hazzam, H.; Benscher, D.; et al. Molecular linkage map for an intraspecific recombinant inbred population of durum wheat (*Triticum turgidum* L. var. *durum*). *Theor. Appl. Genet.* **2001**, *102*, 177–186. [CrossRef]
29. Annicchiarico, P.; Pecetti, L. Developing a tall durum wheat plant type for semi-arid, Mediterranean cereal–livestock farming systems. *F Crop. Res.* **2003**, *80*, 157–164. [CrossRef]
30. Katerji, N.; Van Hoorn, J.W.; Fares, C.; Hamdy, A.; Mastrorilli, M.; Oweis, T. Salinity effect on grain quality of two durum wheat varieties differing in salt tolerance. *Agric. Water Manag.* **2005**, *75*, 85–91. [CrossRef]
31. Katerji, N.; Van Hoorn, J.W.; Hamdy, A.; Mastrorilli, M.; Oweis, T. Salt tolerance analysis of chickpea, faba bean and durum wheat varieties: I. Chickpea and faba bean. *Agric. Water Manag.* **2005**, *72*, 177–194. [CrossRef]
32. Mondini, L.; Nachit, M.M.; Porceddu, E.; Pagnotta, M.A. HRM technology for the identification and characterization of INDEL and SNPs mutations in genes involved in drought and salt tolerance of durum wheat. *Plant Genet. Resour.* **2011**, *9*, 166–169. [CrossRef]
33. Mondini, L.; Nachit, M.M.; Pagnotta, M.A. Allelic variants in durum wheat (*Triticum turgidum* L. var. *durum*) DREB genes conferring tolerance to abiotic stresses. *Mol. Genet. Genom.* **2015**, *290*, 531–544. [CrossRef] [PubMed]
34. Monneveux, P.; Rekika, D.; Acevedo, E.; Merah, O. Effect of drought on leaf gas exchange, carbon isotope discrimination, transpiration efficiency and productivity in field grown durum wheat genotypes. *Plant Sci.* **2006**, *170*, 867–872. [CrossRef]
35. Pagnotta, M.A.; Luca, B.; Poala, F. Selection of Durum Wheat Lines under Organic Management—Preliminary Results/Zenodo. Available online: https://zenodo.org/record/4675513#.YkGRWFXP3IU (accessed on 28 March 2022).
36. Grewal, K.S.; Buchan, G.D.; Tonkin, P.J. Estimation of field capacity and wilting point of some new zealand soils from their saturation percentages. *N. Z. J. Crop Hortic. Sci.* **1990**, *18*, 241–246. [CrossRef]
37. GitHub—Taiyun/Corrplot: A Visual Exploratory Tool on Correlation Matrix. Available online: https://github.com/taiyun/corrplot (accessed on 23 December 2021).
38. Vincent, Q.V. GitHub—Vqv/Ggbiplot: A Biplot Based on Ggplot2. Available online: https://github.com/vqv/ggbiplot (accessed on 23 December 2021).

39. Reynolds, M.; Langridge, P. Physiological Breeding. *Curr. Opin. Plant Biol.* **2016**, *31*, 162–171. [CrossRef] [PubMed]
40. Alahmad, S.; El Hassouni, K.; Bassi, F.M.; Dinglasan, E.; Youssef, C.; Quarry, G.; Aksoy, A.; Mazzucotelli, E.; Juhász, A.; Able, J.A.; et al. A major root architecture QTL responding to water limitation in durum wheat. *Front. Plant Sci.* **2019**, *10*, 436. [CrossRef] [PubMed]
41. Nagel, K.A.; Putz, A.; Gilmer, F.; Heinz, K.; Fischbach, A.; Pfeifer, J.; Faget, M.; Blossfeld, S.; Ernst, M.; Dimaki, C.; et al. GROWSCREEN-Rhizo is a novel phenotyping robot enabling simultaneous measurements of root and shoot growth for plants grown in soil-filled rhizotrons. *Funct. Plant Biol.* **2012**, *39*, 891–904. [CrossRef]
42. Gregory, P.J.; Bengough, A.G.; Grinev, D.; Schmidt, S.; Thomas, W.T.B.; Wojciechowski, T.; Young, I.M. Root phenomics of crops: Opportunities and challenges. *Funct. Plant Biol.* **2009**, *36*, 922–929. [CrossRef] [PubMed]
43. Nakhforoosh, A.; Nagel, K.A.; Fiorani, F.; Bodner, G. Deep soil exploration vs. topsoil exploitation: Distinctive rooting strategies between wheat landraces and wild relatives. *Plant Soil* **2021**, *459*, 397–421. [CrossRef] [PubMed]
44. Lynch, J.P. Steep, cheap and deep: An ideotype to optimize water and N acquisition by maize root systems. *Ann. Bot.* **2013**, *112*, 347–357. [CrossRef]
45. Bengough, A.G.; Gordon, D.C.; Al-Menaie, H.; Ellis, R.P.; Allan, D.; Keith, R.; Thomas, W.T.B.; Forster, B.P. Gel observation chamber for rapid screening of root traits in cereal seedlings. *Plant Soil* **2004**, *262*, 63–70. [CrossRef]
46. York, L.; Slack, S.; Bennett, M.; Foulkes, M.J. Wheat shovelomics I: A field phenotyping approach for characterising the structure and function of root systems in tillering species. *bioRxiv* **2018**, *280875*. [CrossRef]
47. Tracy, S.R.; Nagel, K.A.; Postma, J.A.; Fassbender, H.; Wasson, A.; Watt, M. Crop Improvement from Phenotyping Roots: Highlights Reveal Expanding Opportunities. *Trends Plant Sci.* **2020**, *25*, 105–118. [CrossRef]
48. Roselló, M.; Royo, C.; Sanchez-Garcia, M.; Soriano, J.M. Genetic Dissection of the Seminal Root System Architecture in Mediterranean Durum Wheat Landraces by Genome-Wide Association Study. *Agronomy* **2019**, *9*, 364. [CrossRef]
49. Mengistu, D.K.; Kiros, A.Y.; Pè, M.E. Phenotypic diversity in Ethiopian durum wheat (*Triticum turgidum* var. *durum*) landraces. *Crop J.* **2015**, *3*, 190–199. [CrossRef]
50. Alemu, A.; Feyissa, T.; Maccaferri, M.; Sciara, G.; Tuberosa, R.; Ammar, K.; Badebo, A.; Acevedo, M.; Letta, T.; Abeyo, B. Genome-wide association analysis unveils novel QTLs for seminal root system architecture traits in Ethiopian durum wheat. *BMC Genom.* **2021**, *22*, 20. [CrossRef] [PubMed]
51. Ober, E.S.; Alahmad, S.; Cockram, J.; Forestan, C.; Hickey, L.T.; Kant, J.; Maccaferri, M.; Marr, E.; Milner, M.; Pinto, F.; et al. Wheat root systems as a breeding target for climate resilience. *Theor. Appl. Genet.* **2021**, *134*, 1645–1662. [CrossRef]
52. Puccio, G.; Ingraffia, R.; Giambalvo, D.; Amato, G.; Frenda, A.S. Morphological and Physiological Root Traits and Their Relationship with Nitrogen Uptake in Wheat Varieties Released from 1915 to 2013. *Agronomy* **2021**, *11*, 1149. [CrossRef]
53. Gowda, V.R.P.; Henry, A.; Yamauchi, A.; Shashidhar, H.E.; Serraj, R. Root biology and genetic improvement for drought avoidance in rice. *Field Crops Res.* **2011**, *122*, 1–13. [CrossRef]
54. Fang, Y.; Du, Y.; Wang, J.; Wu, A.; Qiao, S.; Xu, B.; Zhang, S.; Siddique, K.H.M.; Chen, Y. Moderate drought stress affected root growth and grain yield in old, modern and newly released cultivars of winter wheat. *Front. Plant Sci.* **2017**, *8*, 672. [CrossRef] [PubMed]

agronomy

MDPI

Article

A Simple and Accurate Method Based on a Water-Consumption Model for Phenotyping Soybean Genotypes under Hydric Deficit Conditions

Sebastián Simondi [1,†], Esteban Casaretto [2,†], Gastón Quero [2], Sergio Ceretta [3], Victoria Bonnecarrère [4] and Omar Borsani [2,*]

[1] Área de Matemática, Facultad de Ciencias Exactas y Naturales, Universidad Nacional de Cuyo (FCEN-UNCuyo), Padre Contreras 1300, Mendoza M5502JMA, Argentina; ssimondi@uncu.edu.ar
[2] Departamento de Biología Vegetal, Facultad de Agronomía, Universidad de la República, Garzón 809, Montevideo 12900, Uruguay; ecasaretto@fagro.edu.uy (E.C.); gastonquero@fagro.edu.uy (G.Q.)
[3] Instituto Nacional de Investigación Agropecuaria (INIA), Programa de Cultivo de Secano, Estación Experimental INIA La Estanzuela, Ruta 50, Km 11, Colonia 70000, Uruguay; sceretta@inia.org.uy
[4] Instituto Nacional de Investigación Agropecuaria (INIA), Unidad de Biotecnología, Estación Experimental Wilson Ferreira Aldunate, Ruta 48, Km 10, Rincón del Colorado, Canelones 90200, Uruguay; vbonnecarrere@inia.org.uy
* Correspondence: oborsani@fagro.edu.uy
† These authors contributed equally to this work.

Abstract: Drought limits crop productivity and reduces yield stability. Drought tolerance as a selection criterion in breeding programs requires the development of high-throughput, precise, and low-cost phenotyping strategies. We developed a mathematical model, based on biological approaches, for evaluating soybean plants' response to drought under controlled growth conditions. The model describes the kinetics of water consumption of a plant pot substrate system (PPS) with low sampling requirements. The model generated two parameters, $t_{0.5}$ (time necessary for the PPS to reach half of the maximum amount of evapotranspirable water) and $G_w(t_{0.5})$ (stomatal conductance [G_w] at $t_{0.5}$), which determined the water- consumption curve of each genotype. An analysis of the kinetics of water consumption in response to a progressive water deficit in a biparental and breeding population was performed as a preliminary test of the model. A correspondence analysis between the $t_{0.5}$ and $G_w(t_{0.5})$ parameters with the genetic structure of the populations shows a genetic association. The phenotyping methodology presented in this work and drought susceptibility in field conditions are discussed based on previous results. This work could be useful for improving the selection of soybean genotypes in relation to their performance under drought conditions.

Keywords: drought; stomatal conductance; mathematical modeling; crop breeding

Citation: Simondi, S.; Casaretto, E.; Quero, G.; Ceretta, S.; Bonnecarrère, V.; Borsani, O. A Simple and Accurate Method Based on a Water-Consumption Model for Phenotyping Soybean Genotypes under Hydric Deficit Conditions. *Agronomy* **2022**, *12*, 575. https://doi.org/10.3390/agronomy12030575

Academic Editor: Yuan-Ming Zhang

Received: 14 December 2021
Accepted: 24 February 2022
Published: 25 February 2022

Publisher's Note: MDPI stays neutral with regard to jurisdictional claims in published maps and institutional affiliations.

1. Introduction

Despite increases in soybean (*Glycine max* L.) crop yields achieved by breeding and better agricultural practices, its productivity and yield stability are especially susceptible to drought events [1,2]. Drought stress affects both the vegetative and reproductive stages in soybean, reducing leaf area, increasing flower and pod abortion, and diminishing pod and seed size [3]. Several works relate the response in the vegetative stage to drought tolerance during reproductive stages [4–6]. According to Kron et al. (2008) [4], there is a "developmental window" in the V4 stage, in which plants subjected to water stress improve the subsequent drought-stress tolerance in the reproductive stage. Furthermore, Sinclair et al. (2010) [6], using a simulation model, determined that water conservation through an early decrease in stomatal conductance and reduced transpiration rate explains the increase in soybean yield throughout 70 years with drought events.

Soybean breeding programs are mostly focused on increasing total yield and yield stability, which are mainly affected by water shortage during the crop cycle. These programs, particularly public breeding programs, do not use drought response as selection criteria due to the high costs of the massive phenotyping equipment currently available. Most of the high-throughput phenotyping platforms are based on the diagnosis of changes in the plant's physiology, such as leaf conductance, leaf area, and root system development, using a complex and expensive system of analysis [7]. Hence, including drought tolerance as a selection criterion in breeding programs requires the development of high-throughput, precise, and low-cost phenotyping strategies [8]. However, as Valdez et al. (2013) [9] indicate, strategies could include the quantification of two crucial aspects of the plant's water budget: (i) the ability to capture more water, and (ii) the ability to conserve and use captured water more efficiently. Nowadays, it is well documented that water extraction during the key crop stages greatly informs crop performance in water restricted scenarios [9–11].

Water balance in a land crop is greatly dependent on evapotranspiration phenomena. This is the combination of two independent processes involved in water losses from the soil. One is the evaporation of the water content in the soil surface, and the second is the loss of water contained in the plant tissues by transpiration [12]. In crops, the transpiration increases as the plants grow due to the increase of leaf area. At the same time, this causes a contrary effect on the evaporation, which decreases progressively. Evapotranspiration can be determined by measuring several components of water balance in the soil. Specifically, in a controlled close system with no run-off or percolation, the total water content of the system within a certain period can be quantified as weight.

At the whole plant level and under constant water demand, water uptake or water use depends mainly on root-system development, leaf area, transpiration rate, and leaf conductance [9]. Under conditions with increasing water restriction, each of the abovementioned parameters has a different role on water use. However, there is a consensus within the scientific community that transpiration changes are a critical component in contexts where the available water content in the soil is changing [13–16]. Transpiration is determined by the demand and controlled by stomata; during soil desiccation, the water extractable by the roots is continuously reduced, which determines that full transpiration demand cannot be supported. In this situation, the plants respond with stomatal closure to avoid shoot desiccation. In addition, stomatal opening is quite sensitive to the evaporative demand, and a high vapor-pressure deficit (VPD) reduces stomatal opening to restrict water losses. Soil water content and atmospheric demand ultimately determine the dynamic of water consumption [17]. Therefore, leaf conductance measurements using gravimetric methods have been explored as robust parameters for breeding programs [18–21]. Moreover, the variation of stomatal conductance in soybeans in response to VPD and soil water content has been shown to be dependent on the genotype [22,23].

The increase in the number of new genotypes evaluated by breeding programs calls for a high capacity for data acquisition and processing, yield prediction, and responses under different environmental conditions. To this end, the use of tools such as sensors, data analysis programs, algorithms, and models is essential [24]. However, the evaluation of the models with experimental data is necessary for their improvement and adjustment [25,26]. Under optimal development conditions, mathematical models can predict plants' responses more accurately. Under stress conditions, however, a greater range of responses are generated and the prediction accuracy of the models becomes weaker [24]. The results of models focused on the water absorption of crops [27–29] show a great variation among themselves as a consequence of the many factors involved. For example, the substrate information must be improved, as well as the plant-soil interaction [24]. It is necessary to have a better theoretical and empirical basis and an appropriate knowledge of the environmental conditions where the model will be applied. Another aspect that stands out that is not always taken into account when determining the performance of the model is the competence of its users. Therefore, it is necessary to simplify and facilitate the input data models. Most of the relevant mathematical models that we know of are designed to explain crop behavior in response to field environmental changes. Moreover, the time variable is not included in these models, which would not account for the physiological changes at the plant level.

In this study, we developed a simple and informative model based on biological and mathematical approaches for evaluating, under controlled growth conditions, the response of soybean plants to water restriction through the quantification of water consumption in the vegetative stage. To find a physiological explanation of the variables generated by the model, they were correlated with the stomatal conductance dynamic. Also, a preliminary test of the model was performed separately in a biparental segregating population and in a breeding population. Results show that the approach proposed is a valid option to be included in plant phenotyping protocols with limited manpower and infrastructure.

2. Materials and Methods

2.1. Phenotyping Method Based on Water Consumption under Controlled Environmental Conditions

For the development of the mathematical model, we performed an experiment with a soybean genotype (GENESIS 5601). Plants were grown in a 0.5 L plastic bottle (pot) filled with a mix of sand:vermiculite (1:1). This combination of plant, pot, and substrate was defined as a Plant Pot Substrate system (PPS) (Figure S1). Plants were grown in an environment defined by day/night cycle temperatures of 30/20 °C, respectively, and a light/darkness photoperiod of 16/8 h, respectively. Relative humidity (RH) was controlled at 35/40% during the entire growth period. Three seeds per pot were sown, and only one seedling remained after the cotyledon expanded. The homogeneity of the plants was carefully analyzed to avoid any interference related to developmental phenotype. During the first 16 days after sowing (developmental stage V2–3), soybean seedlings were grown without water restriction, and substrate was kept at field capacity with Rigaud and Puppo (1975) [30] medium supplemented with KNO_3 (10 mM final concentration). Since day 17, watering was suspended, and water substrate content was measured daily by gravimetry (water gravimetric content) during the next 10 days of water deficit (dwd) (Figure 1). Stomatal conductance was measured simultaneously with PPS on the abaxial leaf surface with a Porometer Model SC-1 (Decagon Device), as instructed by the manufacturer, since the suspension of watering. Five biological replicates (n = 5) were used for determinations, consisting of five independent PPSs.

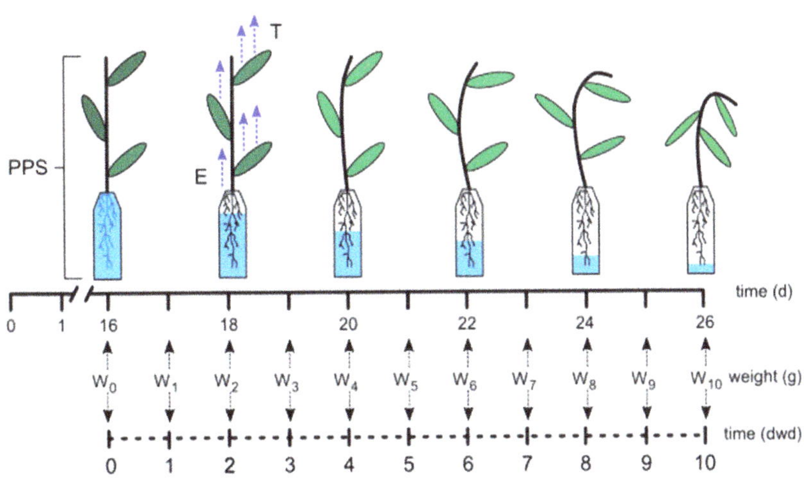

Figure 1. Schematic illustration of the experiment for determining water consumption. Plants were maintained at field-capacity condition until day 16, after which the water supply was suspended. The Plant Pot Substrate system (PPS) was weighted daily for 10 days (t = 0 until t = 10). Evaporation: E; transpiration: T; days of water deficit: dwd.

2.2. Phenotyping Strategy Applied to an F3 Segregating Population

An F3 segregating population of 177 genotypes derived from the crossing of parental lines SO7.6557 × DM6.8 were phenotyped using the methodology and the mathematical model developed in this study. The phenotyping experiment was laid out in a randomized incomplete block design, with three replications in each experiment. Growth conditions were the same as described in the Section 2.1. PPS weight and stomatal conductance were measured simultaneously.

2.3. Phenotyping Strategy Applied to a Breeding Population

A local breeding population composed of a set of 89 genotypes [31] was also phenotyped using the methodology described above and the mathematical model developed in this study. Five well-known commercial varieties were included as checks in all phenotyping experiments. In this case, plants were grown in 2.851 L PVC tubes (11 cm in diameter and 30 cm long) with a mix of sand:vermiculite (1:1) under the same environmental conditions as described previously. Plants were grown without any watering restriction for 30 days (developmental stage V4–5), after which watering was suspended and the PPS weight was registered at days 0, 4, and 8 after suspending the water supply. PPS weight and stomatal conductance were measured simultaneously.

2.4. Genotyping by Sequencing and SNP Calling

The F3 segregating population of 177 genotypes was genotyped using a SoySNP6k chip [32] in an Iscan system (Illumina, San Diego, CA, USA). SNP calling was done using the GenomeStudio software (Illumina, San Diego, CA, USA). Genotyping-by-sequencing (GBS) data were obtained for the breeding population according to the methodology proposed by Quero et al. (2021) [31].

2.5. Data Analysis

2.5.1. Nonlinear Models

To model the weight of the plant pot substrate system and its evapotranspiration, a one-phase exponential decay and one-phase association function were used, respectively. In both cases, the initial values and parameters estimation is described by Equations (11) and (12) of Section 3.6. It is worth noting that the nonlinear models were fitted using the *nls* function of the *stats* [33] package of the statistical R software [33].

2.5.2. Multivariate Characterization

The multivariate characterization of the soybean populations was accomplished through some exploratory analyses and visualization tools. First, genotypes were clustered in groups based on the PCA of the genotypic variability using a hierarchical clustering algorithm (HCPC). Genotypes were grouped based on similarity from the Euclidean distance using the Ward method, and the number of groups was determined by the highest relative loss in inertia using the function *HCPC* of the *FactoMineR* [34] package in the statistical software R [33]. Second, a correspondence analysis (CA) was performed based on the contingency tables of the HCPC analysis from the genotypic characterization, and the clustering was performed according to $t_{0.5}$ and $G_w(t_{0.5})$. The correspondence analysis was performed to visualize the relationship between the two grouping strategies (i.e., based on genetic and phenotypic variables). The CA was conducted using the *CA* function of the *FactoMineR* [34] package in the R software [33].

2.5.3. Statistical Model and Adjustment of Phenotypic Means

Best linear unbiased estimators (BLUEs) for each advanced inbred line were obtained with mixed models to include experimental design components using the following model:

$$y_{ij} = \mu + \alpha_i + \beta_j + \varepsilon_{ij}$$

where y_{ij} is the response variable, μ is the overall mean or intercept, α_i is a random variable associated with the ith assay with $\alpha_i \sim N(0, \sigma^2{}_A)$, β_j is the effect of the jth line, and ε_{ij} is the residual error with $\varepsilon_{ij} \sim N(0, \sigma^2\varepsilon)$. The model was adjusted in the R software [33] using the *lme4* package [35], while the BLUEs were estimated using the *emmeans* package [36], also in R [33]. In all cases, BLUEs were further used as genotypic values for the model, PCA, and CA analyses.

3. Results

In this work, we developed an empirical mathematical model for describing the kinetics of water consumption of a PPS (Figure 2) using an important plant such as soybean. This model was applied for developing a phenotyping methodology for water-deficit response in soybean plants under controlled environmental conditions.

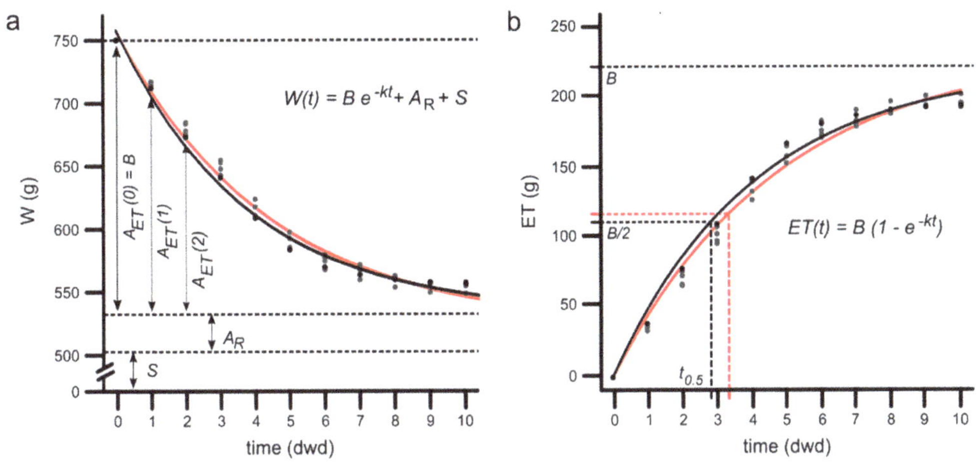

Figure 2. (**a**) Empirical model representing evapotranspiration over time and the empirical model adjustment. Weight (W) of the PPS along the days of water deficit (dwd). Sum of dry weight substrate, pot, and plant weight (S), residual water (A_R), evapotranspirable water (A_{ET}), and potential evapotranspiration (B) of plants at field-capacity conditions. (**b**) Mathematical analysis for modeling $ET(t)$ as a function of time. The parameter $t_{0.5}$ is the time required for the PPS to halve the potentially evapotranspirable water.

3.1. Mathematical Model Development

In the experiment, the PPS system weight (W) is defined as the weight of water (A) plus the rest of the components of the system (S). S is the sum of substrate, pot, and plant weight. The latter was considered constant during the assay because, although it could vary throughout the days, it is insignificant in comparison to the rest of components of S. A is the sum of the transpirable water (A_{ET}) plus a percentage of that which cannot be evapotranspired by the PPS throughout the whole assay. The non-evapotranspirable water is defined as residual water (A_R). The values of A_R depend on the matric potential of the substrate and the transpiratory capacity of the plants (Figure 2a). The amount of potentially evapotranspirable water of the PPS is a function dependent on time (t), and it is named $A_{ET}(t)$. Therefore, $A(t) = A_{ET}(t) + A_R$.

3.2. PPS Weight Modelling over Time

If $W(t)$ is a function of the PPS weight over time (t), then

$$W(t) = A(t) + S = A_{ET}(t) + A_R + S \quad (1)$$

In order to find an algebraic expression for the $W(t)$ function, we assume the following: "The velocity with which the PPS weight varies is directly proportional to the amount of water that can be evapotranspired for the PPS in this fraction of time". This hypothesis can be mathematically expressed through the following differential equation:

$$\frac{dW(t)}{dt} = -kA_{ET}(t) \quad (2)$$

where $k > 0$ is a constant of proportionality. The negative sign of the equation is due to the decrease of the PPS weight over time, as the assay was performed withdrawing watering at $t = 0$.

Since A_R and S are constant magnitudes over time, the derivation of the equality (1) results in $\frac{dW(t)}{dt} = \frac{dA_{ET}(t)}{dt}$, thus, the Equation (2) can be rewritten as follows:

$$\frac{dA_{ET}(t)}{dt} = -kA_{ET}(t) \quad (3)$$

obtaining the first order homogeneous linear differential equation for the function $A_{ET}(t)$. Solving the Equation (3) using separation variables method (Figure S2), the solution can be written as:

$$A_{ET}(t) = Be^{-kt} \quad (4)$$

where B is a positive constant.

By combining the Equations (1) and (4), we obtain the following equation:

$$W(t) = Be^{-kt} + A_R + S \quad (5)$$

The graphic representation and the experimental data adjustment of Equation (5) are shown in Figure 2a.

3.3. *Evapotranspiration Modelling as a Function of Time*

To quantify the evapotranspiration (Figure 2b) ($ET(t)$) of the PPS from the precise moment when watering was suspended ($t = 0$) to time t, is the difference between the PPS weight in both times, that is

$$ET(t) = W(0) - W(t) \quad (6)$$

Combining the Equations (5) and (6), we have that:

$$\begin{aligned} ET(t) &= W(0) - W(t) = A_{ET}(0) + A_R + S - (A_{ET}(t) + A_R + S) \\ &= A_{ET}(0) + A_R + S - A_{ET}(t) - A_R - S \\ &= A_{ET}(0) - A_{ET}(t) = B - Be^{-kt} \\ &= B\left(1 - e^{-kt}\right) \end{aligned}$$

Therefore,

$$ET(t) = B\left(1 - e^{-kt}\right) \quad (7)$$

The graphic representation and the experimental data adjusting of the Equation (7) are shown in Figure 2b.

3.4. *Potential Evapotranspiration Estimated by the Model*

At the moment watering was suspended ($t = 0$), the PPS had the maximum quantity of evapotranspirable water. By definition, this quantity is $A_{ET}(0) = B$, as observed in Figure 2a.

On the other hand, since the constant of proportionality (k) of Equation (7) is positive, we have that

$$\lim_{t\to\infty} ET(t) = \lim_{t\to\infty} B\left(1 - e^{-kt}\right) = B \quad (8)$$

That is, the parameter B is the horizontal asymptote of the function $ET(t)$, and represents the potential evapotranspiration of PPS, as observed in Figure 2b.

3.5. Half-Time of ET

Half-time is, by definition, the time required for the PPS to reduce the potential evapotranspiration in half. As the potential evapotranspiration is B, the half-time, $t_{0.5}$, is expressed as follows:

$$ET(t_{0.5}) = \frac{B}{2} \tag{9}$$

Equation (7) is used to calculate $t_{0.5}$, therefore,

$$t_{0.5} = \frac{1}{k}\ln(2) \tag{10}$$

Thus, the half-time is inversely proportional to the constant k, and independent of B. The graphic representation of $t_{0.5}$ is shown in Figure 2b.

3.6. Parameters of the ET Model Calculated from the Experiment Data

If W_j is the PPS weight registered on the jth day since suspending watering, $\{(j, W_j)\}_{j=0}^{N}$ is the set of data obtained from the assay (Figure 2a).

To determine the ET parameters as a function of time (Figure 2b), with this data set, we can use the least-squares method that requires computational resources or an analytical method from Equation (7). For this last option, we took three specific determinations, W_0, W_n, and W_{2n}, and performed them on days *0*, *n*, and *2n*, respectively. Replacing this data on the Equation (7), we can directly find the parameters, obtaining the following:

$$B = \frac{(W_n - W_0)^2}{W_{2n} - W_n + W_0} \tag{11}$$

$$k = -\frac{1}{n}\ln\left(\frac{W_{2n} - W_0}{W_n - W_0}\right) \tag{12}$$

To develop an empirical model, the PPS weight measurements ($W(t)$) during the whole water-deficit period were used to determine the curve fitting according to the experimental methods described in Figure 2. As shown in Figure 1, 10 days of water restriction determined a curve of PPS weight loss, determined by the water transpired by the plants. As indicated in Figure 2a, W at time t is defined by the parameter B determined by the weight when t is 0 and it represents the potentially evapotranspirable water, A_R indicates the water not extractable, and S indicates the weight of the support and dry substrate.

The evapotranspiration of the PPS over time from the day when watering was withdrawn was determined by Equation (7) and defined by the parameters B and k. It is important to note that B represents the potential evapotranspiration of the PPS. By definition, half-time ($t_{0.5}$) is the time required for the PPS to reach half of the potential evapotranspiration (Figure 2b). These parameters show the kinetics of water consumption of the genotypes and could help in the characterization of soybean genotypes.

3.7. The Model Minimizes Sampling Requirements in Phenotyping Protocols

To simplify the data collection procedure and increase the high throughput capacity, the model parameters were estimated using the minimum sampling. Figure 3a shows the values of K in Equation (7) estimated by the Gauss Newton numeric methods versus the values of K estimated by analytic methods using a sampling of the PPS weight every two days. As observed in Figure 3a, an optimum adjustment was obtained when the values of the PPS weight were sampled on days 4 and 8. The parameters estimated with the weight of these two days are closer to the best model fit. The same results were obtained with parameter B of Equation (7) (Figure 3a), but with an adjustment of 0.96. Both parameters of the model are critical to evaluate the water-consumption curve of a specific genotype. Hence, the PPS weight on days 4 and 8 appears to be enough to describe the kinetics of water consumption throughout the water restriction period. Figure 3b shows that the adjustment curve of Equation (7) with parameters B and K is the same as the one obtained using all the PPS weight data in the Gauss Newton estimation method.

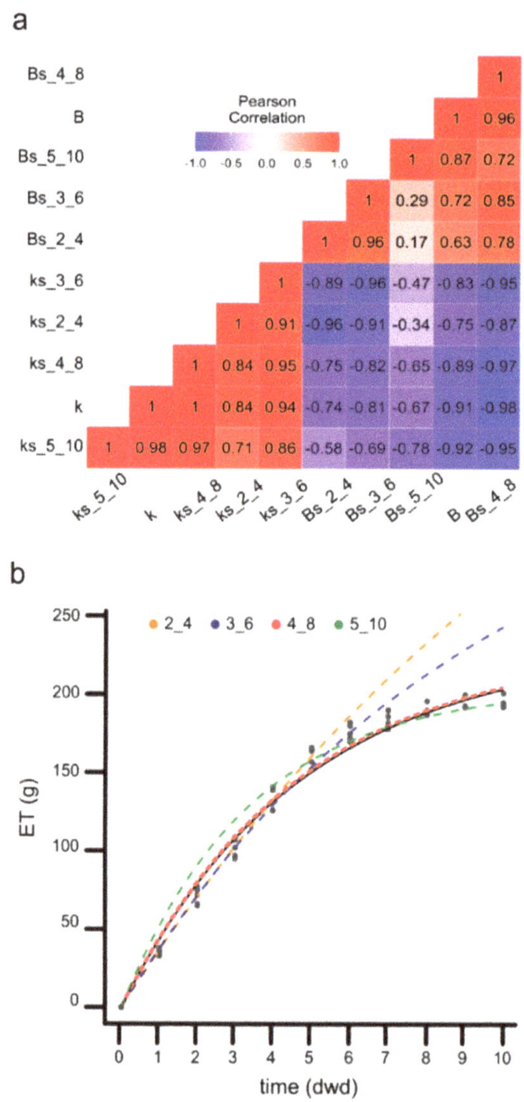

Figure 3. Minimum sampling requirements analysis and adjustment of model. (**a**) Relation between K or B estimated by the Gauss Newton method and the predicted K or B generated by analytic methods with two-time sampling, respectively. Axis indicates the parameters and the days of the sampling. (**b**) ET over time. Graphic representation of Equation (7) adjusted according to the data of the numeric (solid curve) or analytic methods (dashed colored curve). Colors indicate the two specific sampling days used for modeling the curve. Days of water deficit: dwd.

3.8. Stomatal Conductance as a Function of Time

By definition, stomatal conductance (G_w) is the rate of CO_2 entering the substomatal chamber, or the water vapor exiting through the stomata pore of the leaf. Since most of the water lost by the PPS is from stomatal transpiration, it is possible to suppose that the evaporation is negligible. Thus, the variation in weight can be attributed exclusively to transpiration, that is:

$$\frac{dW(t)}{dt} = -\tau G_w(t) \tag{13}$$

where in τ represents a constant of proportion.

By deriving the Equation (5) and replacing in (13), we obtain the following:

$$G_w(t) = \frac{k}{\tau} B e^{-kt} \tag{14}$$

The graphic representation and the experimental data adjustment of the Equation (14) are shown in Figure 4a.

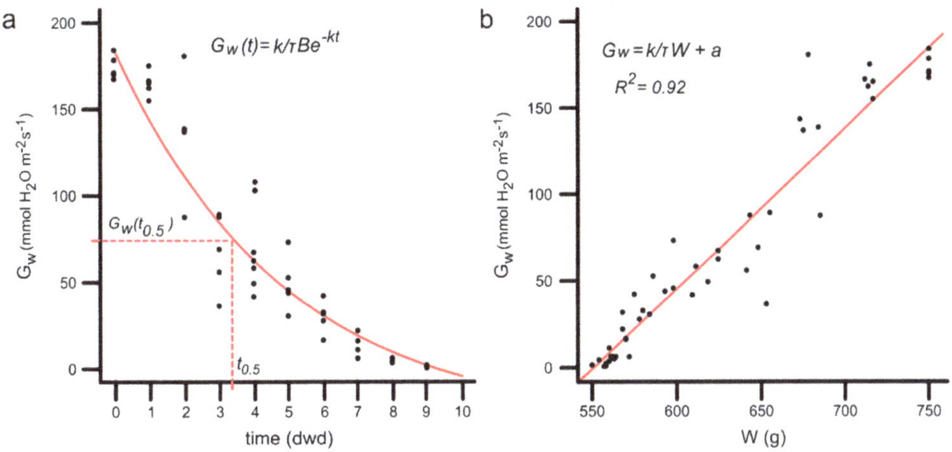

Figure 4. Modeling of stomatal conductance in response to water availability. (**a**) Empirical model representing the conductance over time and the empirical model adjustment. (**b**) Lineal regression between conductance and substrate weight. G_w: stomatal conductance; W: PPS weight; dwd: days of water deficit.

3.9. Conductance as a Function of PPS Weight

By combining the Equation (5) with (14), we obtain the following:

$$G_w(t) = \frac{k}{\tau} B e^{-kt} = \frac{k}{\tau}(W(t) - A_R - S) = \frac{k}{\tau} W(t) + \frac{k}{\tau}(-A_R - S)$$

Thus,

$$G_w = \frac{k}{\tau} W + a \tag{15}$$

where in $a = \frac{k}{\tau}(-A_R - S)$.

Thus, stomata conductance is a linear function of the PPS weight, as shown in Figure 4b.

To understand the physiological component of the kinetics of water consumption dissected by the proposed model, G_w over time was included in the mathematical modeling.

A modeled curve of G_w over the time was performed from Equations (5) and (13) (described in Sections 3.2 and 3.9). Figure 4a shows how the model generated by theoretical tools is adjusted with the experimental data. Using function (4), it is possible to estimate the half-time for $G_w(t_{0.5})$ from the moment watering is suspended. This parameter is relevant as a useful indicator to identify contrasting responses of different genotypes to the hydric deficit. Figure 4b shows the direct theoretical relation between the G_w obtained from Equation (15) and the PPS weight (red curve), and the high correlation with the experimental data ($R^2 = 0.92$).

3.10. Application of the Phenotyping Methodology in Two Breeding Populations at Different Plant Developmental Stages

The methodology was tested in a phenotyping approach in two breeding populations. The parameters of the water-consumption curve defined by the model ($t_{0.5}$ and $G_w(t_{0.5})$) were analyzed and related to the genetic variability in order to improve the understanding of the plant's water-consumption behavior in soybean.

3.10.1. Half Time and Stomatal Conductance

An analysis of the kinetics of water consumption in response to a progressive water deficit was performed in a biparental population. Two parameters, $t_{0.5}$ and $G_w(t_{0.5})$, were selected to characterize the variability of the recombinant genotypes (Figure 5a,b). Values of $t_{0.5}$ between 2 and 6 days showed high variability in the kinetics of water consumption. However, 25% of the genotypes had a $t_{0.5}$ lower than 3 days, and the remaining 75% had a $t_{0.5}$ higher than 3 days. The normal distribution observed confirms that the values of the parameters had a biological behavior inside the population. On the other hand, the distribution of $G_w(t_{0.5})$ ranked between 101.8 and 202.8 mmol H_2O m^2 s^{-1} shows normal distribution that accomplishes the behavior of a water-consumption response (Figure 5b).

When the phenotyping protocol was evaluated in V5 plants of a breeding population, a similar range of $t_{0.5}$ values were obtained in comparison to those obtained in the biparental population when evaluating in V3 plants (Figure 5a,c). However, when $G_w(t_{0.5})$ was analyzed, a wider range of values was obtained (6–300), showing that this parameter is more affected by the developmental stage and genotypic variability (Figure 5b,d).

It is important to point out that $G_w(t_{0.5})$ is the conductance reached at the time when the plant has consumed half of the potentially evapotranspirable water, and not a simple calculation of half G_w at the initial time (t_0). This confirms the idea that parameters identified by the model are biologically relevant and, at least in soybean, that could help in the analysis of plant response to water deficit.

3.10.2. Genetic Structure and Correspondence Analysis

Based on the genetic structure of both breeding populations, we identified three main groups in the biparental population and six groups in the breeding population (Figure 6a,b). The latter shows a higher genetic diversity, which would explain the results of the analysis in the distribution of values for the parameters of the model (Figure 5). The correspondence analysis between HCPC genotypic and HPC phenotypic groups showed the following correspondence: for the $t_{0.5}$ parameter and in the biparental population; tq3 and g2; and tq1 and g3. The relation between tq2 and g1 is not clear. In the case of the breeding population, the correspondence indicated the relation between tq1 and g3; tq2 and g2; and g6, tq3, and g4. It was not possible to find a correspondence for g1 and g5 (Figure 6c,d). When the same correspondence analysis was performed between the genetic groups and the G_w parameters, the results showed correspondences between G_{wq3} and g2, and between G_{wq2} and g1 in the biparental population. It was not possible to find a relation between G_{wq1} and g3 (Figure 6e). In the case of the breeding population, correspondences between G_{wq3}, g3 and g6; G_{wq2} and g1; and g2 and g5 were found (Figure 6f). It was not possible to find a relation between G_{wq1} and g4. This type of analysis indicates that the changes in water consumption under hydric conditions, quantified by the parameters of the model, are related to the genetic components in the genotype.

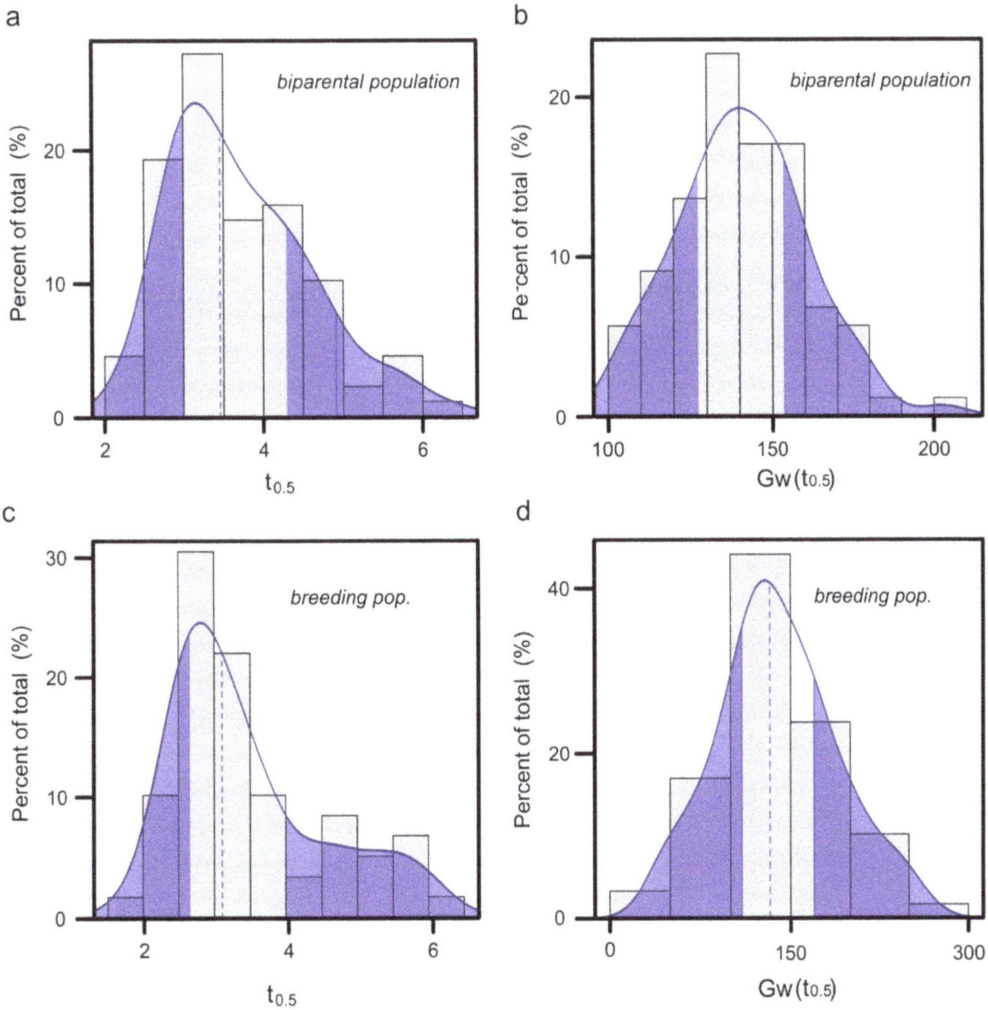

Figure 5. Graphic distribution of $t_{0.5}$ and $G_w(t_{0.5})$ values obtained from two soybean populations. $t_{0.5}$ is expressed in days and $G_w(t_{0.5})$ in mmol H_2O m^2 s^{-1}. A frequency analysis was performed in 177 and 89 genotypes of a biparental and breeding population, respectively. The darker regions represents the 25% of the genotypes with superior and inferior $t_{0.5}$ and $G_w(t_{0.5})$ values of the populations. (**a**,**b**) biparental population; (**c**,**d**) breeding population.

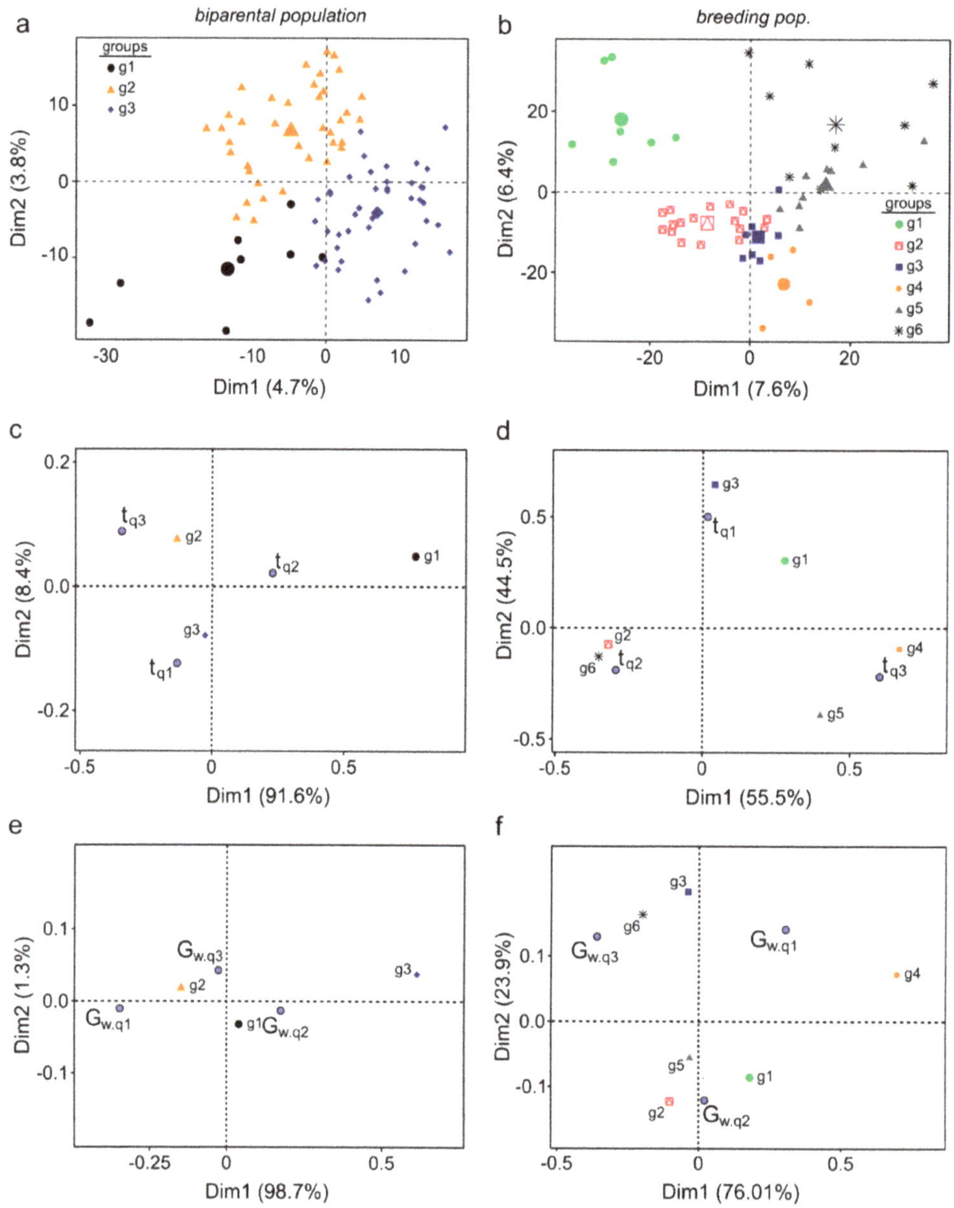

Figure 6. Population structure calculated from genotypic data (molecular markers), phenotypic data (parameters of the model), and the correspondence analysis. The groups were obtained by hierarchical clustering on the principal component (HCPCgenotypic). (**a**) Biparental population: g1, g2, and g3 groups. (**b**) Breeding population: g1, g2, g3, g4, g5, and g6 groups. Correspondence analysis of the contingency table between groups based on the parameters of the models and groups according to genotype. (**c**,**e**) Biparental population. (**d**,**f**) Breeding population. Population structure calculated from parameters $t_{0.5}$ and G_w of the model. tqn and G_{wqn} were obtained by hierarchical clustering on the principal component (HCPCphenotypic).

4. Discussion

In the current climate change scenario, soybean breeders must use effective strategies to develop varieties with a better ability to cope with periods of water shortage. The complexity of drought-tolerance traits has prevented the development of successful and accessible phenotyping strategies for selection, especially in small-scale breeding programs. This situation motivated our work to develop an effective phenotyping strategy. The simplification of phenotyping strategies becomes necessary when a high number of plants must be evaluated at the same time. In this regard, considerable efforts were made to reach this objective. In our case, we set the focus on the development of a model able to characterize and predict the water-consumption curve with low sampling requirements. Since the plant growth system minimizes water losses by evaporation, the water-consumption curve could be related to the transpiration curve, while also including the stomatal response as a parameter of the model. As demonstrated in several studies, water consumption using gravimetric methods correlates with the measurement of the transpiration rate under specific VPD conditions [37], so a fairly high throughput analysis based on that concept could be applied.

The data confirms the strong relation between the kinetics of water consumption and stomatal conductance, which means that the water transpired from the PPS defined in the study is regulated by stomatal conductance, and this last variable is regulated by the water availability in the pot. The model generated in this work contemplates the time variable, therefore it considers the changes in the parameters of water consumption. These types of models included dynamic variables which are scarce and would allow more precise characterization of complex biological phenomena such as plant adaptation to changes in the levels of available water. In addition, the data input of the model is based on a simple variable and easily quantifiable as the weight of PPS.

Moreover, the model can predict the value of water when the conductance is 0 and define the limit of water extraction. This trait has been identified as an important factor, because genotypes with high values of water thresholds begin to partially close their stomata at a relatively high water content, therefore saving soil water [9]. A study using data from different regions and years of the USA has shown, through simulation tools, that this trait would lead to a significant increase in soybean yield, especially in crop seasons classified as dry [6]. An early and accurate screening of the genotypes with specific responses in the water-consumption curve under water deficit seems to be an interesting advantage to be explored in plant breeding programs.

How drought episodes are established in field conditions varies depending on the agro-climatic region, the rain patterns, the soil characteristics, and the atmospheric demand. Under alternating drought conditions, in which there is a frequent period of stress alleviation, genotypes with high evapotranspiration capacity and water extraction (higher B and lower A_R) could be more interesting than those with a contrary response (lower B and high A_R). However, in a drought situation with a low probability of soil water recovery, the selection of genotypes with low B and high A_R could be the main objective for breeding.

The gravimetric measurements of the transpiration-curves trait performed under different VPD could lead to an increase in the adaptation capacity of several crop genotypes to different environments, using the proposed method with a high scale-up potential. Phenotyping protocols have been used in different ways to classify genotypes in response to drought. In this work, we propose the assignment of some parameters of the model as a specific trait of genotypes, so these parameters should be included in drought phenotyping methods. Moreover, associations among the variables of the model increase the possibility of identifying other more informative parameters related to plant-drought tolerance. For example, the model demonstrated that the stomatal conductance could be included as an explicative variable of plant response to progressive water deficit. It is clear that regulating the speed of stomatal closure is a key element in the drought response [9]. In this context, $G_w(t_{0.5})$ appears to be an effective measurement to explain the responsiveness of genotypes to changes in water availability.

Moreover, the phenotype of soybean plants in response to the hydric deficit from two breeding populations could be characterized by the proposed model. Since the phenotype of each genotype is determined by the genome-environment interaction [38,39], the variables generated by the model $t_{0.5}$ and $G_w(t_{0.5})$ were subjected to a correspondence analysis. The analysis confirmed a possible genetic association between the response of genotypes to water restriction and the parameter of the model. As expected, a clearer grouping association was observed in the biparental population than in the breeding population, where the genetic diversity of the germplasm was greater. However, the different number of genotypes included in both analyses could explain the results in the correspondence analysis. Moreover, a genome-wide association analysis could contribute to identifying the specific genomic region and genetic marker associated with the parameters defined by the model.

A recent report has classified the same genotype collection using indexes such as *drought susceptibility index* (DSI) and *yield stability index* (YSI) in relation to the crop cycle group [31], and the authors were able to identify QTLs associated with those traits. A preliminary but not confirmatory correspondence analysis between the parameter of the model $t_{0.5}$ and DSI showed a grouping (Figure S3), indicating that the phenotyping strategy discussed in this work could be useful for improving genotype selection in relation to the performance at field conditions. However, a specific validation assay of water-deficit response in field conditions of a set of genotypes previously characterized by the model represents the next challenge. This point is critical for proposing the phenotyping methodology as a tool to be included in crop breeding programs, especially in those with low-income support.

5. Conclusions

The developed model in the study characterizes and predicts, using gravimetric measurements and with low sampling requirements, the water-consumption curve of soybean plants when watering is withdrawn. The model confirms the strong relationship between the kinetics of water consumption and stomatal conductance. The correspondence analysis between model parameters and the response of the genotypes to water restriction in two different soybean breeding populations confirmed a possible genetic association between parameters of the model and genotypic identity. A preliminary approximation shows that the phenotyping methodology presented in this work could be included in crop breeding programs, especially in those with low-income support.

Supplementary Materials: The following supporting information can be downloaded at: https://www.mdpi.com/article/10.3390/agronomy12030575/s1, Figure S1: Experimental setup and plant growth system. Figure S2: Solution of the first-order linear homogeneous differential Equation (3). Figure S3: Cluster analysis based on drought susceptibility index (DSI) and the parameter of the model $t_{0.5}$.

Author Contributions: S.S. performed the theoretical mathematical model. O.B., V.B., E.C., G.Q. and S.C. were involved in the planning of the work. G.Q. conducted all data analyses of the experiments. E.C. and G.Q. performed the phenotypic evaluation. S.C. generated the breeding populations. O.B. and V.B. wrote the manuscript. All authors corrected the manuscript. All authors have read and agreed to the published version of the manuscript.

Funding: This research was funded by the following projects: CSIC-UdelaR Grupo 418 Estrés abiótico en plantas; Innovagro FSA_1_2013_1_12924, funded by Agencia Nacional de Investigación e Innovación; and Red Nacional de Biotecnología Agrícola, RTS_1_2014_1, funded by Agencia Nacional de Investigación e Innovación, Instituto Nacional de Investigación Agropecuaria, Barraca Erro S.A., Lebu SRL, Fadisol SA, CALMER and COPAGRAN.

Institutional Review Board Statement: Not applicable.

Informed Consent Statement: Not applicable.

Data Availability Statement: Data sharing is not applicable to this article as all new created data is already contained within this article.

Acknowledgments: E.C. would like to thank Comisión Academica de Posgrado (Universidad de la República) for granting him a Doctoral fellowship.

Conflicts of Interest: The authors declare no conflict of interest. The funders had no role in the design of the study; in the collection, analyses, or interpretation of data; in the writing of the manuscript; or in the decision to publish the results.

References

1. Zhang, T.; Lin, X. Assessing future drought impacts on yields based on historical irrigation reaction to drought for four major crops in Kansas. *Sci. Total Environ.* **2016**, *550*, 851–860. [CrossRef] [PubMed]
2. Zipper, S.C.; Qiu, J.; Kucharik, C.J. Drought effects on US maize and soybean production: Spatiotemporal patterns and historical changes. *Environ. Res. Lett.* **2016**, *11*, 094021. [CrossRef]
3. Boyer, J.S. Crop reaction to water and temperature stresses in humid, temperate climate. In *Environmental Stress and Crop Yields*; Raper, C., Jr., Kramer, P., Eds.; Westview Press: Boulder, CO, USA, 1983; pp. 3–7.
4. Kron, A.P.; Souza, G.M.; Ribeiro, R.V. Water deficiency at different developmental stages of glycine max can improve drought tolerance. *Bragantia* **2008**, *67*, 43–49. [CrossRef]
5. He, J.; Du, Y.L.; Wang, T.; Turner, N.C.; Yang, R.P.; Jin, Y.; Xi, Y.; Zhang, C.; Cui, T.; Fang, X.W.; et al. Conserved water use improves the yield performance of soybean (*Glycine max* (L.) Merr.) under drought. *Agric. Water Manag.* **2016**, *179*, 236–245. [CrossRef]
6. Sinclair, T.R.; Messina, C.D.; Beatty, A.; Samples, M. Assessment across the united states of the benefits of altered soybean drought traits. *Agron. J.* **2010**, *102*, 475–482. [CrossRef]
7. Humplík, J.F.; Lazár, D.; Husičková, A.; Spíchal, L. Automated phenotyping of plant shoots using imaging methods for analysis of plant stress responses-A review. *Plant Methods* **2015**, *11*, 29. [CrossRef]
8. Tuberosa, R. Phenotyping for drought tolerance of crops in the genomics era. *Front. Physiol.* **2012**, *3*, 347. [CrossRef]
9. Vadez, V.; Kholova, J.; Zaman-Allah, M.; Belko, N. Water: The most important "molecular" component of water stress tolerance research. *Funct. Plant Biol.* **2013**, *40*, 1310–1322. [CrossRef]
10. Ratnakumar, P.; Vadez, V.; Nigam, S.N.; Krishnamurthy, L. Assessment of transpiration efficiency in peanut (*Arachis hypogaea* L.) under drought using a lysimetric system. *Plant Biol.* **2009**, *11*, 124–130. [CrossRef]
11. Zaman-Allah, M.; Jenkinson, D.M.; Vadez, V. A conservative pattern of water use, rather than deep or profuse rooting, is critical for the terminal drought tolerance of chickpea. *J. Exp. Bot.* **2011**, *62*, 4239–4252. [CrossRef]
12. Consoli, S.; Vanella, D. Mapping crop evapotranspiration by integrating vegetation indices into a soil water balance model. *Agric. Water Manag.* **2014**, *143*, 71–81. [CrossRef]
13. Carter, J.N.; Jensen, M.E.; Traveller, D.J. Effect of Mid- to Late- Season Water Stress on Sugarbeet Growth and Yield 1. *Agron. J.* **1980**, *72*, 806–815. [CrossRef]
14. Meyer, W.S.; Green, G.C. Water Use by Wheat and Plant Indicators of Available Soil Water 1. *Agron. J.* **1980**, *72*, 253–257. [CrossRef]
15. Comstock, J.P. Hydraulic and chemical signalling in the control of stomatal conductance and transpiration. *J. Exp. Bot.* **2002**, *53*, 195–200. [CrossRef]
16. Sinclair, T.R.; Holbrook, N.M.; Zwieniecki, M.A. Daily transpiration rates of woody species on drying soil. *Tree Physiol.* **2005**, *25*, 1469–1472. [CrossRef]
17. Belko, N.; Zaman-allah, M.; Cisse, N.; Diop, N.N.; Zombre, G.; Ehlers, J.D.; Vadez, V. Lower soil moisture threshold for transpiration decline under water deficit correlates with lower canopy conductance and higher transpiration efficiency in drought-tolerant cowpea. *Funct. Plant Biol.* **2012**, *39*, 306–322. [CrossRef]
18. Hanks, R.; Shawcroft, R. An economical lysimeter for evapotranspiration studies 1965. *Agron. J.* **1965**, *57*, 634–636. [CrossRef]
19. Pearcy, R.W.; Schulze, E.; Zimmermann, R. Plant Physiological Ecology. *Plant Physiol. Ecol.* **1989**. [CrossRef]
20. Turner, N.C. Measurement and influence of environmental and plant factors on stomatal conductance in the field. *Agric. For. Meteorol.* **1991**, *54*, 137–154. [CrossRef]
21. Lu, P.; Woo, K.C.; Liu, Z.T. Estimation of whole-plant transpiration of bananas using sap flow measurements. *J. Exp. Bot.* **2002**, *53*, 1771–1779. [CrossRef]
22. Fletcher, A.L.; Sinclair, T.R.; Allen, L.H. Transpiration responses to vapor pressure deficit in well watered "slow-wilting" and commercial soybean. *Environ. Exp. Bot.* **2007**, *61*, 145–151. [CrossRef]
23. Sadok, W.; Sinclair, T.R. Transpiration response of "slow-wilting" and commercial soybean (*Glycine max* (L.) Merr.) genotypes to three aquaporin inhibitors. *J. Exp. Bot.* **2010**, *61*, 821–829. [CrossRef] [PubMed]
24. Stöckle, C.O.; Kemanian, A.R. Can Crop Models Identify Critical Gaps in Genetics, Environment, and Management Interactions? *Front. Plant Sci.* **2020**, *11*, 737. [CrossRef] [PubMed]
25. Basso, B.; Liu, L.; Ritchie, J.T. *A Comprehensive Review of the CERES-Wheat, -Maize and -Rice Models' Performances*; Elsevier Inc.: Amsterdam, The Netherlands, 2016; Volume 136, ISBN 9780128046814.
26. Gaydon, D.S.; Balwinder-Singh; Wang, E.; Poulton, P.L.; Ahmad, B.; Ahmed, F.; Akhter, S.; Ali, I.; Amarasingha, R.; Chaki, A.K.; et al. Evaluation of the APSIM model in cropping systems of Asia. *Field Crops Res.* **2017**, *204*, 52–75. [CrossRef]

27. Van Den Berg, M.; Driessen, P.M.; Rabbinge, R. Water uptake in crop growth models for land use systems analysis: II. Comparison of three simple approaches. *Ecol. Modell.* **2002**, *148*, 233–250. [CrossRef]
28. Wang, E.; Smith, C.J. Modelling the growth and water uptake function of plant root systems: A review. *Aust. J. Agric. Res.* **2004**, *55*, 501–523. [CrossRef]
29. Camargo, G.G.T.; Kemanian, A.R. Six crop models differ in their simulation of water uptake. *Agric. For. Meteorol.* **2016**, *220*, 116–129. [CrossRef]
30. Rigaud, J.; Puppo, A. Indole-3-acetic catabolism by soybean bacteroids. *J. Gen. Microbiol.* **1975**, *88*, 223–228. [CrossRef]
31. Quero, G.; Simondi, S.; Ceretta, S.; Otero, Á.; Garaycochea, S.; Fernández, S.; Borsani, O.; Bonnecarrère, V. An integrative analysis of yield stability for a GWAS in a small soybean breeding population. *Crop Sci.* **2021**, *61*, 1903–1914. [CrossRef]
32. Song, Q.; Hyten, D.L.; Jia, G.; Quigley, C.V.; Fickus, E.W.; Nelson, R.L.; Cregan, P.B. Development and Evaluation of SoySNP50K, a High-Density Genotyping Array for Soybean. *PLoS ONE* **2013**, *8*, e54985. [CrossRef]
33. R Core Team. *R: A Language and Environment for Statistical Computing*; R Foundation for Statistical Computing: Vienna, Austria, 2018.
34. Lê, S.; Josse, J.; Husson, F. FactoMineR: An R Package for Multivariate Analysis. *J. Stat. Softw.* **2008**, *25*, 1–18. [CrossRef]
35. Bates, D.; Mächler, M.; Bolker, B.M.; Walker, S.C. Fitting linear mixed-effects models using lme4. *J. Stat. Softw.* **2015**, *67*. [CrossRef]
36. Russell, L. *Emmeans: Estimated Marginal Means, Aka Leastsquares Means*; R Package Version 1.4.2; R Foundation for Statistical Computing: Vienna, Austria, 2019.
37. Kholová, J.; Nepolean, T.; Tom Hash, C.; Supriya, A.; Rajaram, V.; Senthilvel, S.; Kakkera, A.; Yadav, R.; Vadez, V. Water saving traits co-map with a major terminal drought tolerance quantitative trait locus in pearl millet [*Pennisetum glaucum* (L.) R. Br.]. *Mol. Breed.* **2012**, *30*, 1337–1353. [CrossRef]
38. West-Eberhard, M.J. Phenotypic plasticity and the origins of diversity. *Annu. Rev. Ecol. Syst.* **1989**, *20*, 249–278. [CrossRef]
39. Pigliucci, M. Evolution of phenotypic plasticity: Where are we going now? *Trends Ecol. Evol.* **2005**, *20*, 481–486. [CrossRef]

agronomy

MDPI

Article

Crop Performance Indexes Applied to Legume Used as Summer Cover Crops under Water Deficit Conditions

Verónica Berriel [1], Carlos H. Perdomo [1], Santiago Signorelli [2,3] and Jorge Monza [2,*]

[1] Soil and Water Department, School of Agronomy, Universidad de la República, Montevideo 11000, Uruguay; vberriel@gmail.com (V.B.); chperdomo@gmail.com (C.H.P.)
[2] Plant Biology Department, School of Agronomy, Universidad de la República, Montevideo 11000, Uruguay; ssignorelli@fagro.edu.uy
[3] The School of Molecular Sciences, Faculty of Science, The University of Western Australia, 35 Stirling Highway, Crawley, WA 6009, Australia
* Correspondence: jmonza@fagro.edu.uy

Abstract: Summer legume cover crops (CC) such as *Crotalaria juncea*, *Crotalaria spectabilis*, *Crotalaria ochroleuca*, and *Cajanus cajan* could offer diverse advantages for the environment and productive cropping systems. A low transpiration efficiency (TE) of CC can induce soil water content to levels that present a challenge for the subsequent crop. In a 75-day growth chamber experiment, using the natural abundance of ^{13}C, ^{18}O, and ^{15}N we evaluated the TE and BNF under two soil water conditions. Our results showed that the four species tested are good candidates for their use as CC because they showed good results in terms of productivity parameters, TE, and BFN. *Cajanus cajan* had the highest TE, a high shoot dry matter production, and accumulated more N from BFN in the shoot than *C. spectabilis*, *C. juncea*, and *C. ochroleuca*. $\Delta^{18}O$ increased under moderate water deficit and showed an inversely proportional relationship with the amount of transpired water, supporting the use of this isotopic indicator as a proxy for transpiration and stomatal conductance. For the isotopic parameters no interaction between the factors water regimen and species were found. We propose the mass ratio of nitrogen fixed by the volume of transpired water and the isotopic discrimination of ^{13}C as useful indicators of drought fixing legumes tolerance.

Keywords: legumes; cover crops; drought; biological nitrogen fixation; water use efficiency; nitrogen use efficiency; stable isotopes

Citation: Berriel, V.; Perdomo, C.H.; Signorelli, S.; Monza, J. Crop Performance Indexes Applied to Legume Used as Summer Cover Crops under Water Deficit Conditions. *Agronomy* **2022**, *12*, 443. https://doi.org/10.3390/agronomy12020443

Academic Editor: Rajeev K. Varshney

Received: 18 November 2021
Accepted: 4 February 2022
Published: 10 February 2022

1. Introduction

The annual summer legumes *Crotalaria juncea*, *C. spectabilis*, *C. ochroleuca*, and *Cajanus cajan*, are species that are characterized by their high biomass production and ability for biological nitrogen fixation (BNF) [1–3]. Due to these characteristics, they are commonly used as cover crops (CC) in cropping agriculture rotations in tropical and temperate zones [4,5]. Cover crops could protect the soil surface temporarily or permanently between two commercial crops [6], and their use is a strategy to improve soil quality and reduce nutrient losses, including water shortages. In the context of drought, a species' ability to accumulate dry matter (DM) production and N must be balanced with its water consumption [7].

Transpiration efficiency (TE) also called water-use efficiency (WUE) is defined as the amount of DM produced by water transpired [8] and can be determined on different scales of time (instantaneous or time-integrated measurements) and space (at leaf, shoot or whole plant level). The selection of species based on TE is a key strategy of plants in the acclimation to drought [9]. The reference method to determine TE, at whole plant level and long term is through lysimeters with gravimetric determinations [10], which in practice present some limitations due to labour requirements. Instantaneous WUE or TE can be also determined by measuring the concentration of CO_2 and H_2O vapour, which can be applied at the leaf level and faster than gravimetric determination using lysimeters [11]; however, the robustness of this methodology is more limited than TE determination. In addition, the instantaneous WUE can be estimated by the relationship between the photosynthesis rate (A) and transpiration (T) and the intrinsic WUE by the relationship between A and stomatal conductance (gs) [12].

Another alternative to determine long-term TE, extrapolated to whole plant level that does not require measuring transpiration or DM production, is through isotopic discrimination of ^{13}C ($\Delta^{13}C$) [13,14]. This isotopic indicator is highly heritable in C3 plants and has a low genotype by environment interaction [15,16].

The plant $\Delta^{13}C$ depends on its water status [17] and, therefore, on soil moisture [18]. $\Delta^{13}C$ has a strong negative correlation with rainfall [19] and soil moisture [20]. Moreover, physiological factors such as stomatal conductance and photosynthetic rate also determine $\Delta^{13}C$ [21,22].

Although $\Delta^{13}C$ is widely used as an indicator or proxy for TE, it is not possible to distinguish if under water limiting conditions its variation is due to decreases in stomatal conductance or photosynthetic rate. To differentiate between them, the isotopic discrimination of ^{18}O ($\Delta^{18}O$) can be used, because it does not depend on the photosynthetic rate but on the stomatal conductivity [23]. Thus, the determination of $\Delta^{13}C$ and $\Delta^{18}O$ allows a quick and reliable measurement of TE and the stomatal conductance, respectively [23,24].

The natural abundance of ^{15}N can be used to estimate the BNF. The different composition of nitrogen isotopes in plants grown in the same condition can be attributed to the fact that ^{15}N abundance in the air is lower than in the soil [25]. Therefore, the determination of N isotopes in plants allows estimation of the amount of N obtained through BNF.

Moderated water restrictions have been shown to increase the natural abundance of ^{13}C in plant tissues, including *C. juncea* and *C. spectabilis*, during their growth that allows determining their TE [26]. However, that study was limited to two species and the fixated nitrogen and its relation with the WUE was not evaluated. Therefore, the objective of our work is to evaluate the performance of four species of tropical legumes based on different desirable attributes in CC under water-limiting conditions and also to propose new index parameters related to TE and BNF under water-limiting conditions.

2. Materials and Methods

2.1. Plant Material and Growing Conditions

Seeds of *Crotalaria juncea* L. (Sunnhemp), *C. spectabilis* Roth (Showy rattlepod), *C. ochroleuca* (Slender leaf rattlebox), and *Cajanus cajan* (Pigeon pea) cv. IAPAR 43 were purchased from BRSEEDS Company (Araçatuba, SP, Brazil). Seeds were sown at the rate of one per pot with 4 kg of a typical soil from southern Uruguay (carbon = 11.6 g/kg; clay = 268 g/kg; silt = 487 g/kg; sand = 245 g/kg). The plants were grown in a growth chamber at 30 °C with a relative humidity of approximately 50% and a light intensity of 500 mmol m^{-2} s^{-1} with a 16/8 h light-dark cycle.

The experimental design used was a randomized complete block with two factors, legume species and water regime. Plants from each species, six replicates each, were firstly grown at 80% of field capacity (FC) for 30 days. From day 30, a moderate water deficit was imposed by subjecting the plants to 50% FC for 45 days whereas other plants were kept at 80% FC as control treatment. The daily water volumes of 50% and 80% FC were estimated by gravimetric determination and were calculated considering that the water content at FC (θf, on the basis of mass) was 28.5% (m/m). The amount of transpired water was determined according to Berriel et al. [14], and the TE was calculated at the end of the test as: TE = shoot DM produced/transpired water.

2.2. Isotopic Ratio Mass Spectrometry Determination

To determine the produced shoot biomass expressed as DM, leaves and stems were dried at 60 °C until a constant weight was reached. Plant samples were ground in a fixed and mobile knife mill (Marconi MA-580, Piracicaba, Brazil), achieving a particle size of less than 2 mm and then with a rotary mill (SampleTek 200 vial Rotator, Lawrenceburg,

KY, US) until reaching the required granulometric size for isotopic analysis. One and a half mg of each sample was weighed into tin capsules. The natural abundance of ^{13}C and ^{15}N was determined in a Flash EA 1112 elemental analyser (Milan, Italy) coupled to a Thermo Finnigan DELTAplus mass spectrometer (Bremen, Germany) at the Centre of Nuclear Application in Sustainability Agricultural of School of Agronomy, Uruguay. The isotopic ratio was expressed in delta notation (δ) in parts per thousand or ‰ using the following equation [27]:

$$\delta^{13}C = \left(\frac{R_{sample}}{R_{standard}} - 1\right) \times 1000$$

Carbon isotope discrimination ($\Delta^{13}C$) was calculated using the following equation [28]:

$$\Delta^{13}C = \left(\frac{\delta^{13}C_{air} - \delta^{13}C_{plant}}{1 + \delta^{13}C_{air}/1000} - 1\right) \times 1000$$

The proportion of N fixed from the air (% Ndfa) used the formula of Shearer and Khol [29]:

$$\%Ndfa = \left(\frac{\delta^{15}N_{ref} - \delta^{15}N_{fix}}{\delta^{15}N_{ref} - B}\right) \times 100$$

with % Ndfa the proportion of plant N derived from BNF; Δ^{15}Nref, the $\delta^{15}N$ value of the reference plant (not fixing); $\delta^{15}N_{fix}$, the $\delta^{15}N$ value of the fixing plant; and B, the $\delta^{15}N$ value of a fixing plant growing in a medium without N.

As a reference plant, corn was used, with a value of +9.7‰ of $\delta^{15}N$, determined under the same conditions.

The $^{18}O/^{16}O$ isotopic ratio was determined on the DM of leaves, and the analytical determination was carried out in a Thermo Scientific Delta V mass spectrometer (Bremen GmbH, Germany) with a Conflo IV interface connected to a Costech 4010 elemental analyser (EA) (Milan, Italy) and a high-temperature conversion elemental analyser (CSI laboratory of the University of New Mexico).

2.3. Statistical Analysis

The experimental design consisted of completely randomized blocks with 6 repetitions each. Factors consisted of combinations of four plant species and two soil water regimes (80% FC and 50% FC). The main effects of species and soil water status, as well as their interaction, were analysed by ANOVA and the mean separations were performed with the Tukey's HSD (honestly significant difference) at the 5% significance level using the statistic software InfoStat® version 2020 (Universidad Nacional de Córdoba, Córdoba, Argentina) [30]. The correlation between the variables studied was analysed using the Pearson correlation matrix also using the InfoStat® [30].

3. Results

The legume species *C. juncea*, *C. spectabilis*, *C. ochroleuca*, and *Cajanus cajan* were evaluated according to their shoot DM production, transpired water, TE, and isotopic parameters in two water regimes, moderate water deficit (50% FC) and well-watered (80% FC). For these variables, no interaction between the factors water regimen and species were found, therefore the response of each species to the water regime followed a similar pattern. The main effects of water regime and species were observed on shoot dry matter, transpired water, and TE (Figures 1 and 2).

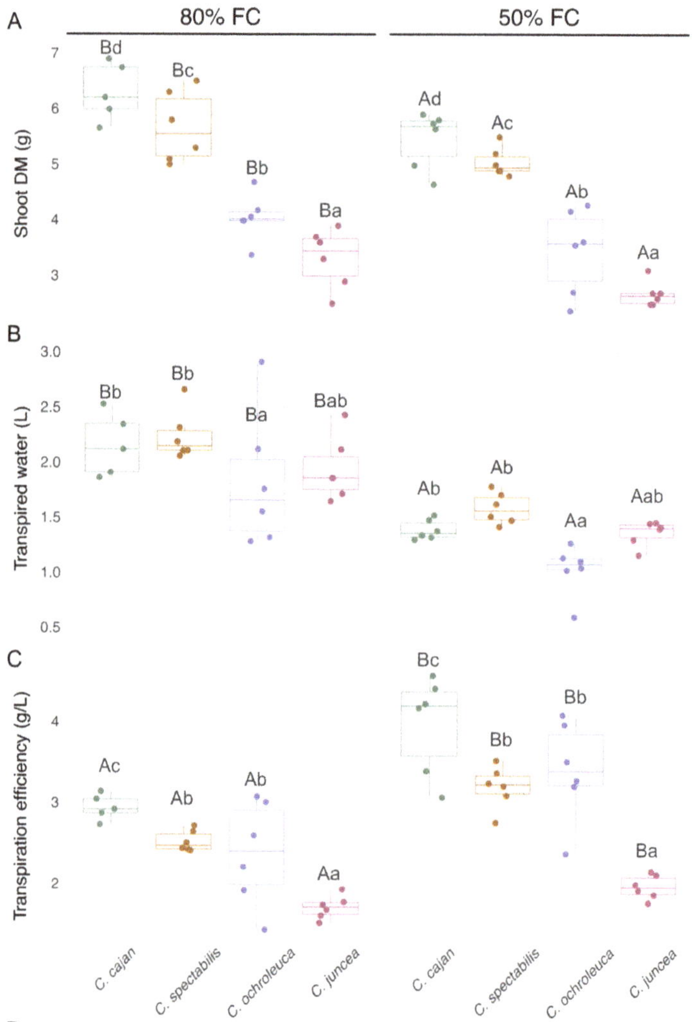

D

Factors	*p*-values		
	Shoot dry matter	Transpired water	TE
Specie	<0.0001	0.0021	<0.0001
Water status	<0.0001	<0.0001	<0.0001
Specie x Water status	NS	NS	NS

Figure 1. Shoot dry matter (DM), transpired water and transpiration efficiency (TE) of legumes used as CC compared with no water deficit (80%) to moderate water deficit (50%). (**A**). DM production expressed in g per plant. (**B**). Transpired water as L of water per plant. (**C**). TE determined as DM produced per transpired water (g of DM/L of water). (**D**). *p*-values of the ANOVA for the species and water status effects and the species × water status interaction. The box plots represent the means, each dot represents independent replicates, and the vertical lines represent the standard deviation. Different letters indicate statistical significance between species (lowercase) and water regimes (capital letters) in a Tukey's HSD (honestly significant difference) post hoc test at 0.05 *p*-value.

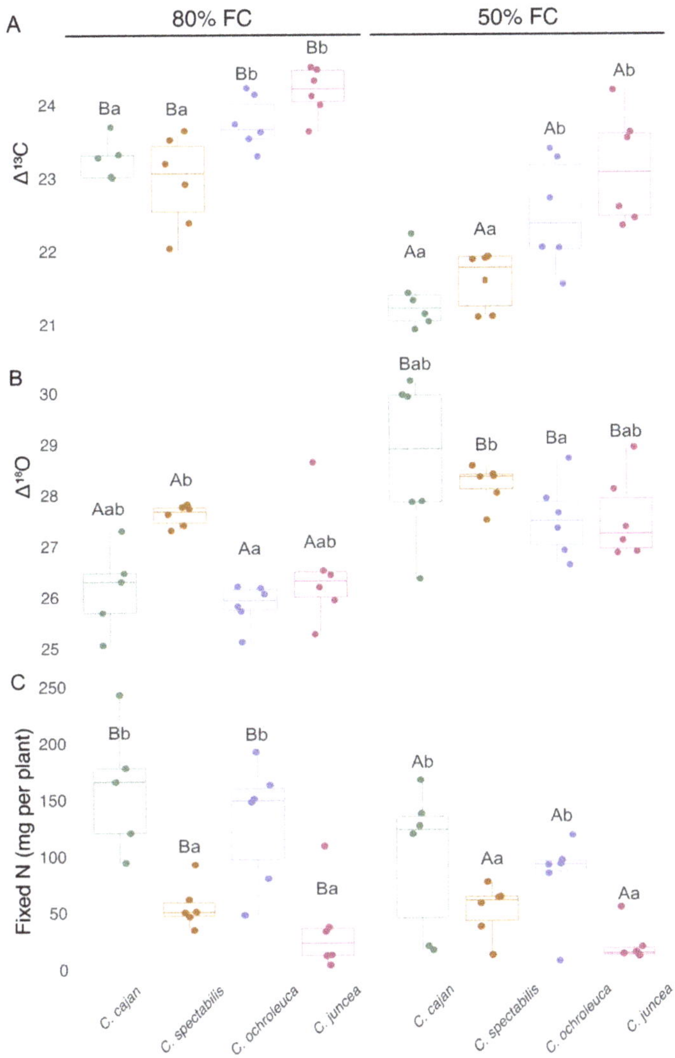

D

Factors	p-values		
	$\Delta^{13}C$	$\Delta^{18}O$	Fixed N
Specie	<0.0001	0.0096	<0.0001
Water status	<0.0001	<0.0001	0.0119
Specie x Water status	NS	NS	NS

Figure 2. $\Delta^{13}C$, $\Delta^{18}O$, and fixed N in legumes used as CC during moderate water deficit (50% FC) and in the absence of water deficit (80% FC). (**A**). Carbon-13 isotope discrimination ($\Delta^{13}C$) (**B**). Oxygen-18 isotope discrimination ($\Delta^{18}O$). (**C**). Fixed nitrogen (mg per plant). (**D**). *p*-values of the ANOVA for the species and water status effects, and the species × water status interaction. The box plots represent the means, each dot represents independent replicates, and the vertical lines represent the standard deviation. Different letters indicate statistical significance between species (lowercase) and water regimes (capital letters) in a Tukey's HDS post hoc test at 0.05 *p*-value.

However, the N fixation did not show differences between species (Figures 1 and 2). In particular, the DM production and transpired water were lower during moderate water deficit (Figure 1). *Cajanus cajan* produced the highest DM and transpiration rate, followed by *C. spectabilis*, *C. ochroleuca*, and *C. juncea* (Figures 1 and 2). In addition, the TE increased in all species under moderate water deficit, being *Cajanus cajan* the most efficient, *C. juncea* the least efficient, and *C. ochroleuca* and *C. spectabilis* showed an intermediate efficiency (Figure 1).

The two factors tested, water regime and species, influenced the isotopic parameters but not their interaction. In general, $\Delta^{13}C$ was lower under moderate water deficit, whereas the $\Delta^{18}O$ increased under moderate water deficit compared to the control condition (Figure 2). Among the species, *Cajanus cajan* and *C. spectabilis* showed a lower $\Delta^{13}C$ than *C. ochroleuca* and *C. juncea* in both control and moderate water deficit (Figure 2). In terms of $\Delta^{18}O$ under moderate water deficit, *only C. ochroleuca* and *C. spectabilis* exhibited differences, being higher for *C. spectabilis* (Figure 2). The correlation analysis, considering all species and water soil content, showed a negative correlation between the variables TE and $\Delta^{13}C$, $\Delta^{13}C$ and $\Delta^{18}O$, and $\Delta^{18}O$ and transpired water (Table 1).

Table 1. Pearson correlation coefficients for the correlations between shoot dry matter (DM), transpired water (T), transpiration efficiency (TE), ^{13}C and ^{18}O isotope discrimination ($\Delta^{13}C$, $\Delta^{18}O$ respectively) from the values obtained for *C. cajan*, *C. spectabilis*, *C. ochroleuca*, and *C. juncea* taken together.

	DM	T	TE	$\Delta^{13}C$	$\Delta^{18}O$
DM	1				
T	0.50 ***	1			
TE	0.49 ***	−0.47 ***	1		
$\Delta^{13}C$	−0.32 ***	0.44 ***	−0.77 ***	1	
$\Delta^{18}O$	0.13 *	−0.44 ***	0.56 ***	−0.69 ***	1

p: * 0.05; *** 0.001.

The %Ndfa was lower under moderate water deficit relative to the control condition, and no differences between the species were found (Figure 3). However, the DM production was different between these species (Table 1) explaining the differences in the amount of N derived from BNF ($p < 0.0001$) in the shoot. The water regime did not affect the amount of total N; thus, when total N was considered irrespectively of the water condition, *C. juncea* and *C. spectabilis* had lower N content (43 and 72 mg, respectively) than *C. ochroleuca* and *Cajanus cajan* (132 and 177 mg, respectively).

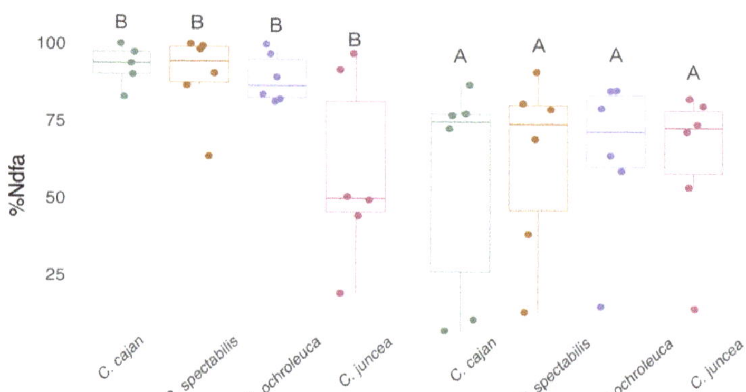

Figure 3. Proportion of nitrogen derived from the air (%Ndfa). The box plots represent the means, each dot represents independent replicates, and the vertical lines represent the standard deviation. The statistical analysis revealed no species effect but a treatment effect. Different letters indicate statistical significance between treatments in a Tukey's HDS post hoc test at 0.05 *p*-value.

Fixed N was affected by the water regime and also by different species, but no statistically interaction was found between water regime and species (Figure 2). In both water conditions, *Cajanus cajan* and *C. ochroleuca* fixed more N than *C. spectabilis* and *C. juncea* (Figure 2). Moreover, in both water regimes *Cajanus cajan* and *C. ochroleuca* had the highest ratio N_{fix}/T and $Nfix/\Delta^{18}O$, while *C. juncea* and *C. spectabilis* had the lowest (Table 2).

Table 2. Performance indexes for legumes that were subjected to moderate water deficit (50% FC) and control (80% FC) conditions. The evaluated indexes related the amount of fixed N from air with the transpired water, $\Delta^{18}O$ and $\Delta^{13}C$ (N_{fix}/T, $N_{fix}/\Delta^{18}O$, and $N_{fix}/\Delta^{13}C$). Different letters indicate statistical significance between species (lowercase) and water regimes (capital letters) in a Tukey's HDS post hoc test at 0.05 *p*-value.

Species	N_{fix}/T		$N_{fix}/\Delta^{13}C$		$N_{fix}/\Delta^{18}O$	
	80% FC	50% FC	80% FC	50% FC	80% FC	50% FC
Cajanus cajan	75 ^{a}	69 ^{a}	6.87 ^{Aa}	4.52 ^{Ba}	6.7 ^{a}	6.3 ^{a}
Crotalaria spectabilis	25 ^{b}	33 ^{b}	2.43 ^{Ab}	2.37 ^{Bb}	2.2 ^{b}	2.3 ^{b}
Crotalaria ochroleuca	77 ^{a}	83 ^{a}	5.49 ^{Aa}	3.60 ^{Ba}	5.7 ^{a}	4.3 ^{a}
Crotalaria juncea	16 ^{b}	14 ^{b}	1.39 ^{Ab}	0.84 ^{Bb}	1.8 ^{b}	1.4 ^{b}
Factor			*p*-value			
Specie	<0.0001		<0.0001		<0.0001	
Water status	NS		0.0325		NS	
Specie × Water status	NS		NS		NS	

In legumes in which $\Delta^{13}C$ correlated more with transpired water than with photosynthetic rate, the $N_{fix}/\Delta^{13}C$ index can be used as an index informing about the nitrogen fixation in relation to water transpired. The moderate water deficit condition implicated a decrease in this index in all four species tested and followed the same trend as the amount of fixed N (Table 2). Finally, strong positive correlations were found between the N_{fix}/T, $N_{fix}/\Delta^{18}O$, and $N_{fix}/\Delta^{13}C$ indexes and between them and the TE when the species and water conditions were grouped (Table 3).

Table 3. Pearson correlation coefficients for the correlation between different indicators of performance, transpiration efficiency (TE), mg of fixed N per L of water (N_{fix}/T), mg of fixed N per ^{13}C isotope discrimination ($N_{fix}/\Delta^{13}C$), and mg of fixed N per ^{18}O isotope discrimination ($N_{fix}/\Delta^{18}O$) from the values obtained for *C. cajan*, *C. spectabilis*, *C. ochroleuca*, and *C. juncea* taken together.

	TE	N_{fix}/T	$N_{fix}/\Delta^{13}C$	$N_{fix}/\Delta^{18}O$
TE	1			
WUEfix	0.45 ***	1		
$N_{fix}/\Delta^{13}C$	0.29 *	0.87 ***	1	
$N_{fix}/\Delta^{18}O$	0.43 ***	0.73 ***	0.85 ***	1

p: * 0.05; *** 0.001.

4. Discussion

The legumes *C. spectabilis*, *C. juncea*, *C. ochroleuca*, and *Cajanus cajan* may be good candidates for cover cropping because of their high ability to fix N [31]. Besides these attributes, high WUE and TE are desirable characteristics for CC in water-limited environments. In this study, we determined these parameters in these four species under different water regimes.

In both water regimes, *Cajanus cajan* had the greatest DM production, TE, and amount of fixed N among the four species tested. Moreover, *Cajanus cajan* had the greatest DM production in relation to transpired water. This shows its potential as a CC in water-limited environments. *Cajanus cajan* can combine these desirable attributes, maximizing the DM produced in relation to water consumed or transpired. These findings show that *Cajanus cajan* has a potential to be used as CC in soils where water restrictions are common.

Most climate change scenarios foresee rainfalls to be decreased around the world and its pattern to be more erratic [32]. Therefore, the redesign of agricultural crop rotations seems to be imminent to mitigate the effect of climate change on natural resources [33–35]. High WUE and TE in drought conditions are desired features that crops, varieties, or genotypes must have to be considered in this redesign [36].

In moderate water deficit, *C. spectabilis, C. juncea, C. ochroleuca,* and *Cajanus cajan* were tolerant to a moderate water deficit, in agreement with previous reports that evaluated the performance of these species under field conditions [37,38]. The strong decrease in transpired water in response to a moderate water deficit, together with a slight decrease in DM production, resulted in an increased TE (Figure 1), which explains their tolerance to water restrictions. Berriel et al. [25] found that moderate drought also negatively impacted DM production and transpired water of *C. spectabilis* and *C. juncea* when plants were studied at whole plant level and in longer term. This suggests that our findings may be translatable for all four species when studied in longer periods of moderate water deficit and no matter the methodological approach used (i.e., whole plant or shoot only).

The decrease in DM production can be a consequence of the reduction in photosynthetic rate, which often relates to a decrease in stomatal conductance. One of the few studies quantifying the impact of water deficit on the variables determining the water–carbon balance in leaves, was carried out in *C. cajan* in which moderate water deficit caused a decrease in transpired water, stomatal conductance, and photosynthetic rate, leading to an increase in instantaneous WUE [39].

In all species, Δ^{13}C decreased in water-limited condition relative to the control condition (Figure 2). Regardless of the water regime, the least discrimination was exhibited by *Cajanus cajan*, the greatest by *C. juncea*, and intermediate discrimination was found in *C. ochroleuca* and *C. spectabilis*. In turn, regardless of the water conditions position (Figure 2). Berriel et al. [25] observed the same pattern for *C. spectabilis* and *C. juncea*, in terms of Δ^{13}C, during water deficit and after rehydration. In our study, when the water deficit prevailed, the decrease in Δ^{13}C indicates that either the decrease in stomatal conductance or the transpiration rate led to a decrease in both transpired water and DM production (Figure 1).

Δ^{13}C was negatively related to TE in all four evaluated species (Figure 1; Table 1). The use of Δ^{13}C as a proxy for TE has been confirmed in different C3 species [40–42] but not in many grain legumes [43]. The relationships we found between Δ^{13}C and TE and those reported by Berriel et al. [14,25], supports the strength of ^{13}C as an isotopic indicator of TE in *Cajanus cajan*, *C. spectabilis*, *C. juncea*, and *C. ochroleuca*.

The variation in Δ^{18}O by the water regimes (Figure 2; Table 1) and the strong inversely proportional relationship between this and transpired water, support the use of this isotopic indicator as a proxy for transpiration and stomatal conductance [44,45]. Determining Δ^{18}O is advantageous compared to determining transpiration and stomatal conductance because it evaluates the transpiration rate in a longer time scale [46].

The negative correlation between Δ^{18}O and Δ^{13}C helped to interpret the differences in TE estimated through the Δ^{13}C proxy (Table 1), indicating that the increase in TE is mainly determined by the decrease in stomatal conductance than by the decrease in photosynthetic rate. This interpretation is consistent with the greater decrease in transpired water than in DM production (Figure 1).

In this study, we defined indexes relating the amount of fixed N (N_{fix}) and transpired water, determined directly, as Δ^{13}C and Δ^{18}O. Based on the relationships established between fixed N and transpired water, we propose three sustainability indexes focused on the amount of water used to fix N and applicable to legumes: N_{fix}/T, $N_{fix}/\Delta^{13}C$, and $N_{fix}/\Delta^{18}O$. All these indexes showed a strong positive correlation among them (Table 2). Methodologically, the $N_{fix}/\Delta^{13}C$ index is the preferred one, given its simplicity to be determined. According to these indexes, *C. cajan* and *C. ochroleuca* are the most promising species as CC in conditions of moderate water deficit. It is worth bearing in mind that the application of rhizobia inoculants would increase the BNF in any of the species, something not tested here, which can cause differences in the ranking produced here. Regardless of this, the information generated in this work, as well as the use of the indexes defined here, can contribute to the study and design of agricultural rotations that allow the generation of ecosystem services and mitigate the impact of climate change on farms. Since this study was completed in controlled conditions, more research is necessary to evaluate the significance of these indexes in field conditions.

Author Contributions: V.B. conceived the study; V.B. performed all the experiments and data analysis; S.S. prepared the figures; C.H.P. and J.M. supervised the study; all the authors contributed to the data interpretation and wrote the manuscript. All authors have read and agreed to the published version of the manuscript.

Funding: This research was funded by Agencia Nacional de Investigación e Innovación from Uruguay, Grant: ANII-FMV_3_2016_1_125492; Doctorado en Biotecnología (Facultad de Ciencias, Universidad de la República) and CAP (Comisión Académica de Postgrado, Universidad de la República).

Institutional Review Board Statement: Not applicable.

Informed Consent Statement: Not applicable.

Data Availability Statement: Not applicable.

Acknowledgments: The authors are active members of the National Research System (SNI).

Conflicts of Interest: The authors declare no conflict of interest.

References

1. Soares, M.B.; Tavanti, R.F.; Rigotti, A.R.; de Lima, J.P.; da Silva Freddi, O.; Petter, F.A. Use of cover crops in the southern Amazon region: What is the impact on soil physical quality? *Geoderma* **2021**, *384*, 114796. [CrossRef]
2. Da Silva, E.C.; Muraoka, T.; Bastos, A.V.S.; Franzin, V.I.; Buzetti, S.; Loureiro Soares, F.A.; Batista Teixeira, M.; Bendassolli, J.A. Biomass and Nutrient Accumulation by Cover Crops and Upland Rice Grown in Succession Under No-Tillage System as Affected by Nitrogen Fertilizer Rate. *J. Crop Sci. Biotechnol.* **2020**, *23*, 117–126. [CrossRef]
3. Pereira Pacheco, L.; Dalla Côrt São Miguel, A.S.; da Silva, R.G.; de Souza, E.D.; André Petter, F.; Kappes, C. Biomass yield in production systems of soybean sown in succession to annual crops and cover crops. *Pesquisa Agropecuária Brasileira* **2017**, *52*, 582–591. [CrossRef]
4. Reddy, P.P. Cover/Green Manure Crops. In *Sustainable Intensification of Crop Production*; Springer: Singapore, 2016; pp. 55–67. [CrossRef]
5. Souza, A.V.S.S.; Souza, T.A.F.; Santos, D.; Rios, E.S.; Souza, G.J.L. Agronomic evaluation of legume cover crops for sustainable agriculture. *Russ. Agric. Sci.* **2018**, *44*, 31–38. [CrossRef]
6. Pound, B.; Anderson, S.; Gundel, S. Species for niches: When and for whom are cover crops appropriate? *Mt. Res. Dev.* **1999**, *19*, 307–312.
7. Sadras, V.; Lake, L.; Li, Y.; Farquharson, E.A.; Sutton, T. Phenotypic plasticity and its genetic regulation for yield, nitrogen fixation and δ13C in chickpea crops under varying water regimes. *J. Exp. Bot.* **2016**, *67*, 4339–4351. [CrossRef]
8. Blankenagel, S.; Yang, Z.; Avramova, V.; Schön, C.-C.; Grill, E. Generating Plants with Improved Water Use Efficiency. *Agronomy* **2018**, *8*, 194. [CrossRef]
9. Ullah, H.; Santiago-Arenas, R.; Ferdous, Z.; Attia, A.; Datta, A. Improving water use efficiency, nitrogen use efficiency, and radiation use efficiency in field crops under drought stress: A review. *Adv. Agron.* **2019**, *156*, 109–157.
10. Vadez, V.; Kholova, J.; Medina, S.; Kakkera, A.; Anderberg, H. Transpiration efficiency: New insights into an old story. *J. Exp. Bot.* **2014**, *65*, 6141–6153. [CrossRef]
11. Polley, H.W. Implications of atmospheric and climatic change for crop yield and water use efficiency. *Crop Sci.* **2002**, *42*, 131–140. [CrossRef]
12. Ali, M.H.; Talukder, M.S.U. Increasing water productivity in crop production—A synthesis. *Agric. Water Manag.* **2008**, *95*, 1201–1213. [CrossRef]
13. Santesteban, L.G.; Miranda, C.; Barbarin, I.; Royo, J.B. Application of the measurement of the natural abundance of stable isotopes in viticulture: A review. *Aust. J. Grape Wine Res.* **2015**, *21*, 157–167. [CrossRef]
14. Berriel, V.; Monza, J.; Perdomo, C.H. Cover Crop Selection by Jointly Optimizing Biomass Productivity, Biological Nitrogen Fixation, and Transpiration Efficiency: Application to Two Crotalaria Species. *Agronomy* **2020**, *10*, 1116. [CrossRef]
15. Elazab, A.; Molero, G.; Serret, M.D.; Araus, J.L. Root traits and δ13C and δ18O of durum wheat under different water regimes. *Funct. Plant Biol.* **2012**, *39*, 379–393. [CrossRef]

16. Yasir, T.A.; Min, D.H.; Chen, X.J.; Condon, A.G.; Hu, Y.G. The association of carbon isotope discrimination (Δ) with gas exchange parameters and yield traits in Chinese bread wheat cultivars under two water regimes. *Agric. Water Manag.* **2013**, *119*, 111–120. [CrossRef]
17. Berriel, V.; Mori, C.; Perdomo, C. Water status and 13C isotopic discrimination of two conventional pastures in Uruguay. *Agrociencia* **2014**, *18*, 1–13. (In Spanish)
18. Hartman, G.; Danin, A. Isotopic values of plants in relation to water availability in the Eastern Mediterranean region. *Oecologia* **2010**, *162*, 837–852. [CrossRef]
19. Stewart, G.R.; Turnbull, M.H.; Schmidt, S.; Erskine, P.D. 13C natural-abundance in plant-communities along a rainfall gradient: A biological integrator of water availability. *Aust. J. Plant Physiol.* **1995**, *22*, 51–55. [CrossRef]
20. Ehleringer, J.R.; Cooper, T.A. Correlations between carbon isotope ratio and microhabitat in desert plants. *Oecologia* **1988**, *76*, 562–566. [CrossRef] [PubMed]
21. Farquhar, G.D.; O'Leary, M.H.; Berry, J.A. On the Relationship Between Carbon Isotope Discrimination and the Intercellular Carbon Dioxide Concentration in Leaves. *Funct. Plant Biol.* **1982**, *9*, 121–137. [CrossRef]
22. Santiago, L.S.; Silvera, K.; Andrade, J.L.; Dawson, T.E. The use of stable isotopes in tropical biology. *Interciencia* **2005**, *30*, 28–35. (In Spanish)
23. Barbour, M.M. Stable oxygen isotope composition of plant tissue: A review. *Funct. Plant Biol.* **2007**, *34*, 83–94. [CrossRef]
24. Hirl, R.T.; Ogée, J.; Ostler, U.; Schäufele, R.; Baca Cabrera, J.; Zhu, J.; Schliep, I.; Wingate, L.; Schnyder, H. Temperature sensitive biochemical 18O-fractionation and humidity-dependent attenuation factor are needed to predict δ18O of cellulose from leaf water in a grassland ecosystem. *N. Phytol.* **2021**, *229*, 3156–3171. [CrossRef] [PubMed]
25. Chalk, P.M.; Craswell, E.T. An overview of the role and signifcance of 15N methodologies in quantifying biological N2 fxation (BNF) and BNF dynamics in agro-ecosystems. *Symbiosis* **2018**, *75*, 1–16. [CrossRef]
26. Berriel, V.; Perdomo, C.; Monza, J. Carbon Isotope Discrimination and Water-Use Efficiency in Crotalaria Cover Crops under Moderate Water Deficit. *J. Soil Sci. Plant Nutr.* **2020**, *20*, 537–545. [CrossRef]
27. Sulzman, E.W. Stable isotope chemistry and measurement: A primer. In *Stable Isotopes in Ecology and Environmental Science*, 2nd ed.; Michener, R., Lajtha, K., Eds.; Blackwell Publishing: Boston, NJ, USA, 2007; pp. 1–21.
28. Farquhar, G.D.; Ehleringer, J.R.; Hubick, K.T. Carbon isotope discrimination and photosynthesis. *Annu. Rev. Plant Biol.* **1989**, *40*, 503–537. [CrossRef]
29. Shearer, G.; Kohl, D.H. N2-fixation in field settings: Estimations based on natural 15N abundance. *Aust. J. Plant Physiol.* **1986**, *13*, 699–756.
30. Di Rienzo, J.A.; Casanoves, F.; Balzarini, M.G.; Gonzalez, L.; Tablada, M.; Robledo, C.W. *InfoStat*; Version 2011; Grupo InfoStat, FCA, Universidad Nacional de Córdoba: Córdoba, Argentina, 2020.
31. Saikia, P.; Nag, A.; Anurag, S.; Chatterjee, S.; Khan, M.L. Tropical Legumes: Status, Distribution, Biology and Importance. In *The Plant Family Fabaceae*; Hasanuzzaman, M., Araújo, S., Gill, S., Eds.; Springer: Singapore, 2020. [CrossRef]
32. Zorrilla-Fontanesi, Y.; Pauwels, L.; Panis, B. Strategies to revise agrosystems and breeding to control Fusarium wilt of banana. *Nat. Food* **2020**, *1*, 599–604. [CrossRef]
33. Condon, A.G. Drying times: Plant traits to improve crop water use efficiency and yield. *J. Exp. Bot.* **2020**, *71*, 2239–2252. [CrossRef]
34. Hernandez-Ochoa, I.M.; Pequeno, D.N.L.; Reynolds, M. Adapting irrigated and rainfed wheat to climate change in semi-arid environments: Management, breeding options and land use change. *Eur. J. Agron.* **2019**, *109*, 125915. [CrossRef]
35. Buto, O.; Galbiati, G.M.; Alekseeva, N.; Bernoux, M. *Climate Finance in the Agriculture and Land Use Sector—Global and Regional Trends between 2000 and 2018*; FAO: Rome, Italy, 2021.
36. Iglesias, A.; Garrote, L. Adaptation strategies for agricultural water management under climate change in Europe. *Agric. Water Manag.* **2015**, *155*, 113–124. [CrossRef]
37. Fageria, N.K. *Maximizing Crop Yields*; Marcel Dekker: New York, NY, USA, 1992.
38. Baligar, V.C.; Fageria, N.K. Agronomy and Physiology of Tropical Cover Crops. *J. Plant Nutr.* **2007**, *30*, 1287–1339. [CrossRef]
39. Sreeharsha, R.V.; Mudalkar, S.; Sengupta, D. Mitigation of drought-induced oxidative damage by enhanced carbon assimilation and an efficient antioxidative metabolism under high CO2 environment in pigeonpea (*Cajanus cajan* L.). *Photosynth. Res.* **2019**, *139*, 425–439. [CrossRef]
40. Maxwell, T.M.; Silva, L.C.R.; Horwath, W.R. Integrating effects of species composition and soil properties to predict shifts in montane forest carbon-water relations. *Proc. Natl. Acad. Sci. USA* **2018**, *115*, E4219–E4226. [CrossRef] [PubMed]
41. Pronger, J.; Campbell, D.I.; Clearwater, M.J.; Mudge, P.L.; Rutledge, S.; Wall, A.M.; Schipper, L.A. Toward optimisation of water use efficiency in dryland pastures using carbon isotope discrimination as a tool to select plant species mixtures. *Sci. Total Environ.* **2019**, *665*, 698–708. [CrossRef] [PubMed]
42. Castillo, A.; Rebuffo, M.; Díaz, P.; García, C.; Monza, J.; Borsani, O. Physiological and biochemical responses to water deficit in *Lotus uliginosus* x L. corniculatus hybrids. *Crop Pasture Sci.* **2017**, *68*, 670–679. [CrossRef]
43. Turner, N.C.; Palta, J.A.; Shrestha, R.; Ludwig, C.; Siddique, K.H.M.; Turner, D.W. Carbon isotope discrimination is not correlated with transpiration efficiency in three cool-season grain legumes (Pulses). *J. Integr. Plant Biol.* **2007**, *49*, 1478–1483. [CrossRef]
44. Farquhar, G.D.; Cernusak, L.A.; Barnes, B. Heavy water fractionation during transpiration. *Plant Physiol.* **2007**, *143*, 11–18. [CrossRef]

45. Ferrio, J.P.; Mateo, M.A.; Bort, J.; Abdalla, O.; Voltas, J.; Araus, J.L. Relationships of grain delta 13C and delta 18O with wheat phenology and yield under water-limited conditions. *Ann. Appl. Biol.* **2007**, *150*, 207–215. [CrossRef]
46. Barbour, M.M.; Cernusak, L.A.; Whitehead, D.; Griffin, K.L.; Turnbull, M.H.; Tissue, D.T.; Farquhar, G.D. Nocturnal stomatal conductance and implications for modelling δ18O of leaf-respired CO2 in temperate tree species. *Funct. Plant Biol.* **2005**, *32*, 1107–1121. [CrossRef]

agronomy

MDPI

Article

Maize and Wheat Response to Drought Stress under Varied Sulphur Fertilisation

Grzegorz Kulczycki [1,*], Elżbieta Sacała [1], Piotr Chohura [2] and Justyna Załuska [1]

1 Institute of Soil Science, Plant Nutrition and Environmental Protection, Wrocław University of Environmental and Life Sciences, Grunwaldzka Str. 53, 50-363 Wrocław, Poland; elzbieta.sacala@upwr.edu.pl (E.S.); 113759@student.upwr.edu.pl (J.Z.)

2 Department of Horticulture, Wrocław University of Environmental and Life Sciences, Grunwaldzki sq 24A, 50-363 Wrocław, Poland; piotr.chohura@upwr.edu.pl

* Correspondence: grzegorz.kulczycki@upwr.edu.pl; Tel.: +48-71-320-5654

Abstract: This study aimed to examine the influence of long-lasting moderate (45% field water capacity—FWC) and severe (30% FWC) water stress and application of sulphur (elemental sulphur or sulphate) on the growth, yield and mineral composition of wheat and maize. Concentrations of macro- and micronutrients were determined in the aboveground parts of the plants. Drought stress caused a marked decrease in the growth parameters of both plants. Under both optimal water conditions (60% FWC) and moderate water stress (45% FWC), grain yields of wheat grown without sulphur application were not significantly different. Applying elemental sulphur caused an increase in grain yield under moderate stress, whereas sulphate was more effective in wheat grown under adequate water supply. Severe water stress significantly lowered wheat yield, regardless of sulphur fertilisation. Increasing water stress resulted in a greater reduction in maize growth, with an average 50% decrease in dry mass under severe water stress. Both crops maintained relatively high levels of macro- (N, P, K, Mg, Ca, S) and microelements (Mn, Fe, Cu, Zn) and did not suffer noticeably from deficiencies in such. Sulphur application did not modify these relationships. In conclusion, sulphur fertilisation may be recommended in wheat cultivation when plants are exposed to moderate water stress.

Keywords: drought; water stress; elemental sulphur; sulphate; macroelements; microelements

Citation: Kulczycki, G.; Sacała, E.; Chohura, P.; Załuska, J. Maize and Wheat Response to Drought Stress under Varied Sulphur Fertilisation. *Agronomy* **2022**, *12*, 1076. https://doi.org/10.3390/agronomy12051076

Academic Editor: Santiago Signorelli

Received: 2 April 2022
Accepted: 27 April 2022
Published: 29 April 2022

1. Introduction

Drought is currently the most important environmental factor that a has huge impact on the growth of plants and their productivity [1,2]. Suboptimal water supply affects plants in a number of interacting manners, and the plant response is dependent on the plant species and stage of plant growth, severity and duration of stress and other environmental factors [1]. Under drought conditions, plants show numerous morphological, physiological, and biochemical changes [2]. Drought disrupts water relations, mineral uptake, photosynthesis efficiency and partitioning of assimilates and ultimately causes a significant reduction in crop yields [1,3,4]. Fahad et al. [1] show that yield losses in maize and wheat caused by drought reach 63–87% and 57%, respectively. Hence, to guarantee successful crop production, it is necessary to find effective ways to mitigate the negative effects of this stress. To achieve this goal, it is very important to use appropriate breeding programs to obtain crop genotypes resistant to suboptimal water supply. Applying a specific mineral fertilisation may help plants to cope with drought stress [2]. Hence, knowledge concerning uptake and accumulation of nutrients in plant tissues is very important from both an agricultural and an ecological perspective. The macroelement sulphur (S) is present in plant tissues in the smallest amount, in comparison to all other essential macronutrients, and yet is considered a limiting element in high-yielding agriculture [5,6]. Research concerning sulphur application in agricultural plant production is therefore very necessary. To ensure proper growth and development, plants require sulphur at a level of 0.1–1.0% on a dry weight basis, and the average concentration of S in plant tissues ranges from 0.2 to 0.5% [7]. Plants take sulphur mainly as sulphate (SO_4^{2-}), but elemental sulphur that is oxidised to sulphate in the soil is another good source of this element [8].

Sulphur plays a crucial role not only in the growth and development of higher plants, but also in stress tolerance and drought tolerance. Data concerning the effect of drought on sulphur nutrition are scant, although some studies have indicated that S nutrition plays a role in stress tolerance and defence mechanisms [1,9,10]. Sulphur, among other things, is a component of glutathione that is an important non-enzymatic antioxidant, being a crucial element in antioxidative mechanisms in plant cells [11]. Sulpholipids containing sulphur are present in chloroplastic membranes where they might protect photosynthetic apparatus under stress conditions. Usmani et al. [12] examined maize grown under drought stress and fertilised with different S fertilisers (K_2SO_4, $FeSO_4$, $CuSO_4$ and Na_2SO_4). They demonstrated that sulphur availability positively influenced some physiological parameters in water-stressed maize and among various S sources, K_2SO_4 application resulted in the maximum increase in plant yield. Lee et al. [10] showed that drought stress induced by PEG (polyethylene glycol) resulted in a reduction in S uptake and significantly decreased the amount of sulphur assimilated into amino acids and proteins. Our earlier study demonstrated that applying elemental sulphur to the soil alleviated the negative effects of stress caused by chromium pollution [13]. Hence, we wished to further investigate if sulphur fertilisation improves plant functioning under drought soil conditions. It is very important to know the effects of S fertilisers on the uptake of other nutrients, particularly nitrogen (N), potassium (K) and phosphorus (P). Relationships between S fertilisers and other minerals under water scarcity conditions are still not clear, and getting to know them will allow for more effective management of crops grown under drought stress.

This study aimed to investigate the reactions of wheat and maize to long-lasting moderate (45% field water capacity—FWC) and severe (30% FWC) drought stress and examine the influence of sulphur fertilisation on yield and mineral composition in the plants.

2. Materials and Methods

2.1. Materials, Setup and Procedure

Research was conducted in the vegetation facilities of the Department of Plant Nutrition of the Wroclaw University of Environmental and Life Sciences in Poland. Experiments were set up in four replicates in Wagner-type pots containing 5 kg of soil. The physical and chemical properties of the soil are described in Table 1. Temperature and light conditions during plant vegetation were natural, while soil moisture was controlled by watering with distilled water and soil moisture was maintained throughout the entire vegetation period of the cultivated plants at 30%, 45% and 60% field capacity (Table 2).

Table 1. Physico-chemical properties of the soil before the experiment.

Agronomic Category of Soil	pH	$C_{organic}$	S_{total}	P	K	Mg	S-SO_4	Zn	Mn	Fe	Cu
	1 M KCl dm^3	g kg^{-1} Soil		mg kg^{-1} Soil Soluble Forms							
Medium	4.80	6.32	0.178	64.0	88.0	48.0	9.26	39.0	110	577	2.94

Initially, the soil had an acidic pH (1 mol dm^{-3} KCl), a medium level of phosphorus according the Egner-Riehm method [14] and low levels of potassium [14] and magnesium according the Schachtschabel method [15]. The amount of overall S and S-SO_4 in the soil classified it as low-fertility soil. There were low levels of the microelement iron present, as well as medium levels of copper and manganese and high levels of zinc according the Rinkis method [16]. Before sowing, calcium was added to the soil (liming) by applying calcium carbonate at a dose calculated for 1Hh (5 g $CaCO_3$). The agricultural plants studied were spring wheat (Tybalt variety) and maize (Mosso variety). Twenty-five grains of wheat were sown into pots and 10 evenly spaced plants were left after thinning, while 12 grains of maize were sown, leaving 6 plants after thinning. The vegetation period of wheat was 115 days, while that of maize was 99 days. Wheat was collected at the full maturity stage and maize was collected at the full bloom stage (BBCH 67). Overall, the experimental design included nine treatments in order to study the interaction of applied sulphur fertilisation and FWC (Table 2).

Table 2. Treatments in the pot experiment.

Field Water Capacity	Form of Sulphur	Dose of S mg kg^{-1}
30%	Without S	0
	S-S^0—elemental	60
	S-SO_4—sulphate (VI)	60
45%	Without S	0
	S-S^0—elemental	60
	S-SO_4—sulphate (VI)	60
60%	Without S	0
	S-S^0—elemental	60
	S-SO_4—sulphate (VI)	60

Sulphur was applied before seeds were sown. Elemental S was ground to an average grain size of less than 0.1 mm to increase the rate of S oxidation in the soil [17]. For both plants, the same dose of nitrogen was applied (1.6 g per pot; NH_4NO_3 in an aqueous solution). Half of the dose was applied before sowing and half during the topdressing stage (spring wheat BBCH 30 and maize BBCH 19). The size of the dose for the remaining macroelements depended on the soil properties. To each 5 kg pot of soil was added 0.6 g phosphorus, 1.5 g potassium and 0.3 g of magnesium. Fertilisation with microelements was applied in standard quantities for pot experiments in compounds that did not contain sulphur. Macro- and microelements were applied before sowing (in an aqueous solution or in solids) and mixed into the entire amount of soil in the pot.

2.2. Methods for Chemical Analysis

Before and after the vegetation experiments, representative soil and plant samples were collected for agricultural and chemical analysis. After conducting the preparations for the soil material, we determined the soil pH of 1 mol dm^{-3} KCl using the potentiometric method, the overall S content (S total) via the Butters–Chenery method [18] and the content of S sulphates (VI) with the Bardsley and Lancaster method [19]. In plant material collected during the study, we determined the overall level of nitrogen (N organic) using the Kjeldahl method and the S total via the Butters–Chenery method. To determine levels of other elements, the plant material was dry mineralised, and then the ash was taken up with nitric acid and measured in solution: phosphorus via the vanadic–molybdate method, potassium and calcium with flame photometry, and magnesium and microelements via atomic absorption spectrophotometry.

2.3. Statistical Methods

The yield sizes and results of the chemical analysis were subjected to a two-way variance analysis. Prior to performing the analysis of variance, tests for homogeneity of variance within groups were performed using the Levene's test and the Shapiro–Wilk test of the correspondence of variables to the normal distribution. The relevance of mean differences was evaluated using the Tukey post hoc test with a significance level of $p = 0.05$. The statistical program R [20] was used for all statistical analyses.

3. Results and Discussion

3.1. Plant Growth and Yield

Drought affects many aspects of plant growth and development, diminishes the germination and establishment of seedlings, reduces cell division and differentiation rates, decreases biomass accumulation and consequently causes dramatically lower crop yields. For major crop plants, average yields can be reduced by more than 50% [1,21].

Our results show that the growth of both wheat and maize markedly dropped under drought conditions (Figures 1a,b, 2a,b, 3a,b and 4a,b), and wheat reacted better than maize to sulphur fertilisation under optimal conditions (60% FWC) and moderate water stress (45% FWC) (Figure 2a).

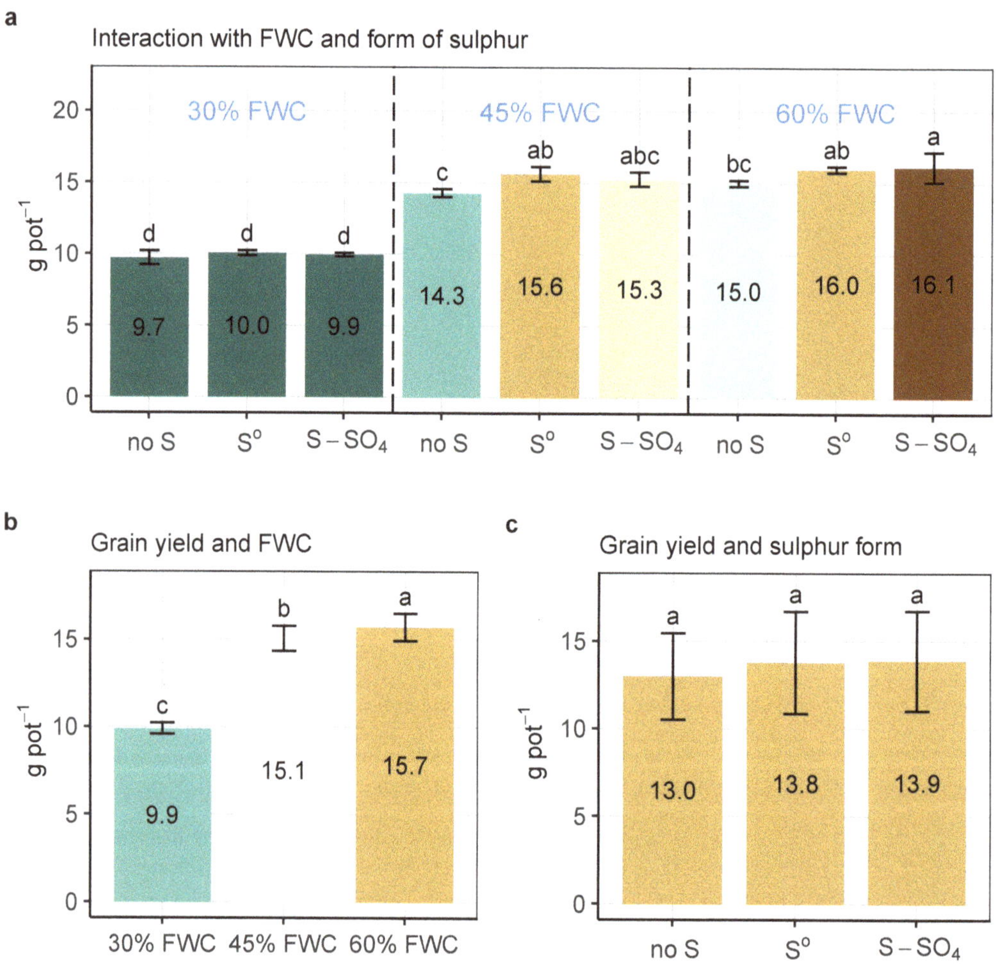

Figure 1. Grain yield of spring wheat. Values indicated by the same letter are not significantly different ($\alpha = 0.05$).

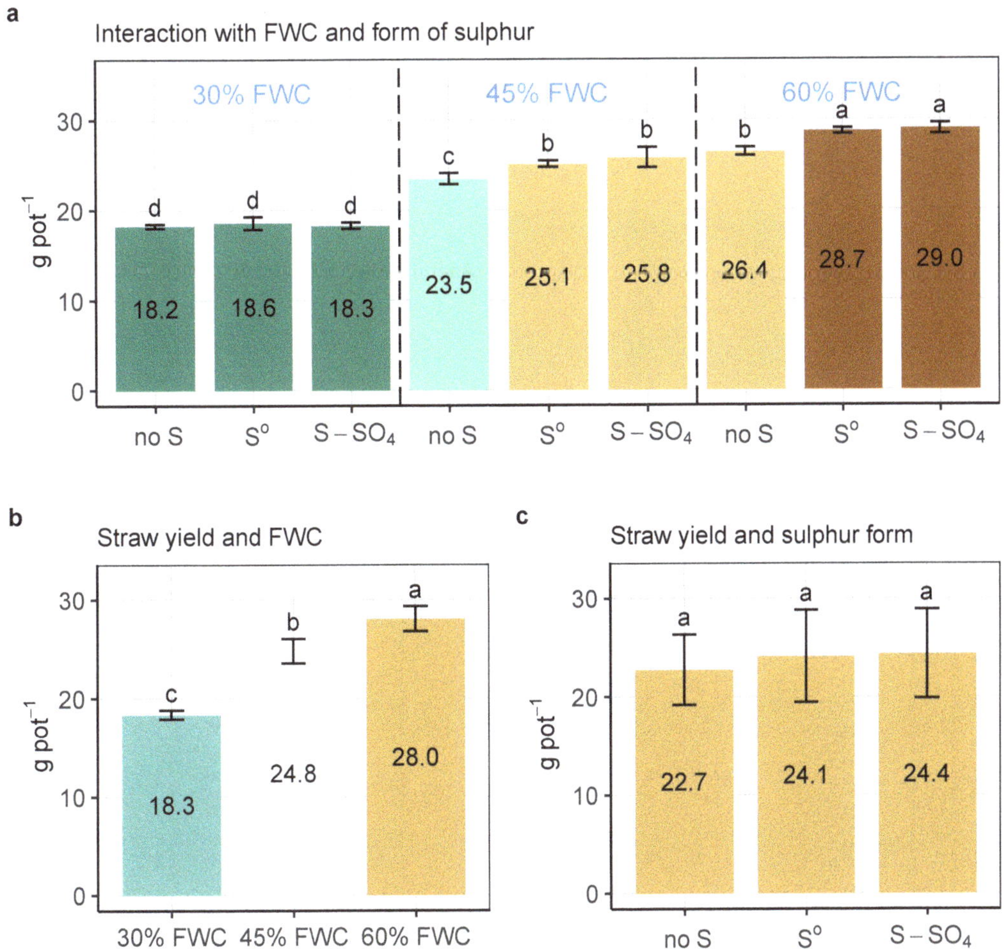

Figure 2. Straw yield of spring wheat. Values indicated by the same letter are not significantly different ($\alpha = 0.05$).

Under both optimal water conditions (60% FWC) and moderate water stress, the grain yields of wheat grown without sulphur application were not significantly different (Figure 1a). Applying elemental sulphur caused an increase in grain yield under moderate stress, whereas sulphate was more effective in wheat grown under adequate water supply (Figure 1a). Sulphur application (Figure 2a)—both elemental and sulphate—significantly improved the yield of wheat straw grown under optimal water conditions and moderate stress, but observed increases did not exceed 10% in comparison to plants grown without sulphur. The severe water stress significantly lowered the yield of grain and straw, regardless of sulphur fertilisation (Figures 1a,b and 2a,b). Under this condition, wheat biomass production (grain, straw) was greatly reduced (Figures 1a,b and 2a,b), but not by more than 40% in comparison to well-watered plants (60% FWC).

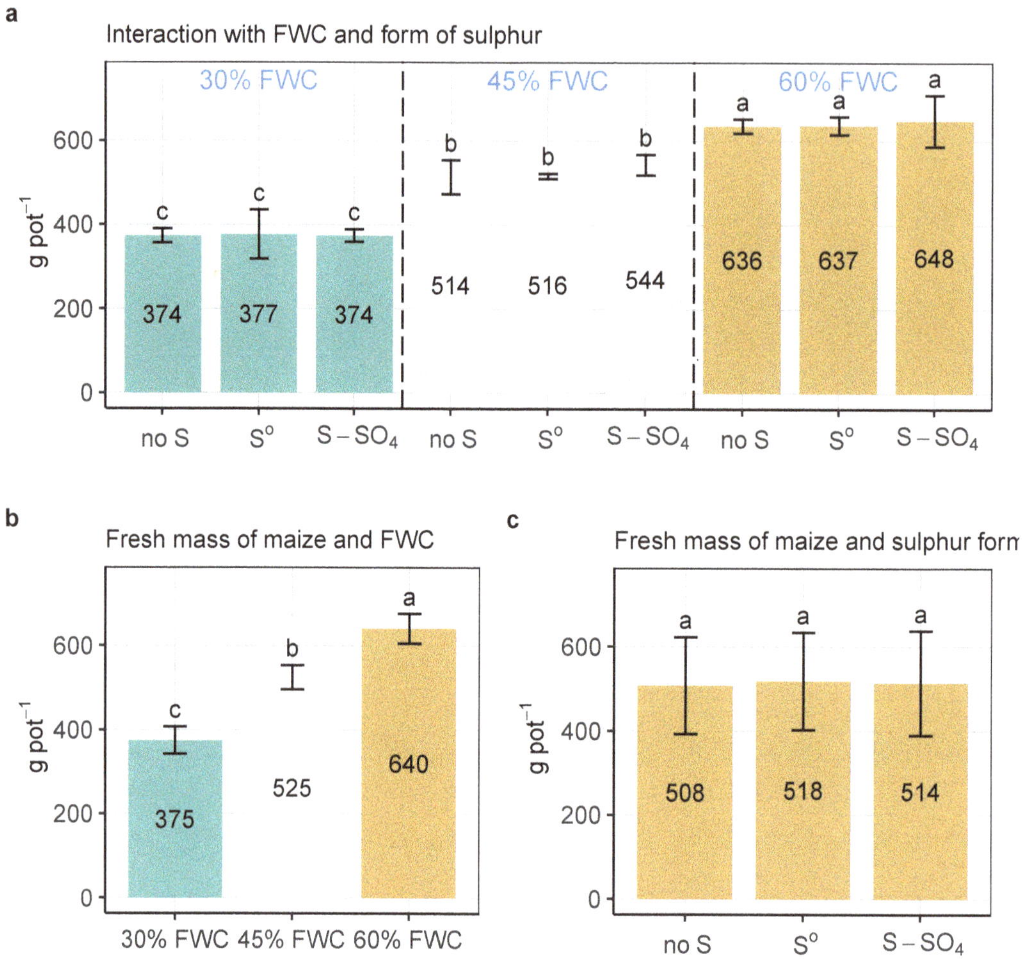

Figure 3. Fresh mass yield of maize. Values indicated by the same letter are not significantly different ($\alpha = 0.05$).

In maize, both stress levels caused a significant reduction in fresh and dry weights of plants (Figures 3a,b and 4a,b). The severe drought stress caused, on average, a 50% reduction in plant dry mass. As a C4 plant, maize uses water very efficiently, but it remains sensitive to water availability. Applying sulphur did not modify maize growth under stress conditions, but it did slightly improve dry matter production in well-watered plants (Figure 4a). Previous studies have shown that drought negatively impacts the yield of crop plants, with the decrease in yield being dependent on the severity of the drought stress and plant growth stage [22–27]. A few studies of sulphur fertilisation indicate that application of this element may help plants better tolerate limited water availability [9,28].

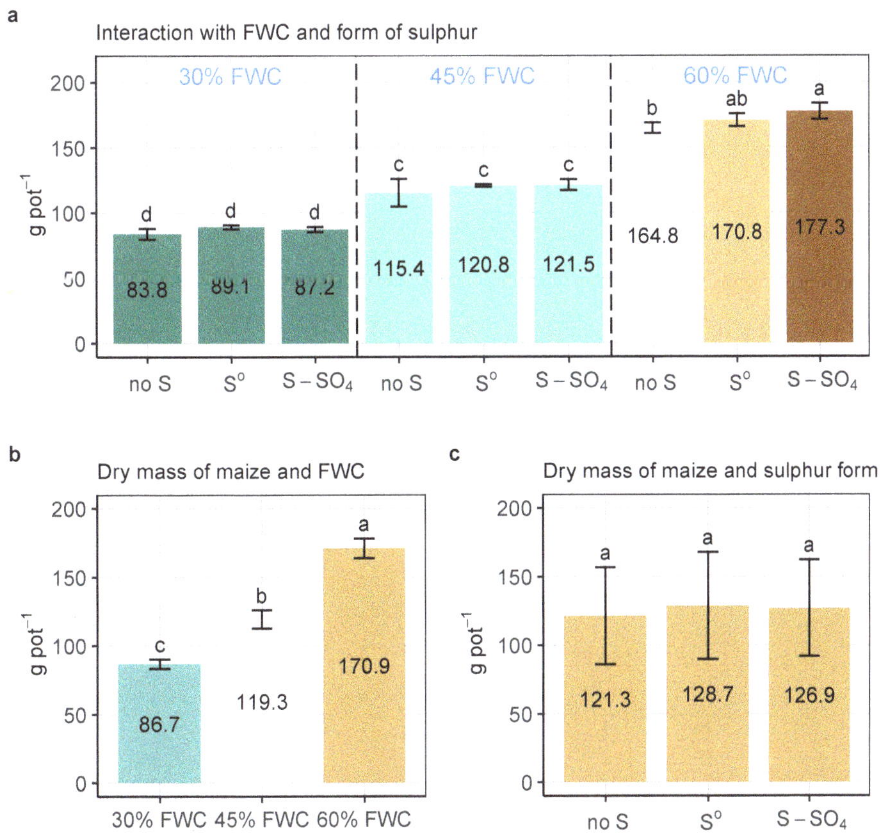

Figure 4. Dry mass yield of maize. Values indicated by the same letter are not significantly different (α = 0.05).

Overall, both maize and wheat were able to survive long-lasting, high-intensity water shortage (30% FWC), and their biomass production was not reduced by more than 50% when compared to well-watered plants. Under severe water stress, applying sulphur did not affect the growth of plants.

3.2. Mineral Nutrition

3.2.1. Macroelements

Plant ability to uptake minerals is a very important factor in determining the quality and quantity of crop yield. Drought stress usually restricts absorption of minerals due to a decreased rate of nutrient diffusion from the soil to the absorbing root surface and lowered translocation within the plant [29,30]. Various studies have shown a decrease in the accumulation of some minerals in plant tissues under water stress, but this response varies across crop species [4,25,29,30]. A lower concentration of particular elements in plant tissues might indicate that mineral uptake is disrupted. Fahad et al. [1] presented the generalisation that under drought conditions, N uptake increases, P uptake declines and K remains unaffected. Our study showed that the concentrations of particular macroelements (S, P, K, Mg, Ca) in wheat grain did not change considerably, either in response to drought conditions or with the addition of sulphur (Tables 3–5).

Table 3. Nitrogen and sulphur content and uptake in cultivated plants. Values indicated by the same letter are not significantly different ($\alpha = 0.05$).

Treatments		Spring Wheat						Maize			
		Grain		Straw		Grain + Straw		Content		Uptake	
FWC	Sulphur	Content g kg^{-1} d.m.				Uptake mg pot^{-1}		g kg^{-1} d.m.		mg pot^{-1}	
Nitrogen											
30%	Without S	20.5	abc	6.64	a	319	c	15.1	a	1 270	c
	S-S^0	18.6	bcd	6.97	a	316	c	15.5	a	1 380	abc
	S-SO$_4$	19.0	abcd	6.89	a	315	c	15.6	a	1 360	bc
45%	Without S	19.2	abcd	6.88	a	436	b	12.3	b	1 420	abc
	S-S^0	17.8	cd	6.37	a	442	b	12.5	b	1 510	ab
	S-SO$_4$	17.3	d	6.63	a	431	b	11.6	b	1 410	abc
60%	Without S	21.2	ab	6.65	a	494	a	8.99	c	1 480	abc
	S-S^0	21.9	a	6.48	a	536	a	8.80	c	1 500	ab
	S-SO$_4$	21.6	ab	6.54	a	537	a	9.02	c	1 600	a
FWC	30%	19.4	b	6.83	a	317	c	15.4	a	1 330	b
	45%	18.1	b	6.63	a	436	b	12.1	b	1 450	a
	60%	21.6	a	6.56	a	522	a	8.94	c	1 530	a
Sulphur	Without S	20.3	a	6.72	a	417	a	12.1	a	1 390	a
	S-S^0	19.4	a	6.60	a	431	a	12.3	a	1 460	a
	S-SO$_4$	19.3	a	6.69	a	428	a	12.1	a	1 460	a
Sulphur											
30%	Without S	1.54	b	1.28	e	38.3	e	0.420	d	35.2	f
	S-S^0	1.60	b	2.08	c	54.7	d	0.640	a	57.1	cd
	S-SO$_4$	1.64	b	1.86	d	50.3	d	0.629	a	54.9	de
45%	Without S	1.50	b	1.47	e	55.9	d	0.396	d	45.9	ef
	S-S^0	1.61	b	2.05	cd	78.1	b	0.550	b	66.5	bc
	S-SO$_4$	1.63	b	2.25	bc	81.2	b	0.519	bc	63.0	cd
60%	Without S	1.64	b	1.44	e	62.5	c	0.316	e	52.1	de
	S-S^0	1.92	a	2.41	ab	99.7	a	0.445	d	76.0	ab
	S-SO$_4$	1.90	a	2.48	a	103	a	0.465	cd	82.4	a
FWC	30%	1.59	b	1.74	a	47.8	c	0.563	a	49.1	b
	45%	1.58	b	1.92	a	71.7	b	0.488	ab	58.5	ab
	60%	1.82	a	2.11	a	88.3	a	0.409	b	70.2	a
Sulphur	Without S	1.56	b	1.40	b	52.2	b	0.378	b	44.4	b
	S-S^0	1.71	a	2.18	a	77.5	a	0.545	a	66.5	a
	S-SO$_4$	1.72	a	2.20	a	78.1	a	0.538	a	66.7	a

Some significant differences were observed in the case of nitrogen (Table 3), which saw a decrease in content in the grains under drought conditions and was not significantly affected by sulphur application. Both sulphur forms applied (S elemental and sulphate) caused a 17% increase in S concentration in the grains of plants grown under optimal conditions (Table 3).

In wheat straw, particular macroelement quantities were more changeable than in grain, with only N levels remaining stable under all conditions (Tables 3–5). Wheat fertilised with sulphur contained, on average, 56% more S than did unfertilised plants. This effect was observed in plants grown under both optimal conditions and drought stress (Table 3).

Water stress promoted S accumulation in maize shoots, with levels increasing with increasing stress intensity (Table 3). Sulphur-fertilised maize accumulated considerably more S than did non-fertilised plants. Under severe water stress, maize shoots contained approximately 50% more S than did non-fertilised plants. A few reports concerning the effect of drought on sulphur nutrition indicate a positive role of sulphur in alleviating the effects of drought stress [9,11]. Fatma et al. [31] demonstrated that an excess S supply improved photosynthesis and growth of mustard grown under salt stress condition.

As a vital constituent of many cellular compounds, sulphur not only plays an important role in the normal functioning of plants, but is also involved in defence mechanisms in stimulating the antioxidative system in cells. Some researchers claim that under stressful conditions, the demand for S is greater and plants increase sulphate uptake compared to other ions [32,33]. Applying sulphur might enhance the efficiency of other essential macronutrients such as N and P [28]. Usmani et al. [12] showed that S availability positively influenced leaf water status, gas exchange characteristics and antioxidative machinery in water-stressed maize plants. In summary, a plant's capacity to acquire S and carry on high sulphur use efficiency plays a significant role in the alleviation of the negative effects of drought stress [10].

In maize tissues, increasing water stress resulted in a greater accumulation of N and P (Tables 3 and 4). Nitrogen concentrations increased by, on average, 36% and 73% for moderate and severe water stress, respectively. Plants require N in large amounts, as it is a constituent of many essential cell compounds and its deficiency rapidly inhibits plant growth. Neither maize nor wheat suffered from nitrogen deficiency, and N concentration in the aboveground parts of stressed plants was similar to or higher than that of well-watered plants.

As N plays a fundamental role in plant growth and productivity, adequate concentration of this element in plant tissues is particularly important to their functioning under stressful conditions. In leaves, most N content is involved in photosynthesis as either enzymes or chlorophyll. Ding et al. [3] contend that photosynthesis and water uptake are the two key traits that enhance crop tolerance to drought. Conversely, however, they also maintain that a high nitrate supply may decrease plant drought tolerance. Nitrogen is also necessary for antioxidative protection as a component of enzymes and osmoprotectants that protects cells from the harmful effects of different abiotic stresses. In contrast, other plant studies have shown that drought affects N metabolism and significantly reduces N concentration [22,34].

Changes in P concentration were relatively small in both plant species (Table 4). Phosphorus is essential in processes connected with the storage and transfer of energy, photosynthesis, regulation of enzyme activity and transport of carbohydrates. Hence, an adequate level of P promotes metabolic processes such as respiration, photosynthesis, cell division and expansion, and the uptake and assimilation of other minerals [35,36]. Several studies indicate that drought stress reduces P uptake, as well as its subsequent transport to the stem, resulting in P deficiency in plant tissues [4,23,37–39]. Despite a slight decrease in P concentration, the examined plants were well supplied with this nutrient (Table 4).

Potassium plays a vital role in the regulation of water status, osmotic adjustment and charge balance in plants. In addition to osmoregulation and stomatal movements, K also regulates enzyme activity and the stability of membranes [7,40,41]. In this study, K content in maize shoots increased by 17% and 48% under moderate and severe water stress, respectively (Table 4). Applying sulphur had no effect on these parameters. Tadayyon et al. [42] obtained similar results with castor bean (*Ricinus communis*), in which K concentration in plant tissue increased with increasing severity of drought stress. Accumulation of K in plant tissues may help plants adjust osmotically and maintain activity of aquaporins involved in water uptake, thus improving drought stress tolerance [43]. The straw of wheat grown under drought stress accumulated significantly less potassium than control plants. This indicates that drought conditions limited potassium uptake and transport within the plant. According to Anschütz et al. [40], in addition to its well-established role as an essential macronutrient, K is also an important signalling agent mediating a wide range of plant adaptive responses to the environment. A disruption of K homeostasis in wheat may impair many biochemical processes and increase a plant's sensitivity to water stress, suggesting that increased K fertilisation could possibly help plants cope better with drought stress. Urbina et al. [44] also demonstrated that severe drought stress decreases K concentration in plants.

Subsequent macroelement calcium regulates any physiological processes, including movement of water and solutes, cell division, cell-wall synthesis, membrane and stomatal functions, and signal transduction. Straw from wheat grown under drought stress contained approximately 70% more calcium than did plants grown under optimal conditions (Table 5). Maize grown under drought stress also accumulated calcium, but the observed increase was lower (41% on average). Alternatively, Nahar and Gretzmacher [45] indicated that a reduction in soil water potential results in reduced calcium uptake. Our results showed that despite the very low mobility of this element, the uptake and distribution of calcium were not disrupted and did not limit plant functioning under stressful conditions. Tadayyon et al. [42] also stated that in *R. communis*, calcium concentration increased as drought stress increased and was lower in control plants.

Little information is available concerning the effect of drought on Mg nutrition in plants. Magnesium plays a vital role in photosynthesis as an essential component of chlorophyll and is also a cofactor for many enzymes and an important agent in protein synthesis. In *R. communis* [42], only very severe drought stress (75% moisture depletion) resulted in a significant decrease in Mg concentration in plant leaves. Nahar and Gretzmacher [45] also found a decrease in Mg concentration in tomato plants under drought stress. Our results showed that the plants were able to take up a sufficient amount of this element under drought conditions (Table 5). Magnesium content in wheat grain was not affected by different treatments, but in wheat straw grown under optimal water conditions, sulphur application resulted in a higher Mg concentration. Sulphate fertilisation was more effective than application of elemental S, with observed increases of 21% and 11% for sulphate and S application, respectively (Table 5). Under drought conditions, applying S did not significantly change Mg levels in wheat straw. In maize shoots, sulphur fertilisation did not modify Mg concentration, although severe water stress caused a considerable increase in Mg content. Under this condition, irrespective of sulphur fertilisation, the mean increase in Mg concentration in maize shoots amounted to 30% more than that of plants grown under optimal water conditions. It can be assumed that photosynthesis was not disrupted by Mg deficiency [46].

In summary, our results show that in plants grown under drought stress, although macronutrient concentrations were somewhat disturbed, relatively high macronutrient levels were maintained overall, and plants did not noticeably suffer from deficiencies.

Table 4. Phosphorus and potassium content and uptake in cultivated plants. Values indicated by the same letter are not significantly different (α = 0.05).

Treatments		Spring Wheat						Maize			
		Grain		Straw		Grain + Straw		Content		Uptake	
FWC	Sulphur	Content g kg^{-1} d.m.				Uptake mg pot^{-1}		g kg^{-1} d.m.		mg pot^{-1}	
Phosphorus											
30%	Without S	2.68	a	1.01	bc	44.4	e	7.92	a	665	c
	S-S^0	2.57	ab	0.888	c	42.3	e	8.07	a	720	c
	S-SO$_4$	2.56	ab	0.922	bc	42.2	e	7.80	a	680	c
45%	Without S	2.27	b	0.970	bc	55.2	d	7.93	a	909	b
	S-S^0	2.38	ab	1.04	abc	64.2	c	7.72	a	934	b
	S-SO$_4$	2.35	ab	1.06	abc	62.6	c	7.62	a	926	b
60%	Without S	2.56	ab	1.14	ab	68.5	bc	8.41	a	1 390	a
	S-S^0	2.61	a	1.25	a	77.6	a	8.30	a	1 420	a
	S-SO$_4$	2.58	ab	1.16	ab	75.2	ab	8.61	a	1 530	a
FWC	30%	2.60	a	0.942	b	43.0	c	7.93	ab	688	c
	45%	2.33	b	1.03	b	60.7	b	7.76	b	923	b
	60%	2.58	a	1.18	a	73.8	a	8.44	a	1 440	a
Sulphur	Without S	2.50	a	1.04	a	56.0	a	8.09	a	986	a
	S-S^0	2.52	a	1.06	a	61.4	a	8.03	a	1 020	a
	S-SO$_4$	2.50	a	1.05	a	60.0	a	8.01	a	1 040	a
Potassium											
30%	Without S	3.94	a	9.56	cd	213	d	68.7	ab	5 750	c
	S-S^0	3.98	a	9.36	d	214	d	74.9	a	6 680	bc
	S-SO$_4$	3.84	a	9.80	bcd	217	d	69.4	ab	6 060	c
45%	Without S	3.64	a	10.2	bcd	292	c	58.3	bc	6 710	bc
	S-S^0	3.50	a	10.8	b	334	b	56.6	bcd	6 840	bc
	S-SO$_4$	3.57	a	10.7	bc	323	bc	54.7	cd	6 640	bc
60%	Without S	3.84	a	12.0	a	375	a	50.2	cd	8 280	a
	S-S^0	4.04	a	10.8	ab	376	a	49.5	cd	8 460	a
	S-SO$_4$	4.03	a	10.6	bc	371	a	44.9	d	7 950	ab
FWC	30%	3.92	a	9.57	b	214	c	71.0	a	6 160	b
	45%	3.57	b	10.6	a	316	b	56.6	b	6 730	b
	60%	3.97	a	11.1	a	374	a	48.2	c	8 230	a
Sulphur	Without S	3.80	a	10.6	a	293	a	59.1	a	6 910	a
	S-S^0	3.84	a	10.3	a	308	a	60.4	a	7 330	a
	S-SO$_4$	3.81	a	10.4	a	304	a	56.3	a	6 880	a

Table 5. Magnesium and calcium content and uptake in cultivated plants. Values indicated by the same letter are not significantly different ($\alpha = 0.05$).

Treatments		Spring Wheat						Maize			
		Grain		Straw		Grain + Straw		Content		Uptake	
FWC	Sulphur	Content g kg^{-1} d.m.				Uptake mg pot^{-1}		g kg^{-1} d.m.		mg pot^{-1}	
Magnesium											
30%	Without S	0.803	a	1.32	ab	31.8	d	6.17	a	517	d
	S-S^0	0.794	a	1.30	ab	32.1	d	5.89	a	526	d
	S-SO$_4$	0.772	a	1.27	abc	30.9	d	5.65	ab	493	d
45%	Without S	0.767	a	1.37	a	43.2	bc	4.97	ab	573	cd
	S-S^0	0.765	a	1.16	bc	42.0	c	5.01	ab	605	bcd
	S-SO$_4$	0.786	a	1.29	abc	44.4	bc	5.10	ab	622	bcd
60%	Without S	0.825	a	1.08	c	41.0	c	4.42	b	729	abc
	S-S^0	0.834	a	1.20	abc	47.7	ab	4.47	b	764	ab
	S-SO$_4$	0.812	a	1.31	ab	51.0	a	4.79	ab	849	a
FWC	30%	0.790	b	1.30	a	31.6	c	5.90	a	512	c
	45%	0.773	b	1.28	a	43.2	b	5.03	b	600	b
	60%	0.824	a	1.20	a	46.6	a	4.56	b	781	a
Sulphur	Without S	0.798	a	1.26	a	38.7	a	5.19	a	606	a
	S-S^0	0.798	a	1.22	a	40.6	a	5.12	a	631	a
	S-SO$_4$	0.790	a	1.29	a	42.1	a	5.18	a	655	a
Calcium											
30%	Without S	0.177	a	6.84	a	127	abcd	6.21	a	520	b
	S-S^0	0.180	a	6.70	a	126	abcd	5.88	a	524	b
	S-SO$_4$	0.189	a	6.62	a	122	bcd	5.89	a	514	b
45%	Without S	0.166	a	6.76	a	161	ab	5.15	ab	595	b
	S-S^0	0.179	a	6.04	ab	159	ab	4.88	abc	590	b
	S-SO$_4$	0.172	a	6.56	a	167	a	4.92	abc	600	b
60%	Without S	0.183	a	3.38	c	92.1	d	3.69	cd	608	b
	S-S^0	0.186	a	3.73	c	110	cd	3.52	d	601	b
	S-SO$_4$	0.192	a	4.63	bc	137	abc	4.47	bcd	793	a
FWC	30%	0.182	ab	6.72	a	125	b	5.99	a	519	b
	45%	0.172	b	6.45	a	162	a	4.99	b	595	ab
	60%	0.187	a	3.91	b	113	b	3.89	c	667	a
Sulphur	Without S	0.176	a	5.66	a	126	a	5.02	a	574	a
	S-S^0	0.181	a	5.49	a	132	a	4.76	a	571	a
	S-SO$_4$	0.185	a	5.94	a	142	a	5.09	a	636	a

3.2.2. Microelements

Because plants require much smaller amounts of microelements than macronutrients, little attention has been given to studying the effects of drought on micronutrient requirements. Indeed, low moisture in the soil could disturb their uptake and induce deficiency in plant tissues. In wheat grain, levels of the examined microelements (Cu, Fe, Mn, Zn) were relatively stable under the tested conditions (Tables 6 and 7). Statistically significant differences were observed for Fe and Mn. Greater differences in Fe concentration were found in wheat straw, with an observed decrease of up to 25% in comparison to control plants. Tadayyon et al. [42] also demonstrated that drought stress decreased Fe content in *R. communis*, although the maximum decline under severe water stress (75% moisture depletion) was only 11%. In this study, the observed reduction in Fe in maize shoots was not statistically significant (Table 6), and sulphur fertilisation had no effect on these parameters. In wheat straw, drought stress decreased not only Fe, but also Cu concentration, although it did not affect Zn or Mn content (Table 7). Applying elemental sulphur caused an increase in Mn content in wheat straw, but no positive effect of sulphur application was observed in maize. Nevertheless, increasing water stress resulted in a higher content of this element in plant tissues. Manganese plays a crucial role in photosynthesis, respiration, antioxidative metabolism and the activation of some enzymes, so a high Mn concentration in plant tissues may be crucial for protecting cells against the harmful effects of reactive oxygen species generated under drought stress.

Samarah et al. [27] found that drought stress increased concentrations of Zn and Cu in soybean seeds and that the increase in mineral concentration was not due to the reduction in dry matter accumulation.

Drought stress in maize shoots caused an increase in the concentration of both Cu and Zn, which was particularly evident under severe water stress (Table 7). Generally, the observed changes exceeded 50% in relation to plants grown in optimal conditions (Table 7). Applying sulphur did not change these relationships. It is possible that higher concentrations of these microelements allow plants to scavenge reactive oxygen species more effectively, ultimately leading to better adaptation to stress conditions. It is noteworthy that the Zn nutritional status of plants is essential for crop productivity and quality worldwide.

3.2.3. Total Mineral Uptake

Generally, drought considerably limited the total uptake of macroelements, with a greater reduction observed in plants exposed to severe water stress (Tables 3–5). The significant decrease in total N uptake by wheat did not result in an analogous decline in N content in the straw and grain, although some negative symptoms were observed in the latter. Although the severe drought stress caused an important decline in the total uptake of numerous nutrients (N, S, P, K, Mg), they remained present in aboveground tissues at high levels (Tables 3–5). Engels and Marschner [47] claim that translocation of minerals is dependent on external factors and is also internally regulated according to the growth-related demand of shoots. Drought-induced inhibition of plant growth (Figures 1–4) reduced plants' mineral nutrient requirements, allowing the plants to maintain an adequate nutritional status. Total mineral uptake was reduced by more than 40% in comparison to well-watered plants in the case of N and P in both plant species and in the case of S and K in wheat. As expected, sulphur fertilisation had an effect on total S uptake by plants grown under both optimal and stressful conditions (Table 3), although the impact was only statistically significant for P, K and Mg in wheat. Reduction in the uptake of various macroelements under stress has been reported in numerous plant species [4,23,29,39,45]. Similar changes were observed for total uptake of microelements and, in general, sulphur fertilisation did not change these relationships (Table 5). Of all the microelements measured here, iron experienced the greatest reduction in both wheat and maize.

Table 6. Manganese and iron content and uptake in cultivated plants. Values indicated by the same letter are not significantly different ($\alpha = 0.05$).

Treatments		Spring Wheat			Maize	
		Grain	Straw	Grain + Straw	Content	Uptake
FWC	Sulphur	Content mg kg^{-1} d.m.		Uptake mg pot^{-1}	mg kg^{-1} d.m.	mg pot^{-1}
		Manganese				
30%	Without S	37.8 ab	91.8 ab	2.04 e	64.8 ab	5.44 b
	S-S^0	37.6 ab	101 ab	2.25 de	69.8 a	6.23 b
	S-SO$_4$	35.5 b	81.3 b	1.84 e	60.8 abc	5.31 b
45%	Without S	37.3 b	93.9 ab	2.74 cd	57.6 abcd	6.66 ab
	S-S^0	37.6 ab	115 a	3.57 ab	56.7 abcd	6.85 ab
	S-SO$_4$	35.9 b	97.2 ab	2.99 bc	53.8 bcde	6.54 ab
60%	Without S	36.1 b	92.3 ab	2.98 bc	43.2 de	7.12 ab
	S-S^0	42.8 a	118 a	4.07 a	40.4 e	6.89 ab
	S-SO$_4$	39.9 ab	89.9 ab	3.25 bc	46.2 cde	8.16 a
FWC	30%	37.0 a	91.4 a	2.04 b	65.2 a	5.66 b
	45%	36.9 a	102 a	3.10 a	56.1 b	6.68 a
	60%	39.6 a	100 a	3.43 a	43.3 c	7.39 a
Sulphur	Without S	37.1 a	92.7 b	2.59 b	55.2 a	6.41 a
	S-S^0	39.3 a	111 a	3.30 a	55.7 a	6.66 a
	S-SO$_4$	37.1 a	89.4 b	2.69 ab	53.6 a	6.67 a
		Iron				
30%	Without S	48.4 ab	65.4 c	1.66 c	57.9 a	4.86 c
	S-S^0	46.8 ab	65.4 c	1.69 c	69.6 a	6.21 bc
	S-SO$_4$	44.7 b	61.9 c	1.57 c	64.6 a	5.63 bc
45%	Without S	46.0 ab	78.3 abc	2.50 b	67.6 a	7.80 b
	S-S^0	45.3 b	69.9 bc	2.51 b	61.8 a	7.47 b
	S-SO$_4$	51.3 ab	69.5 bc	2.53 b	65.9 a	8.01 b
60%	Without S	51.4 ab	85.2 ab	3.02 a	69.9 a	11.5 a
	S-S^0	53.3 a	89.1 a	3.41 a	68.3 a	11.6 a
	S-SO$_4$	51.6 ab	84.7 ab	3.29 a	73.8 a	13.0 a
FWC	30%	46.6 b	64.3 c	1.64 c	64.0 a	5.57 c
	45%	47.5 b	72.6 b	2.51 b	65.1 a	7.76 b
	60%	52.1 a	86.3 a	3.24 a	70.6 a	12.1 a
Sulphur	Without S	48.6 a	76.3 a	2.39 a	65.1 a	8.06 a
	S-S^0	48.4 a	74.8 a	2.54 a	66.6 a	8.44 a
	S-SO$_4$	49.2 a	72.1 a	2.46 a	68.1 a	8.90 a

Table 7. Copper and zinc content and uptake in cultivated plants. Values indicated by the same letter are not significantly different (α = 0.05).

Treatments		Spring Wheat						Maize			
		Grain		Straw		Grain + Straw		Content		Uptake	
FWC	Sulphur	Content mg kg^{-1} d.m.				Uptake mg pot^{-1}		mg kg^{-1} d.m.		mg pot^{-1}	
Copper											
30%	Without S	4.16	a	5.15	b	0.134	c	5.28	abc	0.441	c
	S-S^0	4.27	a	5.28	b	0.141	c	5.62	ab	0.501	bc
	S-SO_4	3.62	a	5.26	b	0.132	c	5.62	ab	0.491	bc
45%	Without S	3.26	a	5.71	ab	0.181	b	6.23	a	0.711	a
	S-S^0	3.86	a	5.73	ab	0.208	a	5.13	abc	0.620	abc
	S-SO_4	4.37	a	5.44	b	0.203	a	5.27	abc	0.639	abc
60%	Without S	3.42	a	6.08	a	0.212	a	3.21	c	0.529	abc
	S-S^0	3.52	a	5.76	ab	0.222	a	3.35	c	0.572	abc
	S-SO_4	3.60	a	5.75	ab	0.225	a	3.80	bc	0.672	ab
FWC	30%	4.02	a	5.23	b	0.136	c	5.51	a	0.478	b
	45%	3.83	a	5.62	a	0.197	b	5.54	a	0.656	a
	60%	3.51	a	5.87	a	0.220	a	3.45	b	0.591	a
Sulphur	Without S	3.62	a	5.65	a	0.176	a	4.91	a	0.560	a
	S-S^0	3.88	a	5.59	a	0.190	a	4.70	a	0.564	a
	S-SO_4	3.86	a	5.48	a	0.187	a	4.89	a	0.600	a
Zinc											
30%	Without S	41.4	a	24.4	a	0.846	b	11.5	a	0.962	bc
	S-S^0	42.3	a	28.8	a	0.960	b	10.8	ab	0.962	bc
	S-SO_4	41.5	a	27.7	a	0.918	b	11.6	a	1.02	bc
45%	Without S	44.3	a	31.3	a	1.37	a	9.41	abc	1.09	abc
	S-S^0	41.0	a	27.6	a	1.35	a	6.91	c	0.836	c
	S-SO_4	42.5	a	29.1	a	1.38	a	7.70	bc	0.936	bc
60%	Without S	43.7	a	27.6	a	1.38	a	7.03	c	1.16	abc
	S-S^0	44.5	a	24.9	a	1.43	a	7.50	c	1.28	ab
	S-SO_4	47.1	a	25.1	a	1.49	a	7.90	bc	1.40	a
FWC	30%	41.7	a	27.0	a	0.908	b	11.3	a	0.980	b
	45%	42.6	a	29.3	a	1.37	a	8.01	b	0.954	b
	60%	45.1	a	25.9	a	1.43	a	7.48	b	1.28	a
Sulphur	Without S	43.1	a	27.8	a	1.20	a	9.31	a	1.07	a
	S-S^0	42.6	a	27.1	a	1.25	a	8.40	a	1.03	a
	S-SO_4	43.7	a	27.3	a	1.26	a	9.08	a	1.12	a

3.3. Soil Parameters after Cultivation

The soil reaction (pH) after the cultivation of both crops significantly increased (in comparison to the initial value), although to a greater extent following wheat cultivation (Tables 1 and 8). The examined treatments (sulphur fertilisation and drought intensities) had relatively little effect on the magnitude of the observed pH increases.

The results of this study are in agreement with our earlier research [13]. The smallest change in pH was observed in the soil after maize cultivation, fertilised by sulphates and well watered (60% FWC). In soil that was not fertilised by sulphur, sulphate concentrations were low (5.81–9.0 mg kg^{-1} soil). As expected, sulphur application resulted in a significant increase in sulphate concentration in the soil, and differences between the two examined forms of sulphur were relatively small, indicating that despite the decrease in soil moisture, oxidation of elemental sulphur was efficient. Moreover, in line with our earlier research [13], the concentration of sulphates in the soil was considerably higher after wheat cultivation than after maize cultivation.

Table 8. Soil pH and the content of S total, sulphates (VI). Values indicated by the same letter are not significantly different (α = 0.05).

Treatments		pH		S Total		S-SO$_4$		S-SO$_4$ in S Total	
		KCl 1M dm^{-3}		mg kg^{-1}				%	
Spring wheat									
30%	Without S	6.41	ab	172	d	12.3	c	7.18	c
	S-S^0	6.12	c	184	cd	79.3	a	43.1	a
	S-SO$_4$	6.19	bc	198	bc	73.4	a	37.1	ab
45%	Without S	6.49	a	183	cd	11.5	c	6.28	c
	S-S^0	6.14	bc	216	ab	68.5	ab	31.7	ab
	S-SO$_4$	6.32	abc	217	a	63.1	ab	29.0	b
60%	Without S	6.54	a	149	e	10.5	c	7.05	c
	S-S^0	6.39	abc	176	d	48.5	b	27.6	b
	S-SO$_4$	6.55	a	171	d	48.4	b	28.3	b
FWC	30%	6.24	b	185	b	55.0	a	29.1	a
	45%	6.32	b	205	a	47.7	a	22.3	a
	60%	6.49	a	165	c	35.8	a	21.0	a
Sulphur	Without S	6.48	a	168	b	11.4	b	6.84	b
	S-S^0	6.22	b	192	a	65.4	a	34.1	a
	S-SO$_4$	6.35	ab	196	a	61.6	a	31.5	a

Table 8. *Cont.*

Treatments		pH		S Total		S-SO$_4$		S-SO$_4$ in S Total	
		KCl 1M dm^{-3}		mg kg^{-1}				%	
Maize									
30%	Without S	5.97	ab	149	d	6.79	d	4.57	c
	S-S^0	5.58	abc	222	ab	11.8	c	5.32	c
	S-SO$_4$	6.07	a	216	b	25.0	a	11.6	a
45%	Without S	6.06	a	150	d	7.62	cd	5.08	c
	S-S^0	6.04	a	235	a	19.9	b	8.49	b
	S-SO$_4$	5.90	ab	230	ab	24.5	a	10.7	a
60%	Without S	5.47	abc	124	e	5.81	d	4.70	c
	S-S^0	5.34	bc	172	c	10.0	cd	5.83	c
	S-SO$_4$	4.97	c	164	cd	10.2	cd	6.21	c
FWC	30%	5.87	a	196	a	14.6	ab	7.16	a
	45%	6.00	a	205	a	17.4	a	8.09	a
	60%	5.26	b	153	b	8.67	b	5.58	a
Sulphur	Without S	5.83	a	141	b	6.74	c	4.78	b
	S-S^0	5.65	a	210	a	13.9	b	6.55	b
	S-SO$_4$	5.65	a	204	a	19.9	a	9.50	a

4. Conclusions

The results of this study show that drought stress can cause a significant reduction in productivity for both maize and wheat, although both plants are able to sustain their vigour and growth despite long-lasting water shortage. Drought-induced changes in mineral composition (macro- and microelements) indicated that minerals were still effectively acquired from the soil and transported throughout the whole plant. Sulphur application did not modify these relationships. Applying sulphur did, however, improve wheat biomass production in plants that were well-watered and grown under moderate drought stress, indicating that sulphur fertilisation may be recommended in wheat cultivation when plants are exposed to moderate water stress.

Finally, on the basis of our data and other studies, we think that further research should focus on other aspects of plant reaction to sulphur supplementation and drought stress, particularly photosynthesis, stress metabolites that improve plant tolerance to water scarcity, and systems involved in nutrient uptake and transport within a plant [11,23,48].

Author Contributions: Conceptualization, G.K. and E.S.; methodology, G.K. and E.S.; statistical analysis, G.K.; investigation-performed the pot experiments, G.K., E.S., J.Z. and P.C.; data curation-compiled and analyzed the results, G.K., E.S., J.Z. and P.C.; writing-original draft preparation, E.S. and G.K.; review and editing, E.S. and G.K. All authors have read and agreed to the published version of the manuscript.

Funding: The APC is financed by Wroclaw University of Environmental and Life Sciences.

Data Availability Statement: The data presented in this study are available on request from the corresponding author.

Conflicts of Interest: The authors declare no conflict of interest.

References

1. Fahad, S.; Bajwa, A.A.; Nazir, U.; Anjum, S.A.; Farooq, A.; Zohaib, A.; Sadia, S.; Nasim, W.; Adkins, S.; Saud, S.; et al. Crop Production under Drought and Heat Stress: Plant Responses and Management Options. *Front. Plant Sci.* **2017**, *8*, 1147. [CrossRef] [PubMed]
2. Seleiman, M.F.; Al-Suhaibani, N.; Ali, N.; Akmal, M.; Alotaibi, M.; Refay, Y.; Dindaroglu, T.; Abdul-Wajid, H.H.; Battaglia, M.L. Drought Stress Impacts on Plants and Different Approaches to Alleviate Its Adverse Effects. *Plants* **2021**, *10*, 259. [CrossRef] [PubMed]
3. Ding, L.; Lu, Z.; Gao, L.; Guo, S.; Shen, Q. Is Nitrogen a Key Determinant of Water Transport and Photosynthesis in Higher Plants Upon Drought Stress? *Front. Plant Sci.* **2018**, *9*, 1143. [CrossRef] [PubMed]
4. Hu, Y.; Schmidhalter, U. Drought and Salinity: A Comparison of Their Effects on Mineral Nutrition of Plants. *J. Plant Nutr. Soil Sci.* **2005**, *168*, 541–549. [CrossRef]
5. Lewandowska, M.; Sirko, A. Recent Advances in Understanding Plant Response to Sulfur-Deficiency Stress. *Acta Biochim. Pol.* **2008**, *55*, 457–471. [CrossRef]
6. Hawkesford, M.J. Sulfate Uptake and Assimilation - Whole Plant Regulation. In *Sulfur Metabolism in Plants: Mechanisms and Applications to Food Security and Responses to Climate Change*; De Kok, J.L., Tabe, L., Tausz, M., Hawkesford, J.M., Hoefgen, R., McManus, T.M., Norton, M.R., Rennenberg, H., Saito, K., Schnug, E., Eds.; Springer: Dordrecht, The Netherlands, 2012; pp. 11–24. [CrossRef]
7. Marschner, P. *Mineral Nutrition of Higher Plants*; Elsevier Ltd.: Amsterdam, The Netherlands; Academic Press: New York, NY, USA, 2012; Chapter 650. [CrossRef]
8. Kulczycki, G. Chapter Three—The Effect of Elemental Sulfur Fertilization on Plant Yields and Soil Properties. In *Advances in Agronomy*; Sparks, D.L., Ed.; Academic Press: Cambridge, MA, USA, 2021; Volume 167, pp. 105–181. [CrossRef]
9. Heidari, M.; Galavi, M.; Hassani, M. Effect of Sulfur and Iron Fertilizers on Yield, Yield Components and Nutrient Uptake in Sesame (*Sesamum indicum* L.) under Water Stress. *Afr. J. Biotechnol.* **2011**, *10*, 8816–8822. [CrossRef]
10. Lee, B.R.; Zaman, R.; Avice, J.C.; Ourry, A.; Kim, T.H. Sulfur Use Efficiency Is a Significant Determinant of Drought Stress Tolerance in Relation to Photosynthetic Activity in Brassica Napus Cultivars. *Front. Plant Sci.* **2016**, *7*, 459. [CrossRef]
11. Chan, K.X.; Wirtz, M.; Phua, S.Y.; Estavillo, G.M.; Pogson, B.J. Balancing Metabolites in Drought: The Sulfur Assimilation Conundrum. *Trends Plant Sci.* **2013**, *18*, 18–29. [CrossRef]
12. Usmani, M.M.; Nawaz, F.; Majeed, S.; Shehzad, M.A.; Ahmad, K.S.; Akhtar, G.; Aqib, M.; Shabbir, R.N. Sulfate-Mediated Drought Tolerance in Maize Involves Regulation at Physiological and Biochemical Levels. *Sci. Rep.* **2020**, *10*, 1147. [CrossRef]
13. Kulczycki, G.; Sacała, E. Sulfur Application Alleviates Chromium Stress in Maize and Wheat. *Open Chem.* **2020**, *18*, 1093–1104. [CrossRef]
14. Egner, H.; Riehm, H. Doppellaktatmethode. In *Methodenbuch Band I. Die Untersuchung von Boden*; Thun, R., Hersemann, R., Knickmann, E., Eds.; Neumann: Radebeul/Berlin, Germany, 1955; pp. 110–125.
15. Schachtschabel, P. Das Pflanzenverfügbare Magnesium Des Boden Und Seine Bestimmung *Z. Pflanz. Düngung Bodenkd.* **1954**, *67*, 9–23. [CrossRef]
16. Rinkis, G. *Optimalisation of Mineral Nutrition of Plants*; Zinatne: Riga, Latvija, 1972.
17. Germida, J.; Janzen, H. Factors Affecting the Oxidation of Elemental Sulfur in Soils. *Nutr. Cycl. Agroecosyst.* **1993**, *35*, 101–114. [CrossRef]
18. Butters, B.; Chenery, E.M. A Rapid Method for the Determination of Total Sulphur in Soils and Plants. *Analyst* **1959**, *84*, 239–245. [CrossRef]
19. Bardsley, C.E.; Lancaster, J.D. Determination of Reserve Sulfur and Soluble Sulfates in Soils. *Soil Sci. Soc. Am. J.* **1960**, *24*, 265–268. [CrossRef]
20. R Core Team. R: A Language and Environment for Statistical Computing. *R Foundation for Statistical Computing*; R Core Team: Vienna, Austria, 2021.
21. Shinozaki, K.; Uemura, M.; Bailey-Serres, J.; Bray, E.A.; Weretilnyk, E. *Biochemistry and Molecular Biology of Plants*, 2nd ed.; Wiley: Hoboken, NJ, USA, 2015; pp. 1051–1101.
22. Basal, O.; Szabó, A. The Combined Effect of Drought Stress and Nitrogen Fertilization on Soybean. *Agronomy* **2020**, *10*, 384. [CrossRef]
23. Bista, D.R.; Heckathorn, S.A.; Jayawardena, D.M.; Mishra, S.; Boldt, J.K. Effects of Drought on Nutrient Uptake and the Levels of Nutrient-Uptake Proteins in Roots of Drought-Sensitive and -Tolerant Grasses. *Plants* **2018**, *7*, 28. [CrossRef]
24. Kamara, A.Y.; Menkir, A.; Badu-Apraku, B.; Ibikunle, O. The Influence of Drought Stress on Growth, Yield and Yield Components of Selected Maize Genotypes. *J. Agric. Sci.* **2003**, *141*, 43–50. [CrossRef]
25. Kapoor, D.; Bhardwaj, S.; Landi, M.; Sharma, A.; Ramakrishnan, M.; Sharma, A. The Impact of Drought in Plant Metabolism: How to Exploit Tolerance Mechanisms to Increase Crop Production. *Appl. Sci.* **2020**, *10*, 5692. [CrossRef]
26. Queiroz, M.; Oliveira, C.; Steiner, F.; Zuffo, A.; Zoz, T.; Vendruscolo, E.; Menis, V.; Mello, B.; Cabral, R.; Menis, T. Drought Stresses on Seed Germination and Early Growth of Maize and Sorghum. *J. Agric. Sci.* **2019**, *11*, 310–318. [CrossRef]
27. Samarah, N.; Mullen, R.; Cianzio, S. Size Distribution and Mineral Nutrients of Soybean Seeds in Response to Drought Stress. *J. Plant Nutr.* **2004**, *27*, 815–835. [CrossRef]

28. Mazhar, A.A.; Mahgoub, M.H.; El-Aziz, N.G.A. Response of *Schefflera Arboricola* L. To Gypsum and Sulphur Application Irrigated with Different Levels of Saline Water. *Aust. J. Basic Appl. Sci.* **2011**, *5*, 121–129.
29. Hu, Y.; Burucs, Z.; von Tucher, S.; Schmidhalter, U. Short-Term Effects of Drought and Salinity on Mineral Nutrient Distribution along Growing Leaves of Maize Seedlings. *Environ. Exp. Bot.* **2007**, *60*, 268–275. [CrossRef]
30. Silva, E.; Nogueira, R.; da Silva, M.; de Albuquerque, M. Drought Stress and Plant Nutrition. *Plant Stress* **2011**, *5*, 32–41.
31. Fatma, M.; Asgher, M.; Masood, A.; Khan, N.A. Excess Sulfur Supplementation Improves Photosynthesis and Growth in Mustard under Salt Stress through Increased Production of Glutathione. *Environ. Exp. Bot.* **2014**, *107*, 55–63. [CrossRef]
32. Ernst, L.; Goodger, J.Q.D.; Alvarez, S.; Marsh, E.L.; Berla, B.; Lockhart, E.; Jung, J.; Li, P.; Bohnert, H.J.; Schachtman, D.P. Sulphate as a Xylem-Borne Chemical Signal Precedes the Expression of ABA Biosynthetic Genes in Maize Roots. *J. Exp. Bot.* **2010**, *61*, 3395–3405. [CrossRef]
33. Nawaz, F.; Shehzad, M.A.; Majeed, S.; Ahmad, K.S.; Aqib, M.; Usmani, M.M.; Shabbir, R.N. Role of Mineral Nutrition in Improving Drought and Salinity Tolerance in Field Crops. In *Agronomic Crops: Volume 3: Stress Responses and Tolerance*; Hasanuzzaman, M., Ed.; Springer: Singapore, 2020; pp. 129–147. [CrossRef]
34. Sacała, E.; Durbajło, W. The Effect of Sodium Silicate on Maize Growing under Stress Conditions (Oddziaływanie Krzemianu Sodu Na Kukurydzę Rosnącą w Warunkach Stresowych). *Przem. Chem.* **2012**, *91*, 941–951. (In Polish)
35. Pilbeam, D.J.; Cakmak, I.; Marschner, H.; Kirkby, E.A. Effect of Withdrawal of Phosphorus on Nitrate Assimilation and PEP Carboxylase Activity in Tomato. *Plant Soil* **1993**, *154*, 111–117. [CrossRef]
36. Warraich, E.; Ahmed, R.; Ashraf, M. Role of Mineral Nutrition in Alleviation of Drought Stress in Plants. *Aust. J. Crop. Sci.* **2011**, *5*, 764–777.
37. Cramer, M.D.; Hawkins, H.J.; Verboom, G.A. The Importance of Nutritional Regulation of Plant Water Flux. *Oecologia* **2009**, *161*, 15–24. [CrossRef]
38. Goicoechea, N.; Antolín, M.; Sánchez-Díaz, M. Influence of Arbuscular Mycorrhizae and Rhizobium on Nutrient Content and Water Relations in Drought Stressed Alfalfa. *Plant Soil* **1997**, *192*, 261–268. [CrossRef]
39. Radersma, S.; Lusiana, B.; van Noordwijk, M. Simulation of Soil Drying Induced Phosphorus Deficiency and Phosphorus Mobilization as Determinants of Maize Growth near Tree Lines on a Ferralsol. *Field Crop. Res.* **2005**, *91*, 171–184. [CrossRef]
40. Anschütz, U.; Becker, D.; Shabala, S. Going beyond Nutrition: Regulation of Potassium Homoeostasis as a Common Denominator of Plant Adaptive Responses to Environment. *J. Plant Physiol.* **2014**, *171*, 670–687. [CrossRef] [PubMed]
41. Cakmak, I. The Role of Potassium in Alleviating Detrimental Effects of Abiotic Stresses in Plants. *J. Plant Nutr. Soil Sci.* **2005**, *168*, 521–530. [CrossRef]
42. Tadayyon, A.; Nikneshan, P.; Pessarakli, M. Effects of Drought Stress on Concentration of Macro- and Micro-Nutrients in Castor (*Ricinus Communis* L.) Plant. *J. Plant Nutr.* **2018**, *41*, 304–310. [CrossRef]
43. Wang, M.; Zheng, Q.; Shen, Q.; Guo, S. The Critical Role of Potassium in Plant Stress Response. *Int. J. Mol. Sci.* **2013**, *14*, 7370–7390. [CrossRef] [PubMed]
44. Urbina, I.; Sardans, J.; Beierkuhnlein, C.; Jentsch, A.; Backhaus, S.; Grant, K.; Kreyling, J.; Peñuelas, J. Shifts in the Elemental Composition of Plants during a Very Severe Drought. *Environ. Exp. Bot.* **2015**, *111*, 63–73. [CrossRef] [PubMed]
45. Nahar, K.; Gretzmacher, R. Effect of Water Stress on Nutrient Uptake, Yield and Quality of Tomato (Lycopersicon Esculentum Mill.) under Subtropical Conditions. *Die Bodenkult.* **2002**, *53*, 45–51.
46. Tränkner, M.; Tavakol, E.; Jákli, B. Functioning of Potassium and Magnesium in Photosynthesis, Photosynthate Translocation and Photoprotection. *Physiol. Plant.* **2018**, *163*, 414–431. [CrossRef]
47. Engels, C.; Marschner, H. Effects of Suboptimal Root Zone Temperatures and Shoot Demand on Net Translocation of Micronutrients from the Roots to the Shoot of Maize. *Plant Soil* **1996**, *186*, 311–320. [CrossRef]
48. Li, Y.; Sun, C.; Huang, Z.; Pan, J.; Wang, L.; Fan, X. Mechanisms of Progressive Water Deficit Tolerance and Growth Recovery of Chinese Maize Foundation Genotypes Huangzao 4 and Chang 7-2, Which Are Proposed on the Basis of Comparison of Physiological and Transcriptomic Responses. *Plant Cell Physiol.* **2009**, *50*, 2092–2111. [CrossRef]

agronomy

MDPI

Article

Soil Water Content Directly Affects Bud Burst Rate in Single-Node Cuttings of Perennial Plants

Santiago Signorelli [1,2,*], Juwita R. Dewi [3] and Michael J. Considine [2,4]

[1] Laboratorio de Bioquímica, Departamento de Biología Vegetal, Facultad de Agronomía, Universidad de la República, Av. Garzón 780, Sayago CP, Montevideo 12900, Uruguay
[2] The Centre of Excellence in Plant Energy Biology, School of Molecular Sciences, The University of Western Australia, 35 Stirling Highway, Crawley, WA 6009, Australia; michael.considine@uwa.edu.au
[3] School of Agriculture and Environment, The University of Western Australia, 35 Stirling Highway, Crawley, WA 6009, Australia; juwita.dewi@research.uwa.edu.au
[4] Horticulture and Irrigated Agriculture, The Department of Primary Industries and Regional Development, Perth, WA 6000, Australia
* Correspondence: ssignorelli@fagro.edu.uy; Tel.: +598-2354-0229

Abstract: The use of single-node cuttings of shoots as explants to study bud dormancy and its physiology under controlled conditions is a common practice in grapevine (*Vitis vinifera* L.) or other perennial plant research. In particular, this method has been extensively used to understand the effect of different chemicals on bud dormancy and bud burst. However, the soil water content in those experiments is usually not reported and its relevance is often neglected. Here, we observed that an unevenly distributed soil water content in a tray containing multiple explants results in an uneven pattern of bud burst within the same treatment. Thus, we hypothesised that soil water content can dramatically affect bud burst. To investigate this, we first established that fresh single-node cuttings were able to transport water into the buds. We then tested the rate of bud burst at different water treatments (35%, 55%, 70%, 85%, and 100% of field capacity; FC). We observed a clear dependence of bud burst on water, in which, at very low levels of water, bud burst does not occur; after 35% FC, bud burst rate increases with water content until around 85% FC; and, from 85% FC, bud burst rate becomes independent of water content. These data highlight the critical importance of monitoring soil water content in any bud burst assay in perennials. Finally, we provide a detailed protocol for determining and controlling field capacity and other soil water content indicators.

Keywords: bud; bud burst; development; dormancy; explants; field capacity; gravimetric water content; grapevine; perennial plants; water

Citation: Signorelli, S.; Dewi, J.R.; Considine, M.J. Soil Water Content Directly Affects Bud Burst Rate in Single-Node Cuttings of Perennial Plants. *Agronomy* **2022**, *12*, 360. https://doi.org/10.3390/agronomy12020360

Academic Editor: Anita Ierna

Received: 16 December 2021
Accepted: 27 January 2022
Published: 31 January 2022

1. Introduction

The use of explants of perennial plants is a common practice to evaluate kinetics of bud burst, for instance, to evaluate the depth of dormancy, the effect of environmental clues, or applied chemicals on bud burst [1–4]. The use of explants enables the use of controlled conditions, which is a major limitation when working with perennial plants, as well as removing the influence of dominance influences of the corpus and competing sinks. Despite the wide use of this methodology, the influence of water availability on bud burst has not been reported. This is particularly important because the application of chemicals usually implies adding water to the single-node cuttings, as water is the most common solvent used. In addition, the water content of soil in single-node cuttings experiments is commonly not reported. We hypothesise that soil water content can explain much of the stochastic variability observed in bud burst experiments and, thus, that determining a fixed water content within the experiments is critical to underpin the validity and reproducibility of data reported.

Water content in the soil is usually expressed as volumetric water content (VWC), gravimetric water content (GWC), or a percentage of field capacity (FC), which simply refers to the maximum amount of water that a soil can retain. VWC (Θ_v) is calculated as:

$$\Theta_v = (V_{wet} - V_{dry})/V_{dry}$$

V_{wet} being the volume of wet soil and V_{dry} the volume of dry soil. GWC (Θ_g) is calculated as:

$$\Theta_g = (W_{wet} - W_{dry})/W_{dry} \quad (1)$$

W_{wet} being the weight (mass) of wet soil and W_{dry} the weight of dry soil. Given that measuring mass is simpler than volume, GWC is generally preferred over VWC.

The objectives of this work were to evaluate the effect of soil water content on bud burst of grapevine single-node cuttings and to provide a protocol to eliminate water content as a variable in single-node cutting experiments. For this purpose, we determined whether uneven water content affects bud burst, whether dormant buds are able to take up water, and the kinetics of bud burst at different % of FC.

2. Materials and Methods

2.1. Preparation of Pots and Calculation of Field Capacity

The substrate of potting mix at pH~6.0, containing fine composted pine bark, coco peat, and brown river sand in the ratio of 2.5:1:1.5 (*w*/*w*), was homogenously mixed and distributed among 19 seedling trays on trays, each adjusted to a mass of 2500 g (Figure 3A). As the potting mix was not completely dry, three additional samples of around 30 g were used to determine the initial moisture content and, hence, initial dry weight. The 30 g samples were incubated in an oven at 105 °C for two days to determine the dry weight (Figure 3B). Once the moisture was determined in the subsamples, the dry weight of the soil in each tray was calculated (Figure 3C). Out of the 19 trays, 15 were used for planting three replicates of the five treatments described below. The remaining four trays were used to determine FC. The FC trays were saturated with water and left in the washing area where they were able to drain (Figure 3D). After 3 h, the trays were placed on a stand with paper below to monitor more accurately the presence of dripping from the trays. Replacing the paper once wet and inspecting every 30 min, we determined the weight of the pots at field capacity when no more drops came out from the tray for 1 h (Figure 3E,F). Once 100% FC was established, the weight to achieve the different FC was calculated (Figure 3G).

2.2. Plant Material and Growth Conditions

Grapevine (*Vitis vinifera* L.) cv. Merlot canes containing mature, dormant buds from node 3 to 10 were collected from a commercial vineyard in Margaret River, Australia (34° S, 115° E) on the second week of June 2017 (winter in the south hemisphere). At that time, the local photoperiod was 10/14 h light/dark (sunrise 07:23 am, sunset 5:17 pm). On the collection day, the temperature was 18 °C/10 °C (high/low), the average temperature for the month was 13 °C, and the highest and lowest in that month were 23 °C (11 June) and 3 °C (9 June), respectively. The highest humidity was 96% (1 June) and 25% (9 June). The canes were wrapped on newspaper, externally moistened by spraying water, placed on black bags sealed with tape to be transported to the lab, and stored at 4 °C for 10 days. Single-node cuttings (explants) were cut about 4–5 cm below the node until 2–3 cm above the node, and immediately planted on the trays containing the potting mix described above. A total of 90 single-node cuttings were planted per treatment divided in three different trays (replicates). Before planting, water was added to each tray until reaching the exact weight for its corresponding treatment of field capacity %. Once the explants were planted, the new weight for each tray was recorded to know the new weight for the treatment tray considering the weight of the plant material. Explants were grown at 20 °C under a photoperiod of 12:12 h, illuminated with a photosynthetic photon flux density of 100 $\mu mol \cdot m^{-2} \cdot s^{-1}$, for 49 days. Explants were watered every day on a set of scales until reaching the corresponding weight of the tray, including the initial weight of the plant material, for each particular water treatment (Figure 3H).

2.3. Water Treatments

To assess the effect of soil water availability on bud burst, we defined five water treatments corresponding to 35%, 55%, 70%, 85%, and 100% of FC. Values below 35% of FC were not considered, as it is unlikely to have growth in such conditions when using an explant system (absence of roots).

2.4. Micro-Computed Tomographies

To investigate the buds' water uptake capacity, buds were dissected from the cane and incubated overnight with caesium iodide (CsI) 10% at 4 °C or without CsI (control). 3D imaging using μCT was performed on each bud (BBCH scale: 00-01 [5]). The micro-computed tomographies (μCT) were performed with a nanofocus CT system (Phoenix Nanotom, General Electric, Heidelberg, Germany) after incubation. Buds were mounted on a rotation stage by means of a Parafilm wrap. A total of 2400 projection images per scan were taken with 0.15° angular steps for a full 360° rotation. Capture time for each image was 500 ms. Settings were 55 kV/182 μA for control samples and 60 kV/167 μA for CsI samples. Image pixel resolution was 2.50–3.25 μm. Slice reconstruction was performed by Octopus Reconstruction version 8.9.0.9 (XRE, Ghent, Belgium) using the filtered back-projection method.

3. Results

To determine if uneven distribution of water affects bud burst, we placed single-node cuttings on seedling trays on concave plastic trays, in which the water was preferably distributed towards the centre of the seedling trays (Figure 1). After 35 days, we observed that the bud burst was not spatially uniform, whereby bud burst of explants in the centre of the tray (dashed rectangle in Figure 1A) greatly exceeded those in the periphery (Figure 1). This was consistent with the hypothesis that water availability was an important variable affecting bud burst kinetics in single-node cutting experiments.

Figure 1. Irregular bud burst on a seedling tray with uneven distribution of water. (**A**) Representative picture of single-node cuttings after 35 days when water was not homogenously distributed across the tray. (**B**) Top view representation of the most hydrated area of the tray. (**C**) Lateral view representation of the tray showing the bending of the seedling tray and the water flux generated towards the centre.

It is well documented that quiescent buds are isolated to prevent dehydration during winter, particularly through the restriction of aquaporin activity, plasmodesmata conductance, and apoplastic conductivity [6–10]. For water to have a direct effect on bud burst, it needs to be transported to the bud after transferring the single-node cutting to the seedling tray. Therefore, we tested if quiescent buds were able to transport water immediately after dissecting them from the cane. Using a contrast agent, and analysing μCT data, we observed that water was effectively transported through the bud 24 h after incubation (Figure 2).

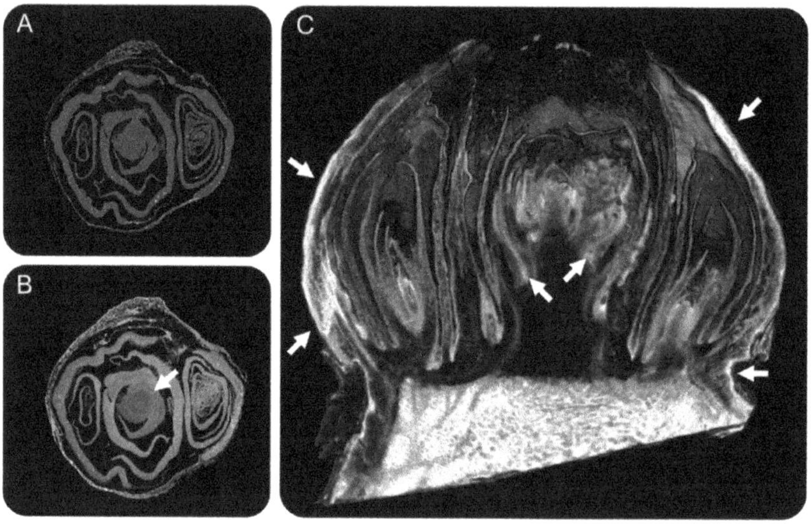

Figure 2. Micro-computed tomographies (μCT) of buds 24 h after being dissected from the cane and incubated with or without contrast agents (CsI). (**A**) Top view section of a control bud without contrast agent. (**B**) Top view section of a bud incubated with the contrast agent. (**C**) Lateral view of a bud incubated with the contrast agent. The arrows indicate the contrast agent, which is observed as white colour. *n* = 3.

Given the evidence presented by Figures 1 and 2, we set up an experiment controlling the water content of the pots every day (Figure 3) for 47 days to see if different kinetics of bud burst are observed. Figure 3 explains all the steps required to control the water content in soil. In our case, we chose 35%, 55%, 70%, 85%, and 100% of FC to test if there is a correlation between bud burst and soil water content.

We observed a strong correlation between bud burst and water treatments (Figure 4). Explants grown on 100% and 85% FC had the greatest rate of bud burst, reaching 50% bud burst at about 24 d, followed by those at 70% FC, which reached 50% bud burst at 30 d. Bud burst of explants grown on 55% FC was considerably delayed, reaching 50% bud burst at 44 d, revealing how small variations in water content can make important differences in bud burst rates. Finally, explants grown on 35% FC did not reach 50% bud burst within the time evaluated (Figure 4A); only 8% of the buds had burst at the end of the experiment. The curves for 100% and 85% FC were almost indistinguishable, being only significantly different between day 25 and 33 (Figure 4A), where the curve for 100% FC reached higher values. Conversely, clear and significant differences were observed between all the other curves since the start of bud burst for each curve. Overall, the graph of Figure 4A suggests a triphasic behaviour between bud burst rate and water content of soil, in which, at very low levels of water (e.g., <35% FC), there is no correlation, as buds are not going to burst (i.e., a zero order kinetic for water), a second phase (e.g., 35–85% FC), in which there is a positive correlation between bud burst rate and water content, and a third phase (e.g.,

>85% FC), in which bud burst rate became, again, independent of water content (i.e., a zero-order kinetic for water), most likely because water was no longer a limiting factor and bud burst is determined by other factors.

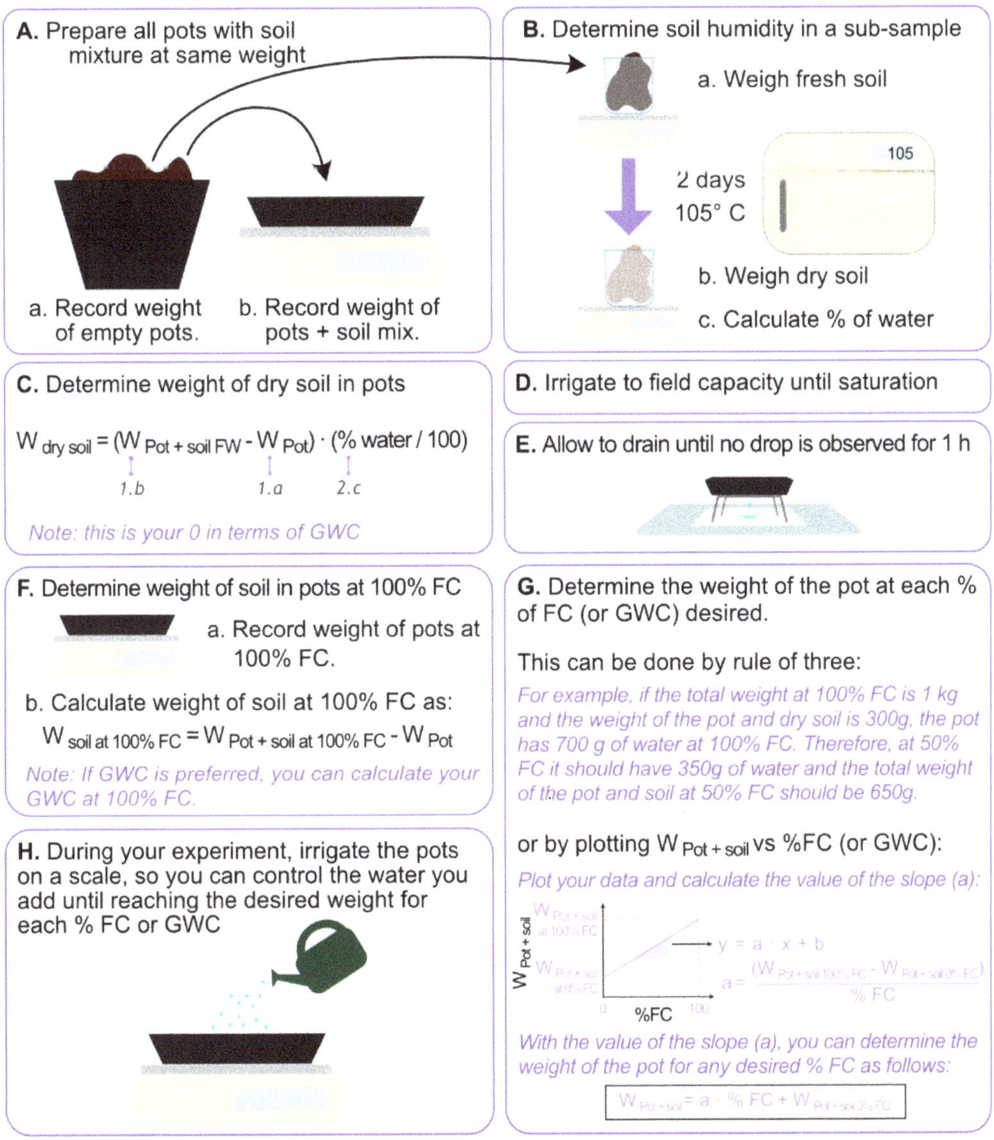

Figure 3. Procedure to determine and control the field capacity in soil and the control of in-soil water levels during the experiment. (**A–H**) describe the successive steps of this protocol, being (**A**) step 1 and (**H**) step 8. The values shown in panel (**G**) are only for exemplification, and are not real data.

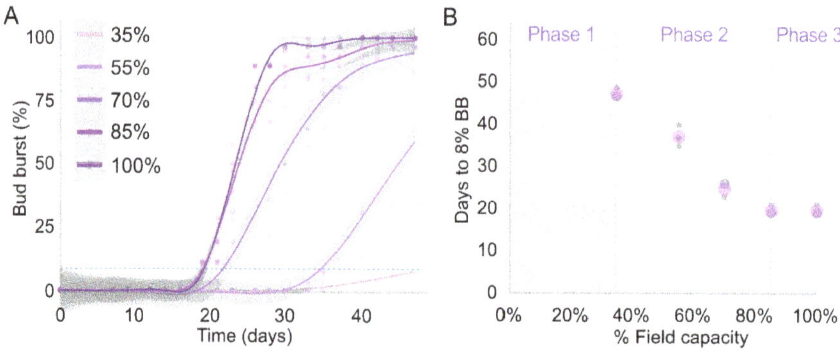

Figure 4. Kinetics of bud burst at different water contents in soil expressed as % of FC. (**A**) Bud burst percentages for each water treatment. The curve represents a regression determined with the generalised additive model, with its 95% confidence level intervals for the prediction presented as the surrounding grey area. The dashed light-blue line marks the 8% bud burst threshold used to determine the days plotted in B. (**B**) Three kinetic behaviours of bud burst rate in relation to water content. The purple circles indicate the means and the grey circles indicate the independent replicates; the bigger the circle, the more replicates giving the same result (n = 3). The first phase indicates a range of water content in which bud burst was unlikely to happen due to water restrictions, where no data (ND) was collected.

To explore this triphasic behaviour further and investigate whether the positive correlation observed is linear, we plotted the days to reach 8% bud burst in relation to the % of FC (Figure 4B). In this case, the three kinetic phases are clearly identified and a linear correlation was observed for the second phase (Figure 4B), suggesting that the kinetic is of the first order (i.e., bud burst rate increases linearly with water content). After 85% FC, the time to reach 8% bud burst is the same as at 100%, confirming that the initiation of bud burst was not affected by water contents of soil above 85% of FC. It is important to note, however, that, if values as high as 90% bud burst are used, there may be an effect of water, as we observed that the buds at 100% of FC were the first to reach the plateau of nearly 100% bud burst (Figure 4A).

4. Discussion

This study aimed to test the often overlooked importance of water availability on bud burst kinetics in explant studies, which have become the main morphometric test for the depth of dormancy in perennating buds. The buds used in this study were collected towards the end of winter, and further stored at 4 °C prior to experimentation, in order to eliminate dormancy as a factor. Our results demonstrated that quiescent buds of explants can transport water as soon as they are subjected to water treatments (Figure 2) and that water availability can directly and strongly influence the rate of bud burst when provided within the range of 35 to 85% FC (Figure 4). For instance, a reduction of 15% FC can delay bud burst by ~50% time (i.e., from 30 days to 44 days to reach 50% BB). This demonstrates that a tight control of water content in soil is required when determining bud burst rate kinetics for comparing different treatments, as these differences observed by water can be much greater (i.e., ~10 days) than those sometimes observed by different exogenous treatments of molecules (e.g., 2–4 days). Therefore, we provided here an illustrative and detailed protocol (Figure 3) explaining how to determine the different percentages of FC when setting up the experiment and how to follow up the experiment. As FC here is expressed as % of mass of water by mass of substrate, a set of scales is required for these calculations. However, it is also possible to use volumetric units and express as volume of water per volume of substrate. We understand that the determination of volume is usually more complex in most laboratories. In our work, we presented all the data as %

FC; however, it is also possible to express the data as GWC or VWC using the equations described in introduction. Alternatively, if the number of treatments represent a major constraint to control daily the water levels of the experiment, we suggest working always above 80–85% FC, a range in which, in our experimental conditions, bud burst takes place and becomes independent of water levels. Thus, we define this as a "safety zone" of water contents when comparing bud burst rates under different treatments, particularly important when the treatments of chemicals implicate the use of water. However, this "safety zone" should be ideally determined in each lab depending on the substrate used, because different substrates have a different water retention capacity, resulting in significant differences in water content relative to soil (i.e., GWC or VWC). In terms of GWC, 85% of FC in our system represented 0.91 $g_{water} \cdot g_{soil}^{-1}$ and 100% of FC was 1.11 $g_{water} \cdot g_{soil}^{-1}$.

Several earlier studies have illustrated the regulated changes in free and bound water [10], in the activity of aquaporin transporters [6,11], the role of plasmodesmata in dormancy transitions of perennating buds [7], the expansion of apoplastic connectivity upon bud burst [9], and the change dynamics in moisture content in the bud across the season [11]. The fact that water has a strong effect on bud burst in explant further supports the importance of those biological processes controlling water content in the bud, such as plasmodesmata regulation, aquaporins activity, etc. A deeper understanding of the signals controlling these mechanisms could allow scientists to develop managing practices to manipulate bud burst that can be used by farmers to synchronize bud burst better. Besides the biochemical mechanism regulating water transport, environmental clues can also affect water availability and transport to the bud. In grapevine, shade [12] and water deficit [13] were shown to reduce water transport capacity in field conditions, whereas high temperatures increase it [14]. The xylem network varies between grapevine cultivars; therefore, some cultivars have greater hydraulic conductivity than others (e.g., Thompson Seedless (high) vs. Merlot (low)) [15,16], as well as the xylem hydraulic response to environmental clues [15]. Likewise, in other perennials, several studies showed that frosting or cold temperatures affect water uptake and transport through the vascular system [17,18]. Therefore, it is expected that environmental clues modulate bud water content in perennials grown in the field. This aspect has not been less explored in research where xylem conductivity was investigated. However, a study conducted on apple trees, where soil frost was prolonged by soil insulation, has shown that the prolonged frost not only reduced water conductivity, but also buds' water content, and this correlated with a delayed bud burst [17]. Hence, the conclusion of our study about the effect of water content on bud burst using an explant system is likely to be translatable to what occurs in field conditions.

5. Conclusions

Single-node cuttings containing quiescent buds are able to transport water, and water has a strong effect on the rate of bud burst. Therefore, we conclude that statements of water content must be included in experiments on bud burst in explants. Soil water content must be controlled in these experiments to avoid the variation in water contents that confounds the differences caused by the specific treatment by chemicals or environmental conditions in bud burst rates, and to avoid false positive effects.

Author Contributions: Conceptualisation, S.S.; methodology, S.S. and M.J.C.; formal analysis, S.S.; investigation, S.S. and J.R.D.; resources, M.J.C.; writing—original draft preparation, S.S.; writing—review and editing, S.S. and M.J.C.; funding acquisition, M.J.C. All authors have read and agreed to the published version of the manuscript.

Funding: This research was funded by the Australian Research Council Discovery Project DP150103211.

Acknowledgments: We thank Jeremy Shaw, Pieter Verboven, and Zi Wang, who contributed to the generation of microCT data. SS is a grateful member of the Uruguayan system of researchers (Sistema Nacional de Investigadores, SNI, Uruguay).

Conflicts of Interest: The authors declare no conflict of interest.

References

1. Vergara, R.; Parada, F.; Rubio, S.; Pérez, F.J. Hypoxia induces H_2O_2 production and activates antioxidant defence system in grapevine buds through mediation of H_2O_2 and ethylene. *J. Exp. Bot.* **2012**, *63*, 4123–4131. [CrossRef] [PubMed]
2. Meitha, K.; Agudelo-Romero, P.; Signorelli, S.; Gibbs, D.J.; Considine, J.A.; Foyer, C.H.; Considine, M.J. Developmental control of hypoxia during bud burst in grapevine. *Plant Cell Environ.* **2018**, *41*, 1154–1170. [CrossRef] [PubMed]
3. Antolín, M.C.; Santesteban, H.; Ayari, M.; Aguirreolea, J.; Sánchez-Díaz, M. Grapevine Fruiting Cuttings: An Experimental System to Study Grapevine Physiology under Water Deficit Conditions. In *Methodologies and Results in Grapevine Research*; Springer: Dordrecht, The Netherlands, 2010; pp. 151–163. [CrossRef]
4. Mullins, M.G.; Bouquet, A.; Williams, L.E. Biology of the grapevine. In *The Biology of Horticultural Crops*; Cambridge University Press: Cambridge, UK, 1992; p. 239. ISBN 9780521305075.
5. Coombe, B.G. Growth Stages of the Grapevine: Adoption of a system for identifying grapevine growth stages. *Aust. J. Grape Wine Res.* **1995**, *1*, 104–110. [CrossRef]
6. Yooyongwech, S.; Horigane, A.K.; Yoshida, M.; Yamaguchi, M.; Sekozawa, Y.; Sugaya, S.; Gemma, H. Changes in aquaporin gene expression and magnetic resonance imaging of water status in peach tree flower buds during dormancy. *Physiol. Plant.* **2008**, *134*, 522–533. [CrossRef] [PubMed]
7. Rinne, P.L.H.; Kaikuranta, P.M.; Van Schoot, C. Der The shoot apical meristem restores its symplasmic organization during chilling-induced release from dormancy. *Plant J.* **2001**, *26*, 249–264. [CrossRef] [PubMed]
8. Beauvieux, R.; Wenden, B.; Dirlewanger, E. Bud dormancy in perennial fruit tree species: A pivotal role for oxidative cues. *Front. Plant Sci.* **2018**, *9*, 657. [CrossRef] [PubMed]
9. Signorelli, S.; Shaw, J.; Hermawaty, D.; Wang, Z.; Verboven, P.; Considine, J.A.; Considine, M.J. The initiation of bud burst in grapevine features dynamic regulation of the apoplastic pore size. *J. Exp. Bot.* **2020**, *71*, 719–729. [CrossRef] [PubMed]
10. Yooyongwech, S.; Horigane, A.K.; Yoshida, M.; Sekozawa, Y.; Sugaya, S.; Cha-Um, S.; Gemma, H. Hydrogen cyanamide enhances MRI-measured water status in flower buds of peach (*Prunus persica* L.) during winter. *Plant Omics* **2012**, *5*, 400–404.
11. Velappan, Y.; Chabikwa, T.G.; Considine, J.A.; Agudelo-Romero, P.; Foyer, C.H.; Signorelli, S.; Considine, M.J. The bud dormancy disconnect: Latent buds of grapevine are dormant during summer despite a high metabolic rate. *J. Exp. Bot.* **2022**. [CrossRef] [PubMed]
12. Schultz, H.R.; Matthews, M.A. Xylem development and hydraulic conductance in sun and shade shoots of grapevine (*Vitis vinifera* L.): Evidence that low light uncouples water transport capacity from leaf area. *Planta* **1993**, *190*, 393–406. [CrossRef]
13. Lovisolo, C.; Schubert, A. Effects of water stress on vessel size and xylem hydraulic conductivity in *Vitis vinifera* L. *J. Exp. Bot.* **1998**, *49*, 693–700. [CrossRef]
14. Galat Giorgi, E.; Keller, M.; Sadras, V.; Roig, F.A.; Perez Peña, J. High temperature during the budswell phase of grapevines increases shoot water transport capacity. *Agric. For. Meteorol.* **2020**, *295*, 108173. [CrossRef]
15. Pouzoulet, J.; Pivovaroff, A.L.; Scudiero, E.; de Guzman, M.E.; Rolshausen, P.E.; Santiago, L.S. Contrasting adaptation of xylem to dehydration in two *Vitis vinifera* L. sub-species. *Vitis-J. Grapevine Res.* **2020**, *59*, 53–61. [CrossRef]
16. Brodersen, C.R.; Choat, B.; Chatelet, D.S.; Shackel, K.A.; Matthews, M.A.; McElrone, A.J. Xylem vessel relays contribute to radial connectivity in grapevine stems (*Vitis vinifera* and *V. Arizonica*; Vitaceae). *Am. J. Bot.* **2013**, *100*, 314–321. [CrossRef] [PubMed]
17. Beikircher, B.; Mittmann, C.; Mayr, S. Prolonged soil frost affects hydraulics and phenology of apple trees. *Front. Plant Sci.* **2016**, *7*, 867. [CrossRef] [PubMed]
18. Christensen-Dalsgaard, K.K.; Tyree, M.T. Frost fatigue and spring recovery of xylem vessels in three diffuse-porous trees in situ. *Plant Cell Environ.* **2014**, *37*, 1074–1085. [CrossRef] [PubMed]

agronomy

MDPI

Article

The Impact of Irrigation on Olive Fruit Yield and Oil Quality in a Humid Climate

Paula Conde-Innamorato [1,*], Claudio García [1], Juan José Villamil [1], Facundo Ibáñez [1], Roberto Zoppolo [1], Mercedes Arias-Sibillotte [2], Inés Ponce De León [3], Omar Borsani [4] and Georgina Paula García-Inza [1]

1 Estación Experimental INIA Las Brujas, Programa Nacional de Investigación en Producción Frutícola, Instituto Nacional de Investigación Agropecuaria (INIA), Canelones 90200, Uruguay; cgarcia@inia.org.uy (C.G.); jjvillamil@inia.org.uy (J.J.V.); fibanez@inia.org.uy (F.I.); rzoppolo@inia.org.uy (R.Z.); ggarciainza@inia.org.uy (G.P.G.-I.)
2 Unidad de Ecofisiología de Frutales, Departamento de Producción Vegetal, Facultad de Agronomía, Universidad de la República (UDELAR), Garzón 780, Montevideo 12900, Uruguay; marias@fagro.edu.uy
3 Departamento de Biología Molecular, Instituto de Investigaciones Biológicas Clemente Estable, Montevideo 11600, Uruguay; iponce@iibce.edu.uy
4 Laboratorio de Bioquímica, Departamento de Biología Vegetal, Facultad de Agronomía, Universidad de la República, Montevideo 12900, Uruguay; oborsani@fagro.edu.uy
* Correspondence: pconde@inia.org.uy; Tel.: +598-99120423

Abstract: The expansion of olive orchards into regions with no tradition of olive production and humid climates, such as Uruguay, with more than 1200 mm of annual rainfall, calls into question the need for irrigation. In these regions, however, years with water deficit during summers are quite common. The vapor pressure deficit during summer is lower than in countries with a Mediterranean climate. The high variability in interannual water availability in the current context of climate change, with a growing tendency for extreme events to occur, emphasizes the need to evaluate the production response of olive trees to irrigation. To achieve this, three irrigation treatments were applied to Arbequina and Frantoio cultivars according to the value of the maximum crop evapotranspiration: a first treatment applying 100% ETc, corresponding to being fully irrigated; a second treatment applying 50% ETc; and a third treatment in which neither irrigation nor rain inputs occurred from the end of the pit hardening period until harvest. Results show the possibility of an increasing fruit weight and pulp/pit ratio through irrigation in the local environmental conditions. The oil content in response to irrigation was different within cultivars. Water restriction conditions did not affect the oil content of olives in Arbequina, while in Frantoio it increased it. Polyphenols in fruit increased under water stress for both cultivars. The technological applicability of the results obtained must be accompanied by an economic analysis. The results obtained highlight the need for better use of irrigation water during the growth and ripening phase of the olive fruit under a humid climate.

Keywords: *Olea europaea* L.; drought stress; stem water potential; fruit growth; oil content; polyphenols

Citation: Conde-Innamorato, P.; García, C.; Villamil, J.J.; Ibáñez, F.; Zoppolo, R.; Arias-Sibillotte, M.; Ponce De León, I.; Borsani, O.; García-Inza, G.P. The Impact of Irrigation on Olive Fruit Yield and Oil Quality in a Humid Climate. *Agronomy* **2022**, *12*, 313. https://doi.org/10.3390/agronomy12020313

Academic Editor: José Casanova Gascón

Received: 15 December 2021
Accepted: 11 January 2022
Published: 26 January 2022

1. Introduction

The expansion of olive trees into new climate areas where temperature and precipitation regimes are different from those of the Mediterranean basin generates uncertainty regarding their ecophysiological response and represents challenges in crop management [1]. In temperate humid regions such as Uruguay, where annual precipitation is around 1200 mm, the need for irrigation is questioned. However, Uruguay presents high interannual climate variability and an irregular rainfall distribution throughout the year, which generates periods of water deficit [2,3]. In addition, these extreme events are expected to be more frequent [4], affecting productive behavior.

The importance of local evaluations has been highlighted by several authors who place emphasis on vapor pressure deficit (VPD) conditions [1,5,6], which is lower in Uruguay than in the Mediterranean region. Lower VPD values are associated with lower tree transpiration and, consequently, lower water consumption. In this context of variability of water supply and environmental characteristics, it is necessary to evaluate the productive response of olive trees under different water status conditions, identifying the best balance between yield, oil quality and water productivity.

Olive has a high capacity to grow under water scarcity conditions due to its morphological characteristics and physiological mechanisms, related with the escape, avoidance and tolerance components of stress resistance [7,8]. However, there is a lot of studies that confirm that this crop positively responds to irrigation. Irrigation increases vegetative shoot growth, as well as final fruit size and yield [9–15].

Olive fruit growth (expressed as fresh weight) follows a double sigmoid curve [16]. Previous reports have identified two periods during fruit growth that are particularly sensitive to water restriction: an initial one during cell division and the expansion phase, from flowering until the end of fruit set, where deficit irrigation can reduce the final fruit number [15,17]; and a second one, during cell expansion and the lipogenesis phase, after pit hardening until harvest, when fruit growth increases sharply as mesocarp cells expand. A deficit in irrigation during this period can reduce the final fruit weight and oil content, and it can affect the oil composition, such as the polyphenol content [18–20].

There are several studies on the effect of water restriction on olive trees in arid regions, but there is a knowledge gap on the response of olive trees to irrigation management in humid temperate climates. The aim of this work was to quantify the impact of different irrigation regimes on fruit growth development and oil quality in two olive cultivars in a humid climate region.

2. Materials and Methods

2.1. Plant Material and Experimental Design

The experiment was conducted at the INIA "Las Brujas" Experimental Station in southern Uruguay (34°40′ S; 56°20′ W; 32 m above mean sea level) using the Arbequina and Frantoio cultivars. The olive trees were planted in 2006 at a density of 416 trees per hectare (4 m between trees and 6 m between rows) and were trained as single-trunk vase shapes, with three to four main branches. The orchard was managed as a commercial farm. Pest management was performed according to the Integrated Pest Management guidelines [21]. For each cultivar, a randomized complete design with three irrigation treatments and four replicates was used. The experimental unit is the tree and there are four trees per treatment, for each cultivar. Three irrigation treatments were applied according to the value of maximum crop evapotranspiration (ETc) (Penman–Monteith equation): a first treatment applying 100% ETc, corresponding to being fully irrigated; a second treatment applying 50% ETc; and a third treatment in which neither irrigation nor rain inputs occurred (non-irrigated) from the end of the pit hardening period until harvest. The experiment was repeated in two years, during the 2018/2019 and 2020/2021 seasons, with a different randomization of the experimental units. The assays were specifically made in years of high fruit load.

2.2. Soil, Irrigation and Tree Water Status

The soil at this site has a fine textured A horizon, with a maximum depth of 30 cm, 2.9% organic matter and pH 6.3, corresponding to a Typic-Vertic Argiudolls soil according to the USDA classification [22]. The soil water curve retention was characterized by measuring water content at tensions of 0.01 and 1.5 MPa (field capacity and permanent wilting point, respectively), using the Richards and Weaver methods [23]. Undisturbed soil sample were used for soil water extraction from different depths up to 0.50 m. Soil moisture was monitored throughout the experiment using three FDR sensors installed at three different depths (15, 30 and 45 cm deep), using an EM50 Digital/Analog Data Logger (Decagon Devices, Inc., Pullman, WA, USA). The total amount of applied water was 190 mm and 410 mm in the first season, and 240 mm and 540 mm in the second season, for 50% ETc and

100% ETc, respectively. Prior to the installation of the experiment, the crop was irrigated according to the value of maximum crop evapotranspiration, so that once the treatments were started, the soil was at field capacity. To avoid the effect of rainfall, after pit hardening the soil around the trees (24 m^2/tree) was covered with nylon (bilayer) in all treatments. The plastic was placed to prevent the entry of rain into the soil and it was not removed until harvest, so that the rain from January to May was not available for the plants and therefore did not affect the treatments. A complete drip irrigation system was used to supply the irrigation water. The system consisted of a 16-diameter (13.6 mm) lateral pipe PE (0.25 MPa) with 0.20 m of emitter spacing. The flow of the self-compensating emitters was 4 L h^{-1}. Therefore, the system was designed to apply 7 mm h^{-1} of water with an average pressure in the lateral pipe of 100 kPa.

The irrigation schedule for the 100% ETc treatment was accomplished daily using the simplified water balance method for the root zone of the crop [24,25], according to the following Equation (1):

$$Dri = ETci - Pei - Ii + DPi + Dri - 1 \quad (1)$$

where *Dr* stands for root zone depletion (mm); *ETc* for maximum crop evapotranspiration (mm), computed as $ET_0{*}Kc$; *Pe* for effective precipitation (mm); *I* for irrigation depth (mm); *DP* for deep percolation outside the root zone (mm); *i* for the current day; and $i - 1$ for the day before. The value used for *Kc* was 0.65 at the beginning of the season and then 0.70 during the mid-season and end-season stages.

The potential crop evapotranspiration was determined as $ET_0 \times kc$ (*kc* values used to calculate the water balance, according to [26]). The irrigation schedule for the 50% ETc treatment was carried out using the same methodology as used for the 100% ETc treatment. The effective precipitation (Pe) used in the soil water balance equation was 0 during pit hardening until harvest. Irrigation depth was computed so that the depletion-water root zone was between the field capacity and readily available water [24]. The daily water balance was calculated for each irrigation treatment with the FAO 56 method [27], and it was used to apply the irrigation during both seasons.

Tree water status was assessed by measuring the stem water potential (SWP) using a Scholander-type pressure chamber (Soil Moisture Equipment Corp., Santa Barbara, CA, USA) every 15 days from the end of pit hardening until harvest. Two hours before the measurement, the shoot was enclosed in a plastic bag, allowing the leaf water potential to balance with the stem water potential for a more stable value than that of the leaf water potential. Measurements on each tree were made between 12h00 and 14h00 on mature leaves exposed to the sun from the middle of the branches [28,29]. The measured SWP values were accumulated over the irrigation period and the cumulative leaf water potential (CLWP) was calculated to compare the level of water stress throughout the entire experiment [30].

2.3. Climate

The climate is temperate humid with an average annual rainfall of 1200 mm unequally distributed throughout the year. The rainfall, mean air temperature, relative air humidity, total day radiation and wind speed were obtained from a Campbell automatic weather station located near the experiment (approximately 0.5 km) (Figure 1). Considering the average historical data (1974–2020), the ET_0 values are as follows: 456.3 mm in summer (maximum values in December and January), 205.2 mm in autumn, 99 mm in winter (minimum values in June and July) and 310.5 mm in spring, with an annual total of 1071 mm.

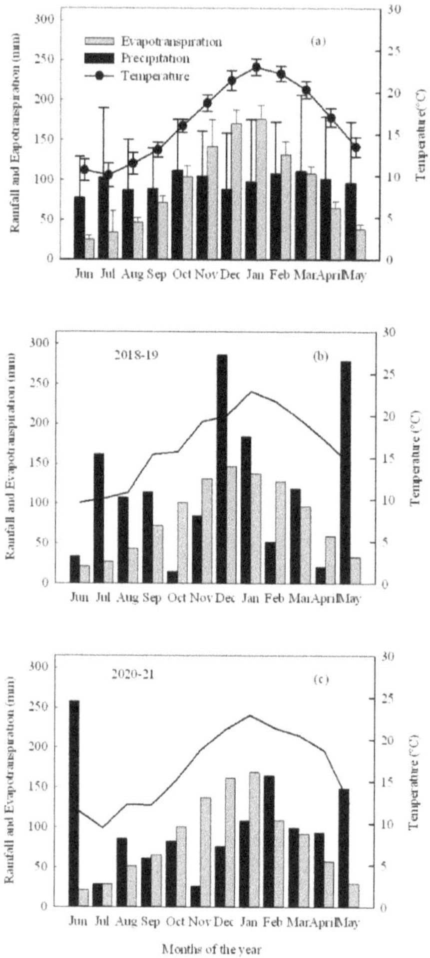

Figure 1. Average monthly values of mean air temperature (continuous line) (°C), evapotranspiration (gray bars) (ET_0, mm × 10) and precipitation (black bars) (mm) at the experimental site in INIA Las Brujas from 1973 to 2018 (**a**), for the 2018/2019 season (**b**), and for the 2020/2021 season (**c**). Vertical bars in (**a**) indicate the standard deviation. Data were recorded at an automatic weather station and is available at http://www.inia.uy/gras/Clima/Banco-datos-agroclimatico 11 January 2022.

Historical data for precipitation and ET_0 show that during spring and summer evapotranspiration exceeds precipitation (Figure 1a), which causes a water deficit of approximately 250 mm during that period. The two seasons in which the experiments were carried out showed differences in rainfall and evapotranspiration. In the first season, high precipitations during December and January were recorded, which by far exceeded the evapotranspiration, and a water deficit occurred in October, November and February (Figure 1b). This generated differences in temperature and relative humidity and, consequently, a vapor pressure deficit that was lower compared to the second season of the experiment during January and higher during February (Figure 2). In the second season (Figure 1c), evapotranspiration exceeded rainfall by approximately 300 mm between the months of September and February, after which the values recorded were similar to what has occurred historically (Figure 1a). In order to characterize our climate region, we compared the VPD of Uruguay and Spain (Córdoba), a main traditional olive cultivation region, considering the average temperature (24 h) and average relative humidity for the 2009–2020 (Uruguay) and 2001–2020 (Spain) periods (Supplementary Figure S1).

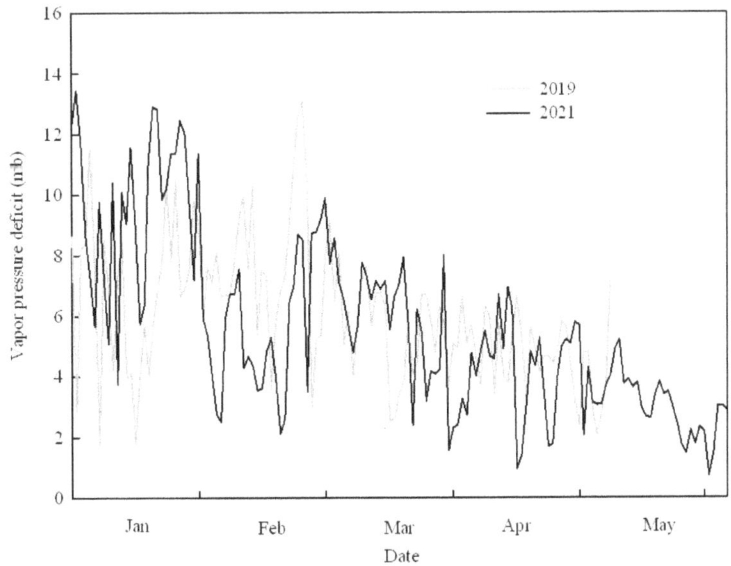

Figure 2. Daily vapor pressure deficit (mb) in Uruguay for the 2019 and 2021 seasons from 1 January to 30 April. Data were recorded at an automatic weather station at the experimental site in INIA Las Brujas (available at http://www.inia.uy/gras/Clima/Banco-datos-agroclimatico 11 January 2022).

2.4. Productive Parameters

The fresh weight of the fruit (g), pit weight (g), and pulp/pit ratio were recorded monthly in 20 olive fruits per tree from the end of pit hardening to harvesting time. Each tree was harvested individually with a trunk vibrator machine at the end of April in both seasons, and fruit maturity index (MI) was recorded based on a 0–7 scale [31]. Fruit yield (kg/tree) was recorded and fruit number per tree was calculated from fruit yield and mean fresh weight of the fruit.

Oil content (%) was measured monthly from the end of pit hardening to harvest time on a sample of 200 g of olives per tree. To determine the fruit moisture content, each sample was ground with a hammer mill and dried at 105 °C for 48 h, after which the dried sample was grinded with a mortar and the oil content was determined following the Soxhlet method [32]. Olive oil from each tree was obtained in an Abencor mill (Mc2 Ingenieria y Sistemas, Sevilla, Spain) for oil composition analysis. Water productivity (WPf) was calculated as kg of fruit per unit of water applied (m^3) throughout the experiment [14].

2.5. Oil Chemical Composition

Acidity: the free fatty acid content was determined following the official analytical method described in [33] and expressed as acidity percentage.

Oil pigments: the content of total chlorophylls and total carotenoids was determined in a spectrophotometer by dissolving 7.5 g of each oil in cyclohexane and measured at 670 and 470 nm, respectively, according to Minguez-Mosquera et al. [34], and expressed as mg kg^{-1} EVOO.

Fatty acids profile: fatty acid methyl esters were prepared from trans-esterification reactions with a cold methanolic solution of potassium hydroxide and analyzed by gas chromatography as described by Feippe et al. [35], with some modifications. Briefly, 0.1 g of the olive oil sample were dissolved in 2 mL of heptane and vortexed for 1 min at 20 °C, then centrifuged at 12,000 rpm for 5 min. The supernatant was trans-esterified by adding 1 mL 4 N KOH in methanol and stirring manually for 1 min. The solution was dried with sodium sulfate powder, centrifugated at 12,000 rpm for 5 min, filtered and injected into a gas chromatograph with a flame ionization detector (GC-FID, Shimadzu model 2010-Plus, Kyoto, Japan). The column used was an Agilent DB-WAX (30 m × 0.25 mm ID, 0.25 μm). The injection temperature was 250 °C and the FID detector temperature was set at 300 °C. The oven temperature was set at 160 °C, increased to 200 °C after 13 min and maintained for 22 min, after which the temperature was increased to 240 °C for the final 25 min. The sample injection volume was 1 μL, and the mobile phase used was nitrogen at 30 mL/min. The hydrogen flow was set at 40 mL/min and the air flow at 400 mL/min. A standard certified Fatty Acids Methyl Esters (FAME) mix (Sigma-Aldrich, St Louis, MO, USA) was used to identify the peaks according to retention times and expressed as percentages.

Total phenolics in olive fruits: total phenolics (TP) were determined according to the method adapted from Sánchez-Rangel et al. [36] with a Folin–Ciocalteu reagent. Two grams of olives were homogenized in an Ultraturrax for 2 min, extracted with 10 mL of 80% methanol, and centrifuged at 10,000 rpm for 4 min at 4 °C. The TP determination was carried out in 96-well microplates, with gallic acid (GA) as the calibration standard; 15 μL of diluted sample extract or GA dilutions were incubated for 15 min in the dark, after the addition of 240 μL of distilled water, 15 μL of Folin–Ciocalteu reagent and 30 μL of 1 N sodium carbonate. The absorbance was read at 760 nm in a Synergy H1 Hybrid Multi-Mode Reader with Gen 5 software (Bio-Tek Inc., Winooski, VT, USA). The results were expressed as mg of gallic acid equivalents (GAE) per kg of fresh olives.

Total phenolic analysis in Virgin Olive Oils (VOO): total phenol content was determined by the Folin–Ciocalteu method described by Gutfinger [37], with some modifications. Briefly, 5 g of olive oil samples were dissolved in 7 mL of $MeOH:H_2O$ (80:20) and vortexed. The mixture was centrifuged for 10 min at 5800 rpm and the procedure was repeated twice. The supernatants were pooled and brought up to a volume of 25 mL with $MeOH:H_2O$ (80:20). An aliquot of 1 mL was transferred to a 10-mL volumetric flask to which 8 mL of distilled water were added followed by 0.5 mL of Folin–Ciocalteu reagent and 0.5 mL of saturated Na_2CO_3. The samples were shaken and left for 15 min in the dark at room temperature. The absorbance was determined spectrophotometrically at 760 nm in a UV–VIS spectrophotometer (Shimadzu™ model UV-3000, Kyoto, Japan). The total amount of TP was calculated and expressed as mg of GAE equivalent per kg of oil by using a calibration curve prepared with pure gallic acid standard solution.

2.6. Statistical Analysis

Since fruit number is a yield component defined during fruit set [38–40] and irrigation treatments are installed after that phase, we analyzed whether there were significant differences between the treatments of fruit number per tree at harvest. As significant differences were detected, productive variables were analyzed with ANCOVA, with fruit number per tree as a covariate. The adjusted model for each cultivar included treatment effect, year effect, and their interaction. The Mixed Models procedure (SAS v.9.4 (SAS Institute, Cary, NC, USA 2013) was used, and the corrected means were contrasted using the Tukey–Kramer test, with a significance level of 5%. Linear functions were fitted to the relationships between the fresh weight of the fruit, fresh weight of the pulp, pulp/pit ratio, maturity index and the CLWP variables. We report those functions that provided the best fits with a significance level of 5%.

3. Results

3.1. Tree Water Status and Fruit Moisture

Plants water status was affected by the levels of irrigation applied during the experiment. The ranges of stem water potential (SWP) during the experiment in both seasons and for both cultivars are presented in Table 1. The values obtained for the non-irrigated treatment were lower than those for irrigated trees. The most negative water potential value reached throughout the experiment was −3.5 MPa, corresponding to the non-irrigated treatment. In the 2021 season, water potential values were more negative than in the 2019 season. During spring and summer of 2021 there were few rain events. The estimated evapotranspiration during those months was higher than the precipitation. The crop water demand exceeded the water supply from the rains. Despite the differences in the ranges between the two seasons, the differences between the treatments in each season were clearly defined and had the same response pattern.

Table 1. Midday stem water potentials (Ψstem) ranges obtained from Arbequina and Frantoio cultivars grown under different water irrigation levels (non-irrigated, 50% ETc and 100% ETc) in two seasons.

Cultivar	2019			2021		
	Non-Irrigated	50% ETc	100% ETc	Non-Irrigated	50% ETc	100% ETc
cv. Arbequina	−1.7 to −2.3	−1.3 to −1.6	−0.8 to −1.3	−2.5 to −3.4	−1.7 to −2.1	−1.3 to −1.8
cv. Frantoio	−1.9 to −2.7	−1.5 to −1.6	−1.1 to −1.5	−3.1 to −3.5	−2.0 to −2.8	−1.9 to −2.4

The values of cumulative leaf water potential (CLWP) of the non-irrigated treatment were lower than those of irrigated trees in both cultivars and seasons (Figure 3). The results show that the generated water deficit was progressive and constant throughout the experiment. The water deficit level reached in the 2021 season was more intense than in 2019 in both cultivars, 'Frantoio' being the cultivar that reached the most negative values. In the 2019 season, values of −248 and −295 MPa in Arbequina and Frantoio, respectively, were recorded in the non-irrigated treatments, while the 100% ETc treatment presented values between −137 and −149 MPa. In the 2021 season, values of −353 and −403 MPa in Arbequina and Frantoio, respectively, were recorded in non-irrigated treatments while the 100% ETc treatment presented values between −211 and −252 MPa.

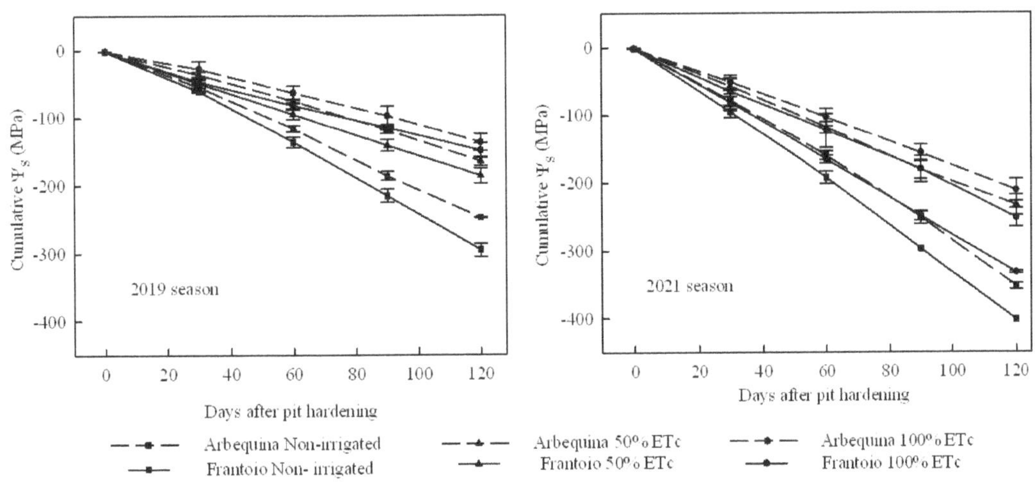

Figure 3. Seasonal evolution of cumulative leaf water potential (CLWP, MPa) of Arbequina (dotted line) and Frantoio (solid line) cultivars from the end of pit hardening to harvest time in the 2019 and 2021 seasons. Treatments included non-irrigated (■), 50% ETc (▲) and 100% ETc (•). Values are the means of four trees.

Fruit moisture content recorded during the experiment also presented differences between treatments, being lower in non-irrigated treatments than in irrigated ones, for both cultivars and both seasons (Figure 4). At the end of the experiment, the 100% ETc treatment presented at least 12% more moisture than the non-irrigated treatment in Arbequina and more than 8% in Frantoio. In 2019, fruit moisture in the non-irrigated treatments was 43 and 48% in Arbequina and Frantoio, respectively, compared to the respective 59 and 56% recorded in the 100% ETc treatments. In 2021, fruit moisture in the non-irrigated treatments was 54 and 49% in Arbequina and Frantoio, respectively, compared to the respective 66 and 63% recorded in the 100% ETc treatments.

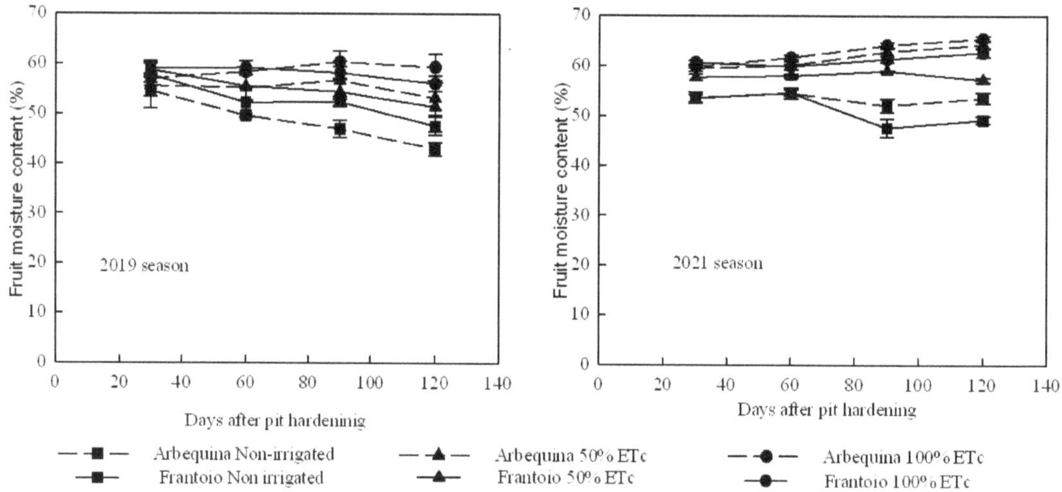

Figure 4. Fruit moisture content (%) in Arbequina (dotted line) and Frantoio (solid line) cultivars from the end of pit hardening to harvest time in the 2019 and 2021 seasons. Treatments included non-irrigated (■), 50% ETc (▲) and 100% ETc (•). Values are the means of four trees.

3.2. Productive Parameters

Significant differences between the treatments of fruit number per tree at harvest were detected, so productive variables were analyzed with ANCOVA, with fruit number per tree as a covariate. Fruit yield (kg/tree) did not show significant differences in Arbequina. In Frantoio, however, fruit yield was significantly higher in the irrigated treatments than in the non-irrigated treatment (Table 2). The maturity index for both cultivars was higher in the non-irrigated treatments than in the irrigated ones. Oil content (% DWB) did not show significant differences between treatments in Arbequina, whereas in Frantoio differences were only observed when comparing the 100% ETc and the non-irrigated treatment, with the oil content being higher in the latter. The fruit yield achieved in both irrigated treatments was similar. Therefore, regardless of the cultivar, WPf was higher in the treatment with 50% ETc than in the one with 100% ETc (Table 2). WPf was not calculated in the non-irrigated treatment, since there was no application of irrigation.

Correlations between the production parameters and CLWP were analyzed (Figure 5). CLWP better represents plant water status when it is closer to zero. Fresh weight of fruit increased with the best water status in both cultivars and in both seasons, as did the fresh weight of the pulp and pulp/pit ratio, presenting significant regressions in all cases. A positive relationship was observed between the fresh weight of fruit and CLWP, as irrigated trees of both cultivars presented a higher fresh weight of fruit than those of the non-irrigated treatment (Figure 5). In Arbequina, an adjustment greater than 0.4 was obtained, while in Frantoio this value was greater than 0.66. Similar results were obtained in the relationship between fresh weight of pulp and CLWP, with an adjustment greater than 0.51 in Arbequina and at 0.65 in Frantoio. A positive relationship was also observed between the pulp/pit ratio and CLWP, as Arbequina and Frantoio exhibited an adjustment greater than 0.68

and 0.71, respectively. The maturity index presented a negative relationship with CLWP, being significantly higher in non-irrigated treatments compared to those with irrigation (Figure 5). Arbequina presented an adjustment of 0.72, with MI ranges that varied between 1 and 3.7, while Frantoio presented an adjustment of 0.40, with MI ranges between 1.1 and 2.1 (Table 2).

Table 2. Fruit yield (kg/tree), maturity index, oil content (%) and water productivity (WPf, kg fruit/m^3 water applied) of Arbequina and Frantoio cultivars grown in the fully irrigated treatment with 100% ETc, in treatment with 50% ETc and in the non-irrigated treatment at INIA Las Brujas. Mean of the two seasons.

Evaluated Parameters	cv. Arbequina			cv. Frantoio		
	Irrigation Treatment					
	Non-Irrigated	**50% ETc**	**100% ETc**	**Non-Irrigated**	**50% ETc**	**100% ETc**
Fruit yield (kg/tree)	35.2 ^{a,*}	42.2 ^{a}	45.2 ^{a}	31.5 ^{b}	45.5 ^{a}	52.4 ^{a}
Maturity index	3.32 ^{a}	2.21 ^{b}	1.91 ^{b}	2.31 ^{a}	1.32 ^{b}	0.96 ^{b}
Oil content (% DWB)	39.6 ^{a}	37.7 ^{a}	37.6 ^{a}	39.2 ^{a}	36.9 ^{a,b}	34.5 ^{b}
WPf (kg fruit/m^3 water applied)	#	19.6	9.5	#	21.2	11.0

* Different letters within the row indicate significant differences for each cultivar separately (HSD Tukey–Kramer $p \leq 0.05$). # Since the non-irrigated treatment did not receive water applications, the WPf was not calculated.

3.3. Polyphenols Content in Fruits

Total phenols in fruit showed a negative relationship with the reduction in CLWP (Figure 6). Arbequina showed a reduction of 2730 and 1180 mg GAE/kg FW in total phenols in fruit, as CLWP was lower in 2019 and 2021, respectively. Frantoio showed a lesser reduction, being 470 and 610 mg GAE/kg FW in total phenols in fruit, as CLWP was lower in 2019 and 2021, respectively.

3.4. Oil Chemical Composition

The fatty acids profile was affected by irrigated treatments in both seasons, mainly for Arbequina. In the 2019 season, palmitoleic (C16:1) and linoleic (C18:2) acids increased in the non-irrigated treatment, whereas stearic (C18:0) and oleic (C18:1) decreased in Arbequina. In Frantoio, only stearic acid showed an increase without irrigation. In the 2021 season, Arbequina showed higher levels of linoleic acid in the non-irrigated treatment, similarly to the first season but in a much higher percentage in all treatments. Arachidic (C20:0) and eicosenoic (C20:1) acids showed lower levels under non-irrigated treatments in the 2021 season in Arbequina. As for Frantoio, linolenic (C18:3) and eicosenoic acids were lower and stearic acid was higher in the non-irrigated treatment. The MUFA/PUFA ratio was modified only in Arbequina in the first season, being significantly higher in the 100% ETc treatment. The oil polyphenol content was significantly higher in the non-irrigated treatment than in the 100% ETc treatment for Arbequina in the 2019 season and for Frantoio in 2021. Total carotenoids were significantly higher in the non-irrigated treatments in both cultivars, except for Arbequina in 2021, when no significant differences were recorded. Total chlorophylls in the non-irrigated treatment were lower in Arbequina in both seasons and higher in Frantoio, which only presented significant differences with the other treatments in the 2021 season. Free acidity was analyzed as the quality control of the extraction process, which in all cases was less than 0.15% (data not shown).

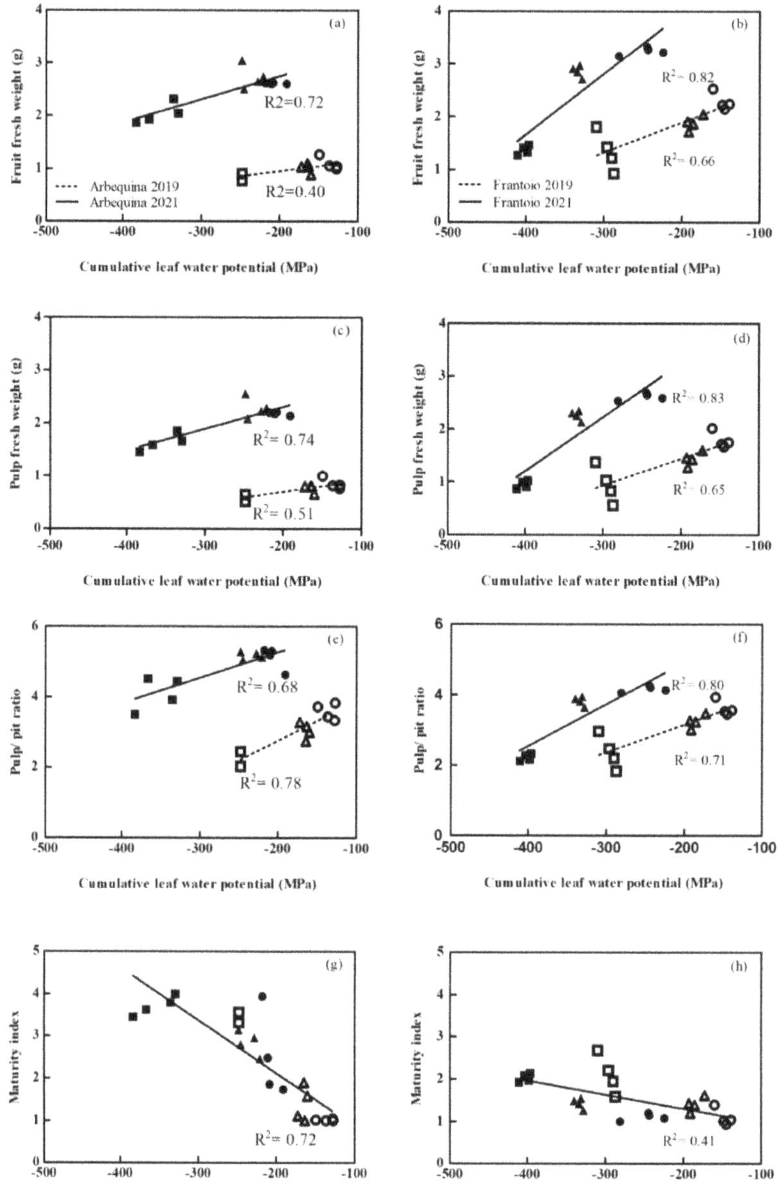

Figure 5. Fresh weight of fruit in grams (**a**,**b**), fresh weight of pulp in grams (**c**,**d**), pulp/pit ratio (in fresh weight basis) (**e**,**f**) and maturity index (**g**,**h**) at harvest time as a function of the cumulative leaf water potential (MPa) in Arbequina (**left**) and Frantoio (**right**) cultivars. Treatments included: non-irrigated (■), 50% ETc (▲) and 100% ETc (•). In panels (g) and (h), the regression was done for the two years together. The empty symbols correspond to the 2019 season and the full symbols to 2021.

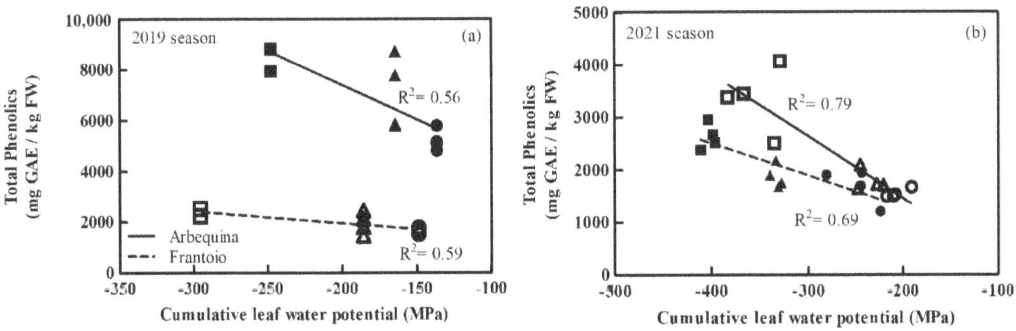

Figure 6. Total phenolics in fruit (mg GAE/kg FW) as a function of the cumulative leaf water potential (MPa) in the 2019 (**a**) and 2021 (**b**) seasons. Treatments included non-irrigated (■), 50% ETc (▲) and 100% ETc (•). Full symbols correspond to cv. Arbequina and empty symbols to cv. Frantoio.

4. Discussion

Olive (*Olea europaea* L.) is a typical tree of the Mediterranean climate that has traditionally been cultivated in rainfed conditions and in regions with high vapor pressure deficit (VPD). VPD is the difference between the saturation vapor pressure and real vapor pressure during a given period [27]. Mairech et al. [5] express the need for calibrating irrigation at a local level given the high dependence of the water requirements with VPD. Many authors report the range of xylem potential reached during irrigation experiments, but few studies refer to VPD [15,41]. We compared the VPD of Uruguay and Spain (Córdoba) (Supplementary Figure S1) and it is observed that Uruguay presents notoriously lower values of VPD. During the summer, VPD values in Spain double those of Uruguay. VPD is directly associated with evapotranspiration (ET_0), as higher VPD would generate higher water consumption. Annual ET_0 values in our conditions are lower than those reported for other olive regions around the world [1,5]. During summer, the period in which the experiment was carried out, the most frequent ET_0 value in Uruguay was 456 mm (Figure 1a), while in the Mediterranean basin it is higher (for instance, 600 mm in Sevilla [1]). These climatic differences can influence olive tree physiology, affecting the productive variables.

Plant water status achieved by the different treatments in both seasons generated a moderate stress in the 50% ETc treatment and a severe one in the non-irrigated treatment according to Fernández et al. [42]. Stem water potential (SWP) in both seasons was in the range of −1.7 and −3.5 MPa in the non-irrigated treatment, −1.3 and −2.8 MPa in the 50% ETc treatment and between −0.8 and −2.4 MPa the in 100% ETc treatment, all within the range measured by other authors [14,30,43]. In Arbequina, the stress during the experiment reached cumulative leaf water potential gradients between treatments from −137 to −248 in 2019 and from −211 to −353 in 2021. In Frantoio, the CLWP gradient was −149 to −295 in 2019 and −251 to −403 in 2021 (Figure 3). These values were similar to those reported by Gucci et al. [30]. This allows us to classify the stress level of treatments and therefore to compare our results with reports made in other sites and cultivars.

Final yield is determined by fruit number, fruit weight and oil content [44]. Fruit number is defined in the flowering–fruit setting phase, prior to establishing the experiment [38–40]. As significant differences in fruit number per tree were detected at harvest between treatments, this parameter was used as a covariate. Therefore, we focused on fruit weight and oil content responses. It was observed that, under the experimental conditions, both cultivars responded positively to irrigation, increasing the fruit weight and the pulp/pit ratio in comparison to the non-irrigated treatment. A similar response was also observed during the same water deficit period by Lavee et al. [45] in cv. Muhasan, where fruits were significantly smaller in the non-irrigated trees than in the irrigated ones. The magnitude of the irrigation effects was different depending on the cultivar. Fruit weight in

Arbequina in the irrigated treatment increased by 40% in comparison to the non-irrigated treatment, while in Frantoio it increased by 68%, which represents a 66% increase in yield per tree. Moreover, a positive linear relationship between the pulp/pit ratio and water plant status was observed in both cultivars (Figure 5b). This is in accordance with previous studies by Gómez-Rico et al. [46] and Lavee et al. [45] who observed that irrigation increases the pulp/pit ratio in comparison to rain-fed trees. However, other authors did not find effects in this ratio under mild water stress [12,45]. In the framework of this experiment, the impact of irrigation on absolute fruit weight was lower in Arbequina, a cultivar with small fruits. In addition, it did not translate into a yield increase (kg/plant). In Frantoio, the increase in fruit weight doubled with irrigation, impacting the yield (Table 2 and Figure 5). This could be due to the genetic characteristic of the fruit size limiting the response to irrigation.

Fruit oil content does not yet show a consensual pattern of response to water restriction. This parameter is highly variable depending on the moment and level of water restriction applied [14,46]. In Arbequina, we found no effect of irrigation on the oil content on a dry basis (Table 2). Similar results were obtained by Hueso et al. [14] in Arbequina under similar water-stress conditions (up to −2.6 MPa) and by Ahumada-Orellana et al. [47] even under severe water stress (up to −6 MPa). However, Iniesta et al. [40] found a higher oil content in water deficit treatments in comparison to irrigated ones, under similar stress conditions as those of our work (−2.9 and −3.6 MPa), also in Arbequina. Oil content is conditioned by the genotype–environment interaction [48]. In this study, a different response in oil content according to cultivars was observed, in agreement with Iniesta et al. [40]. In particular, the oil content in Frantoio was higher in the non-irrigated treatment than in the 100% ETc treatment (Table 2), while in other works a reduction in oil content has been recorded when stem water potential was near −4 MPa [46,49].

Regarding MI, the negative effect of the fruit load on the advancement of maturity has been widely reported. In this sense, it is expected that the treatments with more load have a delayed maturity. Despite having corrected the maturity mean values for fruit number, a negative linear relationship was maintained between MI and water status in both cultivars, with a greater slope in Arbequina (Figure 5). The same pattern was found by Inglese et al. [50] in cv. Carolea when the irrigation in the final phase of fruit development delayed the MI, and by Motilva et al. [51] in Arbequina. However, Iniesta et al. [40] did not find that a deficit in irrigation leads to an earlier ripeness.

Polyphenols have been associated with defense mechanisms against water stress [52]. In our study, total phenols in fruit increased in the non-irrigated treatment in both cultivars. According to Talhaoui et al. [53], the transfer of phenolic compounds from fruits to oil did not surpass 2% in a study with six cultivars that explains qualitative and quantitative changes in phenolic compounds of olive oil during oil extraction in relation to fruits. Our results show that the content of polyphenols in Arbequina and Frantoio oil increased for the non-irrigated treatment compared with the 100% ETc treatment, as also reported by several authors [51,54,55]. During the 2021 season, Arbequina showed the same tendency but the differences were not statistically significant. In the 2019 season, the content of polyphenols in Frantoio increased with irrigation restriction at 50% ETc, reaching the highest level. A similar effect was observed by Tognetti et al. [56], who reported that total phenolics at 66% ETc where higher than in the non-irrigated and 100% ETc treatments. Other works with Frantoio found inconsistent results between years in the phenol content in response to the level of irrigation [57]. Despite the fact the levels found in both seasons are different, they are in concordance with previously reported oils from similar experiments [18,46,51]. In humid climate conditions, the differences between two growing seasons could affect not only the oil content [58] but also the minor oil components, such as phenolic compounds and the fatty acid profile.

The carotenoids content in oil was affected by plant water status, increasing with water restriction between the non-irrigated and 100% ETc treatments in Arbequina and Frantoio, respectively. Only Arbequina showed no differences in season 2021 (Table 3). However, previous works on the effects of irrigation on the carotenoid content found different responses. Sena-Moreno et al. [59] reported an increase in carotenoids with water restriction, consistent with our results, while Tovar et al. [18] did not find differences. The chlorophyll content in Arbequina was reduced by 65% (2019) and 81% (2021) in the non-irrigated treatment compared to the irrigated ones. This response was probably due to the advanced maturity index in the non-irrigated treatment at harvest. For Frantoio, the only difference was found in the 2021 season, where the chlorophyll content increased by 76%. The pigments content (chlorophylls and carotenoids) varies depending on the cultivar, the fruit ripening stage, the weather conditions and the oil-extraction processes [60]. The lipophilic characteristics of these compounds determine the affinity of the oily phase and the migration ratio into the EVOO, playing a role in oxidative stability [61].

Table 3. Physicochemical composition of olive oil in the non-irrigated, 50% ETc and 100% ETc treatments: total phenolics, chlorophylls, total carotenoids, free acidity and fatty acids composition. Olives from the Arbequina and Frantoio cultivars harvested in 2019 and 2021 obtained the corresponding virgin olive oil (VOO).

	2019						2021					
	cv. Arbequina			cv. Frantoio			cv. Arbequina			cv. Frantoio		
	Non-Irrigated	50% ETc	100% ETc	Non-Irrigated	50% ETc	100% ETc	Non-Irrigated	50% ETc	100% ETc	Non-Irrigated	50% ETc	100% ETc
Total Phenolics (mg GAE/kg EVOO)	147.5 [a,*]	138.3 [a,b]	121.7 [b]	343.5 [a,b]	372.4 [a]	306.4 [b]	86.4 [a]	72.6 [a]	80.2 [a]	133.5 [a]	105.9 [b]	104.0 [b]
Totals Carotenoids (mg Car/kg EVOO)	3.54 [a]	3.36 [b]	2.88 [b]	7.29 [a]	5.81 [b]	5.59 [b]	0.68 [a]	0.77 [a]	0.79 [a]	5.19 [a]	3.08 [b]	3.41 [b]
Total Chlorophylls (mg Ch/kg EVOO)	0.69 [b]	1.94 [a]	1.99 [a]	6.81 [a]	5.16 [a]	5.79 [a]	0.14 [b]	0.75 [a]	0.75 [a]	7.47 [a]	3.85 [b]	4.04 [b]
Fatty acid composition:												
Palmitic Acid (%)	14. 82 [a]	15.07 [a]	15.25 [a]	13.54 [a]	14.19 [a]	13.67 [a]	16.69 [b]	17.41 [a]	17.18 [a,b]	14.00 [a]	13.81 [a]	14.64 [a]
Palmitoleic Acid (%)	2.19 [a]	1.99 [a,b]	1.93 [b]	1. 23 [a]	1.45 [a]	1.55 [a]	1.62 [b]	2.16 [a,b]	2.28 [a]	1.07 [a]	0.85 [a]	1.09 [a]
Stearic Acid (%)	1.51 [c]	1.74 [a]	1.65 [b]	2.25 [a]	1.89 [b]	1.58 [b]	1.71 [a]	1.75 [a]	1.69 [a]	2.26 [a]	1.90 [b]	1.70 [c]
Oleic Acid (%)	67.56 [b]	68.73 [a,b]	69. 57 [a]	73.45 [a]	73.05 [a]	74.07 [a]	60.23 [a]	60.52 [a]	61.24 [a]	71.10 [a]	70.28 [a]	71.01 [a]
Linoleic Acid (%)	12.84 [a]	11.41 [b]	10.42 [c]	8.17 [a]	8.14 [a]	7.78 [a]	18.42 [a]	16.66 [a,b]	16.14 [b]	10.11 [b]	11.59 [a]	9.808 [b]
Linolenic Acid (%)	0.50 [a]	0.51 [a]	0.56 [a]	0.75 [a]	0.68 [b]	0.68 [b]	0.60 [a]	0.65 [a]	0.66 [a]	0.63 [c]	0.70 [b]	0.82 [a]
Arachidic Acid (%)	0.30 [a]	0.29 [a]	0.31 [a]	0.31 [a]	0.31 [a]	0.31 [a]	0.34 [b]	0.40 [a]	0.40 [a]	0.41 [a]	0.40 [a]	0.40 [a]
Eicosenoic Acid (%)	0.28 [a]	0.27 [a]	0.31 [a]	0. 30 [a]	0.29 [a]	0.37 [a]	0.27 [b]	0.31 [a]	0.30 [a]	0.31 [c]	0.36 [b]	0.40 [a]
MUFA/PUFA	5.23 [c]	5.94 [b]	6.55 [a]	8.39 [a]	8.49 [a]	9.00 [a]	3.28 [a]	3.66 [a]	3.81 [a]	6.76 [a]	5.84 [b]	6.83 [a]

* Means ($n = 4$) followed by the same letter within a row for each cultivar and each season are not significantly different at $p < 0.05$ (Tukey's test). GAE: Gallic acid equivalent; Ch: total chlorophylls; Car: total carotenoids. MUFA/PUFA: $\sum$ monounsaturated fatty acids (MUFA)/$\sum$ polyunsaturated fatty acids (PUFA) ratios.

The effect of different irrigation strategies on the olive oil fatty acid composition remains unclear [4]. In our study, oleic acid, the main olive oil fatty acid, was significantly lower for Arbequina in the 2019 non-irrigated treatment (Table 3). Severe and prolonged water stress during fruit growth increases fruit temperature [62] and consequently the oleic acid proportion fell [63]. In the same regard, we found a reduction in oleic acid as linoleic acid increases with severe water restriction in Arbequina. The MUFA/PUFA ratio decreased concomitantly for Arbequina in the 2019 season in the non-irrigated treatment, with no changes in Frantoio in either season. Despite slight differences in the percentages, the obtained quality complied with the IOC specifications for EVOOs regarding the fatty acid composition of both cultivars.

It is important to find the best balance between yield, oil quality and water-saving issues [30]. The relationship between fruit production and ETc was shown to be curvilinear by Moriana et al. [10], which means that high production could be reached at lower values than those of the maximum potential ETc. It was demonstrated that under full irrigation olives can achieve high yields (8 t/ha/year) in humid template regions [58]. In our study, if we compare WP (estimated according to kg/tree based on the applied water) between both irrigation treatments, we observe that it was always higher in the 50% ETc treatment than in the 100% ETc one, since yields in kg/tree were similar between both and the water applied was half in the 50% ETc one. Moreover, oil polyphenols content was similar in both irrigated treatments, which may affect oil stability [64]. Fruit moisture at harvest ranged between 43% and 65.5% in Arbequina in the non-irrigated and 100% ETc treatments, respectively. In Frantoio, the range was between 47.6% and 62.7% in the non-irrigated and 100% ETc treatments, respectively. High fruit moisture may have negative effects on the oil extraction process [65,66]. A reduction in irrigation translates into a lower percentage of fruit moisture, which facilitates oil extraction [64–66] and reduces costs at the olive oil mill. The economic valuation of the investment in irrigation must take into account other benefits, such as avoiding severe droughts in the spring or in the years of installation of the crop. Our results provide information on how to manage the irrigation already installed during the fruit growth phase, a phase in which the published results are highly variable and for which there is no evidence in our agroecological conditions.

In summary, this work provided experimental evidence about the productive behavior of the Arbequina and Frantoio cultivars in a temperate humid climate under different water deficit conditions. We demonstrated that irrigation in a low VPD environment increases fruit weight and the pulp/pit ratio, and that it resulted in a significant increase in yield (kg/tree) in Frantoio. The oil content in response to irrigation was different between cultivars. Water restriction conditions did not affect the oil content of olives in Arbequina, while in Frantoio it increased by water restriction in the evaluated range of stem water potential (from −0.8 to −3.5 MPa). The content of polyphenols in fruits and in oil increased under water restriction, with lesser changes in other oil quality parameters. A moderate water restriction (50% ETc) produced the most balanced result between yield, oil quality and WP. Irrigation during the growth and ripening of the fruit also affects the vegetative development and therefore will affect the flowering potential for the next season, in this way it is also intervening in the expression of alternate bearing. For this reason, future studies should address aspects of partition and the relationship of vegetative–reproductive growth to carry out a comprehensive analysis of the benefits of irrigation in our agroclimatic conditions.

Supplementary Materials: The following supporting information can be downloaded at: https://www.mdpi.com/article/10.3390/agronomy12020313/s1, Figure S1: Annual vapor pressure deficit (mb) in Uruguay and in Spain. Data recorded by an automatic weather station at the experimental site in INIA Las Brujas as an average of the 2009–2020 period (available at http://www.inia.uy/gras/Clima/Banco-datos-agroclimatico 11 January 2022) and by the weather station at Córdoba as an average of the 2001–2020 period (available at https://www.juntadeandalucia.es/agriculturaypesca/ifapa/riaweb/web/estacion/14/6 11 January 2022).

Author Contributions: Conceptualization, P.C.-I., C.G., O.B., F.I., R.Z., J.J.V., M.A.-S., I.P.D.L. and G.P.G.-I.; formal analysis, P.C.-I. and G.P.G.-I.; methodology, P.C.-I., O.B., F.I. and C.G.; software, C.G. and G.P.G.-I.; investigation, P.C.-I., J.J.V., G.P.G.-I., C.G., M.A.-S. and F.I.; writing—original draft preparation, P.C.-I., G.P.G.-I. and M.A.-S.; writing—review and editing, P.C.-I., G.P.G.-I., M.A.-S., O.B., F.I., C.G., J.J.V., I.P.D.L. and R.Z.; supervision, O.B., I.P.D.L. and R.Z.; project administration, P.C.-I.; funding acquisition, P.C.-I., C.G., F.I. and R.Z. All authors have read and agreed to the published version of the manuscript.

Funding: This research was funded by The National Institute of Agricultural Research (Instituto Nacional de Investigación Agropecuaria–INIA, Uruguay) (Project INIA FR 22).

Institutional Review Board Statement: Not applicable.

Informed Consent Statement: Not applicable.

Data Availability Statement: Not applicable.

Acknowledgments: We are grateful to David Bianchi, Cecilia Martínez, César Burgos and Sergio Bentancor for their collaboration in this project and to Andrés Coniberti for his contribution in the experimental design.

Conflicts of Interest: The authors declare no conflict of interest.

References

1. Torres, M.; Pierantozzi, P.; Searles, P.; Rousseaux, M.C.; García-Inza, G.; Miserere, A.; Bodoira, R.; Contreras, C.; Maestri, D. Olive cultivation in the Southern Hemisphere: Flowering, water requirements and oil quality responses to new crop environments. *Front. Plant Sci.* **2017**, *8*, 1830. [CrossRef]
2. Tiscornia, G.; Cal, A.; Giménez, A. Análisis y caracterización de la variabilidad climática en algunas regiones de Uruguay. *Rev. Investig. Agropecu.* **2016**, *42*, 66–71.
3. Vaughan, C.; Dessai, S.; Hewitt, C.; Baethgen, W.; Terra, R.; Berterretche, M. Creating an enabling environment for investment in climate services: The case of Uruguay's National Agricultural Information System. *Clim. Serv.* **2017**, *8*, 62–71. [CrossRef]
4. Moretti, S.; Hernández, M.L. Regulation of olive fatty acid desaturation by environmental stresses and culture conditions. *Annu. Plant Rev. Online* **2021**, *4*, 687–712. [CrossRef]
5. Mairech, H.; López-Bernal, Á.; Moriondo, M.; Dibari, C.; Regni, L.; Proietti, P.; Villalobos, F.J.; Testi, L. Is new olive farming sustainable? A spatial comparison of productive and environmental performances between traditional and new olive orchards with the model OliveCan. *Agric. Syst.* **2020**, *181*, 102816. [CrossRef]
6. Khosravi, A.; Zucchini, M.; Giorgi, V.; Mancini, A.; Neri, D. Continuous monitoring of olive fruit growth by automatic extensimeter in response to vapor pressure deficit from pit hardening to harvest. *Horticulturae* **2021**, *7*, 349. [CrossRef]
7. Connor, D.; Fereres, E. The physiology of adaptation and yield expression in olive. In *Horticultural Reviews*; John Wiley & Sons Ltd.: Hoboken, NJ, USA, 2005; Volume 31, pp. 155–229.
8. Fernández, J.E. Understanding olive adaptation to abiotic stresses as a tool to increase crop performance. *Environ. Exp. Bot.* **2014**, *103*, 158–179. [CrossRef]
9. Fernández, J.E.; Moreno, F. Water use by the olive tree. *J. Crop Prod.* **1999**, *2*, 101–162. [CrossRef]
10. Moriana, A.; Orgaz, F.; Pastor, M.; Fereres, E. Yield responses of a mature olive orchard to water deficits. *J. Am. Soc. Hortic. Sci.* **2003**, *128*, 425–431. [CrossRef]
11. Grattan, S.R.; Berenguer, M.J.; Connell, J.H.; Polito, V.S.; Vossen, P.M. Olive oil production as influenced by different quantities of applied water. *Agric. Water Manag.* **2006**, *85*, 133–140. [CrossRef]
12. Gucci, R.; Lodolini, E.M.; Rapoport, H.F. Water deficit-induced changes in mesocarp cellular processes and the relationship between mesocarp and endocarp during olive fruit development. *Tree Physiol.* **2009**, *29*, 1575–1585. [CrossRef] [PubMed]
13. Searles, P.S. EL uso del agua en olivo. El consumo del agua por el cultivo de olivo (*Olea europaea* L.) en el noroeste de Argentina: Una comparación con la Cuenca Mediterránea Sección especial. *Ecol. Austral* **2011**, *21*, 015–028.
14. Hueso, A.; Trentacoste, E.R.; Junquera, P.; Gómez-Miguel, V.; Gómez-del-Campo, M. Differences in stem water potential during oil synthesis determine fruit characteristics and production but not vegetative growth or return bloom in an olive hedgerow orchard (cv. Arbequina). *Agric. Water Manag.* **2019**, *223*, 105589. [CrossRef]
15. Pierantozzi, P.; Torres, M.; Tivani, M.; Contreras, C.; Gentili, L.; Parera, C.; Maestri, D. Spring deficit irrigation in olive (cv. Genovesa) growing under arid continental climate: Effects on vegetative growth and productive parameters. *Agric. Water Manag.* **2020**, *238*, 106212. [CrossRef]
16. Hartmann, H.T. Growth of the olive fruit. *Proc. Am. Soc. Hortic. Sci.* **1949**, *54*, 86–94.
17. Trentacoste, E.R.; Calderón, F.J.; Contreras-Zanessi, O.; Galarza, W.; Banco, A.P.; Puertas, C.M. Effect of regulated deficit irrigation during the vegetative growth period on shoot elongation and oil yield components in olive hedgerows (cv. Arbosana) pruned annually on alternate sides in San Juan, Argentina. *Irrig. Sci.* **2019**, *37*, 533–546. [CrossRef]
18. Tovar, M.J.; Romero, M.P.; Alegre, S.; Girona, J.; Motilva, M.J. Composition and organoleptic characteristics of oil from Arbequina olive (*Olea europaea* L.) trees under deficit irrigation. *J. Sci. Food Agric.* **2002**, *82*, 1755–1763. [CrossRef]
19. Artajo, L.S.; Romero, M.P.; Tovar, M.J.; Motilva, M.J. Effect of irrigation applied to olive trees (*Olea europaea* L.) on phenolic compound transfer during olive oil extraction. *Eur. J. Lipid Sci. Technol.* **2006**, *108*, 19–27. [CrossRef]
20. Ahumada-Orellana, L.E.; Ortega-Farías, S.; Searles, P.S. Olive oil quality response to irrigation cut-off strategies in a super-high density orchard. *Agric. Water Manag.* **2018**, *202*, 81–88. [CrossRef]
21. Malavolta, C.; Perdikis, D. Crop Specific Technical Guidelines for Integrated Production of Olives. IOBC-WPRS Commission IP Guidelines. 2018. Available online: https://www.iobc-wprs.org/ip_integrated_production/IP_practical_guidelines.html (accessed on 10 January 2022).
22. Durán, A.; Califra, A.; Molfino, J.H.; Lynn, W. *Keys to Soil Taxonomy for Uruguay*; USDA, Natural Resources Conservation Service (NRCS): Washington, DC, USA, 2006; 77p.
23. Richards, L.; Weaver, L. Moisture retention by some irrigated soils as related to soil moisture tension. *J. Agric. Res.* **1944**, *69*, 215–234.
24. Allen, R.G.; Pereira, L.S.; Raes, D.; Smith, M. Crop evapotranspiration. In *Guideline for Computing Crop Water Requirements*; FAO Irrigation and Drainage Paper No. 56; FAO: Rome, Italy, 1998.
25. Pereira, L.S.; Perrier, A.; Allen, R.G.; Alves, I. Evapotranspiration: Concepts and future trends. *J. Irrig. Drain. Eng.* **1999**, *125*, 45–51. [CrossRef]
26. Puppo, L.; García, C.; Girona, J.; García-Petillo, M. Determination of young olive-tree water consumption with drainage lysimeters. *J. Water Resour. Prot.* **2014**, *6*, 841–851. [CrossRef]
27. Allen, R.; Pereira, L.; Raes, D.; Smith, M. *Evapotranspiración del Cultivo: Guía para la Determinación de los Requerimientos de Agua de los Cultivos*; Estudio FAO de Riego y Drenaje N°. 56; FAO: Rome, Italy, 2006.

28. Scholander, P.F.; Bradstreet, E.D.; Hemmingsen, E.A.; Hammel, H.T. Sap pressure in vascular plants. *Science* **1965**, *148*, 339–346. [CrossRef]
29. McCutchan, H.; Shackel, K.A. Stem-water potential as a sensitive indicator of water stress in prune trees (*Prunus domestica* L. cv. French). *J. Am. Soc. Hortic. Sci.* **1992**, *117*, 607–611. [CrossRef]
30. Gucci, R.; Caruso, G.; Gennai, C.; Esposto, S.; Urbani, S.; Servili, M. Fruit growth, yield and oil quality changes induced by deficit irrigation at different stages of olive fruit development. *Agric. Water Manag.* **2019**, *212*, 88–98. [CrossRef]
31. Uceda, M.; Frías, L. Épocas de recolección. Evolución del contenido graso del fruto y de la composición y calidad del aceite. In *Proceedings of II Seminario Oleícola International*; IOOC: Córdoba, Spain, 1975.
32. IUPAC. *Standard Methods for the Analysis of Oils, Fats and Derivatives*, International Union of Pure and Applied Chemistry IUPAC, Ed.; 1st Supplement to the 7th ed.; Pergamon Press: Oxford, UK, 1992.
33. European Union Regulation. Commission Regualation (EEC). No 2568/91 of 11 of July 1991 on the Characteristics of Olive oil and Olive-Residue Oil and on the Relevant Methods of Analysis. *Off. J. Eur. Commun.* **1991**, *L248*, 1–83.
34. Minguez-Mosquera, M.I.; Rejano-Navarro, L.; Gandul-Rojas, B.; Sanchez-Gomez, A.H.; Garrido-Fernandez, J. Color-pigment correlation in virgin olive oil. *J. Am. Oil Chem. Soc.* **1991**, *68*, 332–336. [CrossRef]
35. Feippe, A.; Ibáñez, F.; Altier, G.P. Fruit ripening stage effect on the fatty acid profile of "Arbequina" and "Picual" olives in Uruguay. *Acta Hortic.* **2010**, *877*, 1495–1499. [CrossRef]
36. Sánchez-Rangel, J.C.; Benavides, J.; Heredia, J.B.; Cisneros-Zevallos, L.; Jacobo-Velázquez, D.A. The Folin-Ciocalteu assay revisited: Improvement of its specificity for total phenolic content determination. *Anal. Methods* **2013**, *5*, 5990–5999. [CrossRef]
37. Gutfinger, T. Polyphenols in olive oils. *J. Am. Oil Chem. Soc.* **1981**, *58*, 966–968. [CrossRef]
38. Trentacoste, E.R.; Puertas, C.M.; Sadras, V.O. Effect of fruit load on oil yield components and dynamics of fruit growth and oil accumulation in olive (*Olea europaea* L.). *Eur. J. Agron.* **2010**, *32*, 249–254. [CrossRef]
39. Martín-Vertedor, A.I.; Rodríguez, J.M.P.; Losada, H.P.; Castiel, E.F. Interactive responses to water deficits and crop load in olive (*Olea europaea* L., cv. Morisca) I.—Growth and water relations. *Agric. Water Manag.* **2011**, *98*, 941–949. [CrossRef]
40. Iniesta, F.; Testi, L.; Orgaz, F.; Villalobos, F.J. The effects of regulated and continuous deficit irrigation on the water use, growth and yield of olive trees. *Eur. J. Agron.* **2009**, *30*, 258–265. [CrossRef]
41. Villalobos, F.; Orgaz, F.; Testi, L.; Fereres, E. Measurement and modeling of evapotranspiration of olive (*Olea europaea* L.) orchards. *Eur. J. Agron.* **2000**, *13*, 155–163. [CrossRef]
42. Fernández, J.E.; Torres-Ruiz, J.M.; Diaz-Espejo, A.; Montero, A.; Álvarez, R.; Jiménez, M.D.; Cuerva, J.; Cuevas, M.V. Use of maximum trunk diameter measurements to detect water stress in mature "Arbequina" olive trees under deficit irrigation. *Agric. Water Manag.* **2011**, *98*, 1813–1821. [CrossRef]
43. Marra, F.P.; Marino, G.; Marchese, A.; Caruso, T. Effects of different irrigation regimes on a super-high-density olive grove cv. "Arbequina": Vegetative growth, productivity and polyphenol content of the oil. *Irrig. Sci.* **2016**, *34*, 313–325. [CrossRef]
44. Steduto, P.; Hsiao, T.; Fereres, E.; Raes, D. Respuesta del rendimiento de los cultivos al agua. In *FAO: Estudio de Riego y Drenaje No. 66, (rev.)*; Organización de las Naciones Unidas para la Alimentación y la Agricultura: Roma, Italy, 2012; 510p.
45. Lavee, S.; Hanoch, E.; Wodner, M.; Abramowitch, H. The effect of predetermined deficit irrigation on the performance of cv. Muhasan olives (*Olea europaea* L.) in the eastern coastal plain of Israel. *Sci. Hortic.* **2007**, *112*, 156–163. [CrossRef]
46. Gómez-Rico, A.; Salvador, M.D.; Moriana, A.; Pérez, D.; Olmedilla, N.; Ribas, F.; Fregapane, G. Influence of different irrigation strategies in a traditional Cornicabra cv. olive orchard on virgin olive oil composition and quality. *Food Chem.* **2007**, *100*, 568–578. [CrossRef]
47. Ahumada-Orellana, L.E.; Ortega-Farías, S.; Searles, P.S.; Retamales, J.B. Yield and water productivity responses to irrigation cut-off strategies after fruit set using stem water potential thresholds in a super-high density olive orchard. *Front. Plant Sci.* **2017**, *8*, 1280. [CrossRef]
48. Mousavi, S.; de la Rosa, R.; Moukhli, A.; El Riachy, M.; Mariotti, R.; Torres, M.; Pierantozzi, P.; Stanzione, V.; Mastio, V.; Zaher, H.; et al. Plasticity of fruit and oil traits in olive among different environments. *Sci. Rep.* **2019**, *9*, 16968. [CrossRef]
49. Gómez-del-Campo, M. Summer deficit-irrigation strategies in a hedgerow olive orchard cv. "Arbequina": Effect on fruit characteristics and yield. *Irrig. Sci.* **2013**, *31*, 259–269. [CrossRef]
50. Inglese, P.; Barone, E.; Gullo, G. The effect of complementary irrigation on fruit growth, ripening pattern and oil characteristics of olive (*Olea europaea* L.) cv. Carolea. *J. Hortic. Sci. Biotechnol.* **1996**, *71*, 257–263. [CrossRef]
51. Motilva, M.J.; Tovar, M.J.; Romero, M.P.; Alegre, S.; Girona, J. Influence of regulated deficit irrigation strategies applied to olive trees (*Arbequina cultivar*) on oil yield and oil composition during the fruit ripening period. *J. Sci. Food Agric.* **2000**, *80*, 2037–2043. [CrossRef]
52. Šamec, D.; Karalija, E.; Šola, I.; Vujčić Bok, V.; Salopek-Sondi, B. The role of polyphenols in abiotic stress response: The influence of molecular structure. *Plants* **2021**, *10*, 118. [CrossRef]
53. Talhaoui, N.; Gómez-Caravaca, A.M.; León, L.; De La Rosa, R.; Fernández-Gutiérrez, A.; Segura-Carretero, A. From olive fruits to olive Oil: Phenolic compound transfer in six different olive cultivars grown under the same agronomical conditions. *Int. J. Mol. Sci.* **2016**, *17*, 337. [CrossRef] [PubMed]
54. Gómez-Rico, A.; Salvador, M.D.; La Greca, M.; Fregapane, G. Phenolic and volatile compounds of extra virgin olive oil (*Olea europaea* L. cv. Cornicabra) with regard to fruit ripening and irrigation management. *J. Agric. Food Chem.* **2006**, *54*, 7130–7136. [CrossRef]

55. Dag, A.; Ben-Gal, A.; Yermiyahu, U.; Basheer, L.; Nir, Y.; Kerem, Z. The effect of irrigation level and harvest mechanization on virgin olive oil quality in a traditional rain-fed "Souri" olive orchard converted to irrigation. *J. Sci. Food Agric.* **2008**, *88*, 1524–1528. [CrossRef]
56. Tognetti, R.; D'Andria, R.; Sacchi, R.; Lavini, A.; Morelli, G.; Alvino, A. Deficit irrigation affects seasonal changes in leaf physiology and oil quality of *Olea europaea* (cultivars Frantoio and Leccino). *Ann. Appl. Biol.* **2007**, *150*, 169–186. [CrossRef]
57. Caruso, G.; Gucci, R.; Urbani, S.; Esposto, S.; Taticchi, A.; Di Maio, I.; Selvaggini, R.; Servili, M. Effect of different irrigation volumes during fruit development on quality of virgin olive oil of cv. Frantoio. *Agric. Water Manag.* **2014**, *134*, 94–103. [CrossRef]
58. Conde-Innamorato, P.; Arias-Sibillotte, M.; Villamil, J.J.; Bruzzone, J.; Bernaschina, Y.; Ferrari, V.; Zoppolo, R.; Villamil, J.; Leoni, C. It is feasible to produce olive oil in temperate humid climate regions. *Front. Plant Sci.* **2019**, *10*, 1544. [CrossRef]
59. Sena-Moreno, E.; Pérez-Rodríguez, J.M.; De Miguel, C.; Prieto, M.H.; Franco, M.N.; Cabrera-Bañegil, M.; Martín-Vertedor, D. Pigment profile, color and antioxidant capacity of Arbequina virgin olive oils from different irrigation treatments. *JAOCS J. Am. Oil Chem. Soc.* **2017**, *94*, 935–945. [CrossRef]
60. Giuffrida, D.; Salvo, F.; Salvo, A.; Cossignani, L.; Dugo, G. Pigments profile in monovarietal virgin olive oils from various Italian olive varieties. *Food Chem.* **2011**, *124*, 1119–1123. [CrossRef]
61. Criado, M.N.; Romero, M.P.; Casanovas, M.; Motilva, M.J. Pigment profile and colour of monovarietal virgin olive oils from Arbequina cultivar obtained during two consecutive crop seasons. *Food Chem.* **2008**, *110*, 873–880. [CrossRef] [PubMed]
62. García, J.M.; Hueso, A.; Gómez-del-Campo, M. Deficit irrigation during the oil synthesis period affects olive oil quality in high-density orchards (cv. Arbequina). *Agric. Water Manag.* **2020**, *230*, 105858. [CrossRef]
63. García-Inza, G.P.; Castro, D.N.; Hall, A.J.; Rousseaux, M.C. Responses to temperature of fruit dry weight, oil concentration, and oil fatty acid composition in olive (*Olea europaea* L. var. 'arauco'). *Eur. J. Agron.* **2014**, *54*, 107–115. [CrossRef]
64. García, J.M.; Cuevas, M.V.; Fernández, J.E. Production and oil quality in "Arbequina" olive (*Olea europaea*, L.) trees under two deficit irrigation strategies. *Irrig. Sci.* **2013**, *31*, 359–370. [CrossRef]
65. Ellis, A.C.; Gambaro, A. Characterisation of Arbequina extra virgin olive oil from Uruguay. *J. Food Res.* **2018**, *7*, 79–90. [CrossRef]
66. Fernández, J.E.; Diaz-Espejo, A.; Romero, R.; Hernandez-Santana, V.; García, J.M.; Padilla-Díaz, C.M.; Cuevas, M.V. Precision irrigation in olive (*Olea europaea* L.) tree orchards. In *Water Scarcity and Sustainable Agriculture in Semiarid Environment: Tools, Strategies, and Challenges for Woody Crops*; Elsevier: Oxford, UK, 2018; pp. 179–217, ISBN 9780128131640.

agronomy

MDPI

Article

Pre- and Postharvest Management of Sunburn in 'Granny Smith' Apples (*Malus* × *domestica* Borkh) under Neotropical Climate Conditions

Vivian Severino [1,*], Mercedes Arias-Sibillotte [1], Santiago Dogliotti [1], Erna Frins [2], José Antonio Yuri [3] and Jaime González-Talice [1]

[1] Facultad de Agronomía, Universidad de la República (UDELAR), Garzón 780, Montevideo CP 12900, Uruguay; marias@fagro.edu.uy (M.A.-S.); sandog@fagro.edu.uy (S.D.); jgonzalez@fagro.edu.uy (J.G.-T.)

[2] Facultad de Ingeniería, Universidad de la República (UDELAR), Julio Herrera y Reissig 565, Montevideo CP 11300, Uruguay; efrins@fing.edu.uy

[3] Centro de Pomáceas, Universidad de Talca, 1 Poniente 1141, Talca CP 3460000, Chile; ayuri@utalca.cl

* Correspondence: vseverin@fagro.edu.uy; Tel.: +598-99168635

Citation: Severino, V.; Arias-Sibillotte, M.; Dogliotti, S.; Frins, E.; Yuri, J.A.; González-Talice, J. Pre- and Postharvest Management of Sunburn in 'Granny Smith' Apples (*Malus* × *domestica* Borkh) under Neotropical Climate Conditions. *Agronomy* **2021**, *11*, 1618. https://doi.org/10.3390/agronomy11081618

Academic Editor: Santiago Signorelli

Received: 14 July 2021
Accepted: 11 August 2021
Published: 14 August 2021

Publisher's Note: MDPI stays neutral with regard to jurisdictional claims in published maps and institutional affiliations.

Abstract: Sun damage on apples is attributed to the occurrence of high temperatures, incident radiation, and fruit and plant water parameters, all dependent on climatic conditions and management. The development of new production areas and climate changes increase the interest in studying the behavior of the fruit under different conditions. The effect on sun damage of three nets and two chemical protectants was evaluated in a commercial orchard of 'Granny Smith' in a neotropical climate during the 2012–2016 seasons. We recorded the evolution, incidence and severity of sunburn and sunscald. Fruit surface temperature, fruit size, sprouting and return to flowering was also recorded. Incident radiation and air temperature were considered for the assessment of the crop's microclimate. The transmittance in PAR wavelength and the air temperature variation on netting treatments reached 0.4 and 9 °C, respectively. The fruit surface temperature for the White-Net and Black-Net-50% treatments was always lower than 46 °C. For the sun damaged fruits, the Black-Net-50% treatment showed the highest proportion of slightly sunburned fruits (94%) and the lowest proportion of heavy damages, with a sunscald index equal to or less than 2 (on a scale of 1–4) in all the tested conditions. In a neotropical climate, protectant applications did not reduce the incidence of sunburn, but with the use of nets it was possible to reduce both sunburn and sunscald without affecting growth processes dependent on leaf net assimilation.

Keywords: nets; sun chemical protectants; sunscald; climate changes

1. Introduction

Sunburn damage originates during fruit growth and may or may not be visible at this stage. At harvest and packing, fruits showing an irregular, yellow-to-brown spot on the sunburned side are rejected. Damage without apparent symptoms at harvest and/or packing manifests itself after cold storage (*sunscald*) [1–3]. These colorations result from the synthesis of polyphenols and β-carotenoids in the affected portion of the skin as a potential protective mechanism against oxidative stress due to sun exposure [4–6]. Severe damage is characterized by dark-brown to black coloration due to necrosis of the epidermis and subepidermis [3]. The fruit skin's physicochemical properties in each variety, such as homogeneity, thickness and composition of the epicuticular wax, and pigment concentration, modify the refraction of incident light and determine the sensitivity to sunburn [4], being 'Granny Smith' the most sensitive cultivar [7].

High solar incident radiation and high air temperature cause photo-oxidative stress conditions and increase fruit temperature. These are considered the environmental factors that determine sunburn [3,4,8]. Schrader et al. [1] propose a base fruit temperature of 46 to

49 °C for slight damage, and 52 °C for the occurrence of necrosis, the latter irrespective of incident radiation. Considering that the fruit temperature in the side exposed to the sun can be 18 °C higher than the air temperature and 9 °C higher than in the unexposed side [9], the air temperature values at which fruit damage occurs should be higher than 28–32 °C [3]. Although there is consensus on the predisposing environmental conditions for the occurrence of sunburn, recent works have focused on studying the role of the water status of the fruit and the tree. The location of the fruit in the canopy and its degree of acclimatization are associated with the severity of sun damage. Orchard characteristics, such as tree vigor, the presence of windbreaks and row direction, affect the incident radiation and the sensible heat of the fruit. Finally, crop water status plays a role due to its ability to cool the tissues and, therefore, reduce fruit temperature [2,10–13].

Quality losses due to sun damage have led to the development of cultivation practices to mitigate them [14], such as water spraying on foliage and fruits to lower fruit temperature [15], the use of shade nets [16], chemical formulations that reflect or filter UV radiation [17], antitranspirants [18], and canopy management [2]. Although the reduction in sunburn incidence using different types of netting, structures and cover is widely reported, the mechanisms involved are not fully described and results are highly variable depending on local climate, cultivar and management [16,19]. The nets reduce the total incident radiation, altering the crop's microclimate and modifying the gaseous exchange between plant and atmosphere [20,21]. As a result, many of the plant's physiological processes are modified, producing changes in both water use efficiency and net assimilation [22]. The effect of the netting on the crop's photosynthetic capacity depends on the total incident photosynthetically active radiation (PAR) at each location and for each time of the year. When the decrease in radiation due to the nets maintains PAR levels above the canopy PAR saturation, no problems in carbohydrate production are expected [23,24]. Reduced assimilation rates are associated with smaller fruit size and lower new-shoot growth [25].

The decrease in air temperature under the net, coupled with a decrease in wind speed, affects air temperature, relative humidity, and water vapor-pressure deficit. These environmental variables modify the photosynthetic efficiency, in addition to the water status of the plants and, therefore, the cooling capacity of the foliage and the fruit, thus altering the fruit's sensitivity to sun damage. The approaches linked to photosynthetic efficiency and conducted on recently harvested fruits show a decrease from 0.71 to 0 in the Fv/Fm ratio of fruit skin between 39 and 42 °C. Authors who subscribe to these approaches conclude that sunscald is an expression of photo-oxidative stress in the fruit's skin, which is promoted by peel temperatures above 40 °C. Short-term field tolerance is acquired by insolation at sub-injurious temperatures, but this tolerance is rather weak and does not abide the atmospheric conditions that prevail in the Israeli summer [26]. McCaskill et al. [22] point out that reductions of up to 2 °C in fruit temperature can be obtained under nets, which reduce the intensity of the solar beam by interception and scattering while allowing sufficient air flow to enable the transfer of heat from the fruit's surface to the air. On the other hand, nets can modify the quality of incident light, affecting photomorphogenic processes and other developmental processes such as bud differentiation and the return to flowering in the following spring [27,28].

Another reported management is the application of inorganic physical blockers, i.e., particles that block, reflect and scatter solar radiation [18]. These particles have been used in various crops since the 1970s with varied results. Gindaba and Wand [17] report a decrease in fruit temperature and sun damage in apple trees with the application of chemical protectants. The main drawbacks of this practice include the need for several applications to maintain constant coverage and the difficulty for the removal of the applied products at the packing lines [29].

Most of the research related to sunburn has been conducted in arid or semi-arid climates, with just a few studies performed in humid regions or addressing the differences in sensitivity to sunburn between fruits from different growing environments [30]. Research on crop physiology under netting has been carried out in climates with high incidence of hail, related to altitudes and/or latitudes higher than those of our experiment [31], where solar radiation is not always excessive and therefore the risk of sunburn is lower [23]. Changes in the climate of traditional growing areas increase the interest in understanding the response of temperate fruit trees to neotropical climatic conditions, such as those of our study [32].

The objective of this work is to quantify sunburn damage on apples in a neo-tropical climate and to evaluate the effect of black and white netting, as well as the application of sunburn protectants, on sprouting and growth parameters, and on sunburn damage at harvest and after cold storage.

2. Materials and Methods

The experiment was carried out during the 2012/2013 to 2015/2016 seasons (hereafter, seasons 1 to 4) in an apple orchard (*Malus* × *domestica* Borkh) of 'Granny Smith'/M7 pollinized with 'Gala', located in the department of San José, in southern Uruguay (34°38′18″ S; 56°40′06″ W, 28 masl). The orchard was established in 2003 with north–south row orientation and a planting frame of 4 m × 1.5 m. The climate of this regional ecotone is classified by Bernardi et al. (2016) as neotropical. The soil types are mainly Argiudolls and Hapluderts and a drip irrigation system setup with a maximum daily watering capacity of 4.5 mm.

Treatments consisted of netting and sunburn protectant (PRO) applications. Netting was evaluated throughout all 4 seasons, whereas PROs were applied in seasons 2 and 3. The nets used in all cases were monofilament nets with the following characteristics: translucent white net 20% (WN) (only in seasons 3 and 4); black net 35% (BN35); and black net 50% (BN50). The PROs used were kaolin (Surround WP®, 50 K/ha, Tessenderlo Kerley, Inc., Phoenix, AZ, USA) and $CaCO_3$ (Purshade®, 30 l/ha, Tessenderlo Kerley, Inc., Phoenix, AZ, USA), applied 4 and 5 times for seasons 2 and 3, respectively. All treatments were applied each season in mid-December, coinciding with weeks 6 and 9 after full bloom (WAFB), and were maintained until harvest. The flowering dates for cycles 1 to 4 were the following: 27 September (cycle 1), and 28 October, 3 and 14 (cycles 2, 3 and 4, respectively). The PRO treatments were repeated whenever necessary, depending on rainfall levels (after an accumulation of 10 mm) and/or fruit cover levels.

2.1. Experimental Design

2.1.1. Field Experiment

The experimental design consisted of randomized complete blocks with 3 replicates per treatment. The net treatments covered a surface of 600 m^2 (12 × 50 m), including three rows and its inter-rows, whereas PRO and Control treatments each consisted of one row of 15 m. In all cases, measurements were made in the three central trees.

Four fruits were marked in each replicate between 40 and 50 days after full bloom (DAFB) and after fruit drop by carbohydrate balance [33–35], one in each of the following conditions: (a) exposed green fruit with no visible sun damage (GEF); (b) exposed fruit with red coloration (RF); (c) exposed fruit with incipient sunburn (SBF) or sunburn browning, according to Racsko and Schrader [18]; (d) internal green fruit with no visible sun damage (GIF) (Figure 1). Those fruits were marked at a height between 1.5 and 2 m in the internal and external part of the canopy.

At harvest, fruits were visually classified according to the degree of sunburn into the following categories proposed by Torres et al. [33]: HF, healthy fruit; Mild, slight discoloration on the skin or mild sunburned symptoms on less than 25% of the surface; Mod, moderate sunburned fruit with yellowing and browning on 25 to 50% of the skin surface; and Sev, severe sunburned fruit with more than 50% sunburn or dark brown patches over light browning. These categories consider both symptom type and quantification of the surface, with symptoms described by Racsko and Schrader [18] and Torres et al. [36], as a way of following a standard evaluation procedure to reduce the difficulties of visual grading.

Figure 1. Examples of fruit categories at three stages of damage development. At the beginning of treatments (6 WAFB = weeks after full bloom): RF = red fruit, GEF = green external fruit, GIF = green internal fruit, and SBF = sunburned fruit. At harvest: HF = healthy fruit, Mild = slightly sunburned fruit, Mod = moderately sunburned fruit, Sev = heavily sunburned fruit; and Post-harvest: HF = healthy fruit, SSF = slightly sunscalded fruit, MSF = moderately sunscalded fruit, HSF = heavily sunscalded fruit.

2.1.2. Postharvest Experiment

Postharvest experiment was evaluated in seasons 2 and 3, using a randomized $5 \times 2 \times 2$ factorial design in each season. We applied five levels for field treatments, two levels for fruit condition at harvest (HF and Mild) and two levels for antiscaldant application to prevent superficial scald (with or without product application). The experimental unit consisted of a box with an average of 78 fruits, which was replicated 3 times. Each replicate was palletized and stored in commercial cold storage at a temperature of 0–1 °C and 95% RH. Diphenylamine (DPA) treatment was performed in season 2 at 2000 ppm for 1 min at 22 °C. In season 3, 75 mg m^{-3} of 1-MCP was applied in a commercial storage at room temperature.

2.2. Evaluations

2.2.1. Microclimate

The effect of the nets on the radiation affecting the crop was evaluated with an Ocean Optics spectrometer, model S2000, ~1 nm resolution, which covers a wavelength range from 250 to 800 nm. The spectra were evaluated twice in the tested period, on cloudless days and between 12:00 and 15:00 h (local time = GMT − 3 h).

During the 15/JAN-27/FEB period of season 4, the air temperature (AT) of each treatment was recorded every half hour using i-button sensors located inside the canopy of the trees and protected from direct solar radiation.

The maximum fruit surface temperature (FST) was recorded using an IR camera (FLIR E50) on three days at 12 WAFB of seasons 2 and 3, between 13:00 and 16:00 h, with maximum AT (ATmax) between 33 and 35 °C.

2.2.2. Sunburn

From 6 WAFB onwards, the evaluation of sun damage was recorded weekly by visual determination on the marked fruits.

At harvest, the fruits from each treatment were manually sorted into the four above-mentioned categories: HF, Mild, Mod and Sev (Figure 1). Fruits classified as HF and Mild were retained for postharvest evaluation.

2.2.3. Fruit Growth and Sprouting

From 6 WAFB onwards, fruit size development was measured weekly on a random sample of 15 fruits per tree. After harvest, we counted the number of shoots and spurs on one branch per tree and measured the length of the shoots. In the blooms corresponding to seasons 2, 3 and 4 for each treatment, we evaluated the return to flowering on the same branch.

2.2.4. Post-Harvest

The incidence and severity of scald was evaluated three times during each conservation season. In season 2, evaluations were made at 60, 120 and 210 days of cold storage, while in season 3 evaluations were made at 82, 133 and 245 days. Incidence was expressed as the presence or absence of alterations, and severity was classified on a 4-point scale, in which 1 corresponds to healthy fruits (HF), 2 to slightly scalded fruits (SSF, 25% of the fruit's surface), 3 to moderately scalded fruits (MSF, 25–50% of the surface), and 4 to heavily scalded fruits (HSF, greater than 50%) (Figure 1).

2.3. Statistical Analysis

Statistical analyses were performed using the R statistical software. The interpretation of binary data (0/1) was performed with generalized linear models (GLM) and binomial distribution. For continuous variables, normality and homogeneity of variances were tested prior to the analysis using Shapiro and Levene tests, respectively ($p > 0.05$). An analysis of variance (ANOVA) was performed when the assumptions of normality and homogeneity of variances were met, and a non-parametric Kruskal–Wallis test was applied when the ANOVA was not appropriate. The analysis of ordinal variables such as the level of sunburn or scald was performed with a Cumulative Link Model (CLM) and a Type II ANOVA test to establish the significance of the model effects. When required, Tukey or Mann–Whitney–Wilcoxon tests ($p < 0.05$) were performed.

3. Results

3.1. Changes in Microclimate

Incident Radiation and Air and Fruit Temperature

The nets mainly modified the maximum temperatures depending on their color. Daily ATmax showed the greatest differences among treatments, reaching a 9 °C maximum variation in the studied period. The Control treatment presented an ATmax of 43 °C, while the WN treatment reached 47 °C in the evaluated period. The lowest ATmax values were recorded in black net treatments: 39 °C and 38 °C for BN50 and BN35, respectively (Figure 2). Daily ATmed ranged from 23 °C to 25 °C, and ATmin from 9 °C to 10 °C (data not shown).

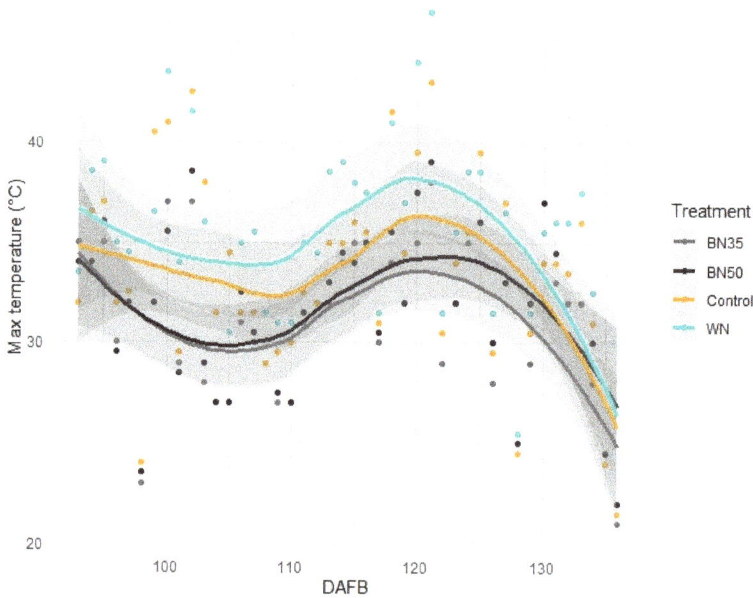

Figure 2. Evaluation of the maximum air temperature in the treatments under netting (BN35 = Black net 35, BN50 = Black net 50, WN = White net) and the Control treatment. DAFB = days after full bloom in season 4. Dots represent the maximum daily temperature. Lines and 0.95 confidence intervals were calculated according to the geom_smooth function (loess method in R).

FST showed differences in relation to treatment and fruit type (SBF, RF, GEF, GIF) (Figures 3 and 4). The highest FST were recorded in the PRO and Control treatments, with 35 and 43% of measurements above 46 °C, respectively. BN35 had 14% of the FST measurements above 46 °C, while WN and BN50 did not present temperatures above this threshold. SBF showed the highest dispersion, with a mean FST of 43 °C, higher than RF and GIF (41 °C) (Figure 3). The distribution of the intrafruit surface varied among treatments (data not shown). The standard deviation of temperature values was 3.1 for the Control treatment to 1.9 for BN50, an example of which is shown in Figure 4.

The transmittance of BN50 and WN in the PAR wavelength range was 0.40 (±0.03) and 0.70 (±0.03), respectively. In the near IR (between 700 and 800 nm), the behavior of both materials varied. The transmittance of WN increased to values close to 100%, whereas in the case of BN50 the transmittance decreased to values of around 0.35 (Figure 5).

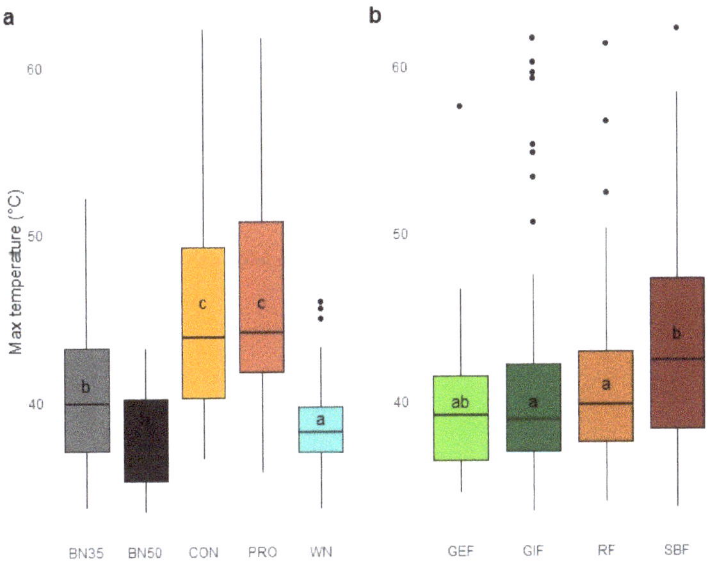

Figure 3. Maximum fruit surface temperature (FST) according to: (**a**). treatment (BN35 = Black net 35, BN50 = Black net 50, WN = White net, CON = control and PRO = protectants), and (**b**). fruit condition (RF = red fruit, GEF = green external fruit, GIF = green internal fruit, and SBF = sunburned fruit). The box represents interquartile range (IQR); the upper (Q3) and lower (Q1) quartiles are the ends. The line in the box is the median and the whiskers are the range of the data, excluding outliers. Outliers are values higher than Q3 + 1.5 IQR or smaller than Q1 − 1.5 IQR. Boxes with different letters indicate significant differences ($\alpha = 0.05$).

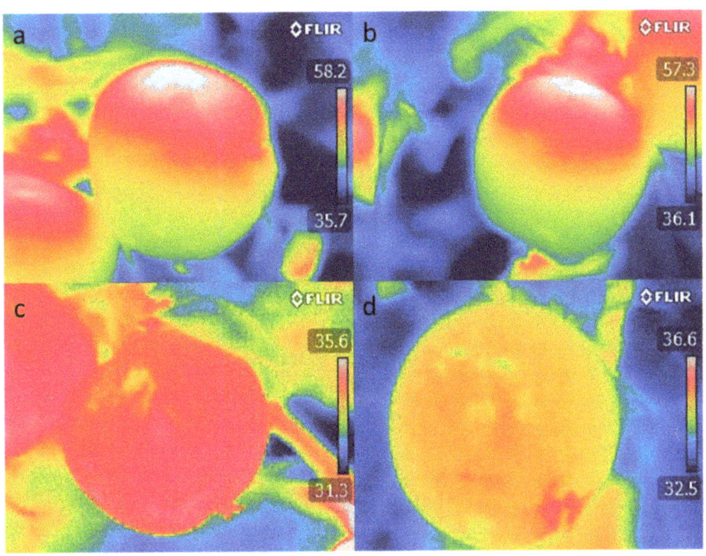

Figure 4. Examples of fruit temperature evaluated with an IR camera (FLIR E50). The scale represents the temperature (°C) in each image. (**a**,**b**) are fruits in the Control treatment (CON), while (**c**,**d**) are fruits in BN50 (Black net 50%).

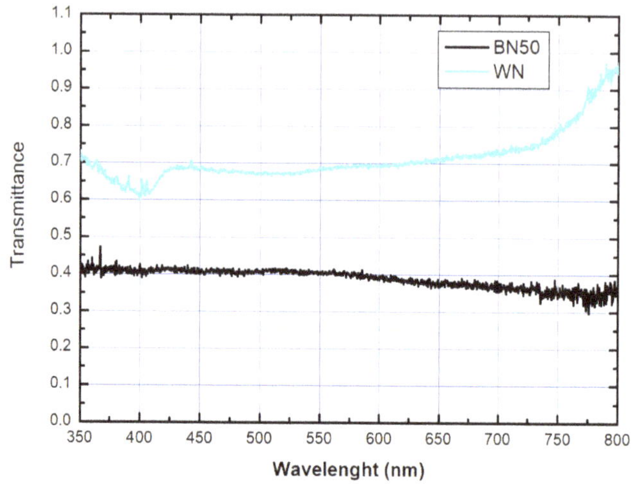

Figure 5. Transmitted radiation proportion between 350 to 800 nm for BN50 and WN treatments (Black Net 50% and White Net, respectively).

3.2. Sunburn

3.2.1. Sunburn Changes during the Season

The incidence of sunburn showed a different evolution according to the season, the initial condition of the fruit, and the protection treatment. In green internal fruits, the percentage of sunburn did not exceed 20%, while in sunburned fruits the final sunburn percentages ranged between 50 and 100% under all treatments in seasons 2 and 3 and under non-mesh treatments in seasons 1 and 4. Under netting treatments, the incidence of sunburn decreased drastically in seasons 1 and 4 in the first weeks of measurement and remained at low values as the season progressed. In green external fruits and red fruits, progressive increases were recorded during the evaluation period, with variations in the evolution between treatments and seasons. For red fruit, the maximum values of sunburn under netting (BN35, BN50, WN) were 75%. Values close to 100% were recorded for the PRO and Control treatments in season 2. For green external fruits, sunburn damage also increased as the season progressed, with results similar to those recorded for red fruits (Figure 6).

The incidence of final sunburn on marked fruits was significant for all the analyzed sources of variation (Figure 7). In the analysis according to season, seasons 2 and 3 showed the highest percentages of sunburned fruits, without reaching statistical differences with season 1 but doing so with season 4 (Figure 7a). In the analysis according to fruit type, Green internal fruits showed a median equal to 0% and lower than all other fruit types. Sunburned fruits and red fruits showed final sunburn values higher than 60% (65 and 62%, respectively) that did not differ statistically from those of green external fruits (Figure 7b). The treatments with the highest values of sunburned fruits were PRO and Control (64 and 50%, respectively), although pairwise comparisons did not reach significant differences (Figure 7c).

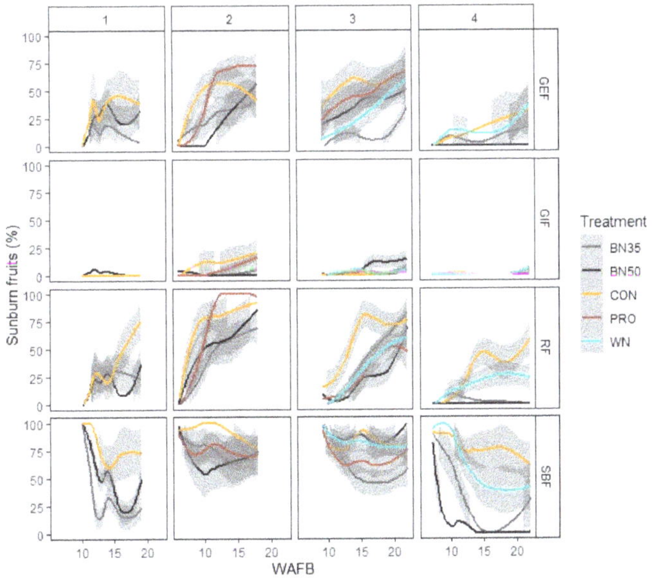

Figure 6. Evolution of the percentage of sunburned fruits by treatment (BN35 = Black net 35, BN50 = Black net 50, WN = White net, CON = control and PRO = protectants) according to weeks after full bloom (WAFB), fruit condition (RF = red fruit, GEF = green external fruit, GIF = green internal fruit, and SBF = sunburned fruit) and season (1 to 4). Lines and 0.95 confidence intervals were calculated according to geom_smooth function (loess method in R).

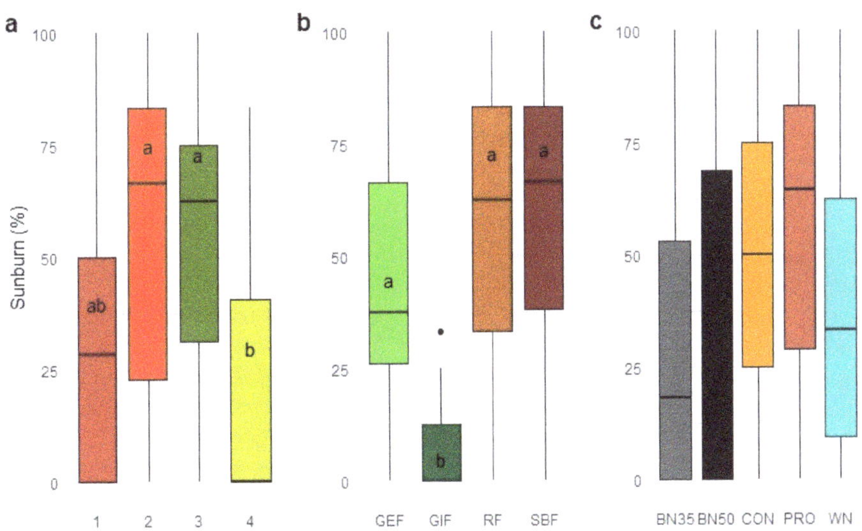

Figure 7. Incidence of final sunburn in evaluated fruits, according to: (**a**). season (1 to 4), (**b**). fruit condition (RF = red fruit, GEF = green external fruit, GIF = green internal fruit, and SBF = sunburned fruit), and (**c**). treatment (BN35 = Black net 35, BN50 = Black net 50, WN = White net, CON = control and PRO = protectants). The box represents interquartile range (IQR); the upper (Q3) and lower (Q1) quartiles are the ends. The line in the box is the median and the whiskers are the range of the data, excluding outliers. Outliers are values higher than Q3 + 1.5 IQR or smaller than Q1 − 1.5 IQR. Boxes with different letters indicate significant differences ($\alpha = 0.05$); the absence of letters indicate non-significant differences.

3.2.2. Sunburn at Harvest

The percentage of sunburned fruits out of the total harvested fruits showed differences between treatments, seasons, and their interaction (Figure 8a). The distribution between sunburn levels was significant only within treatments. The BN50 treatment showed the highest proportion of Mild and the lowest proportion of Mod + Sev, whereas PRO showed the highest proportion of Sev (Figure 8b).

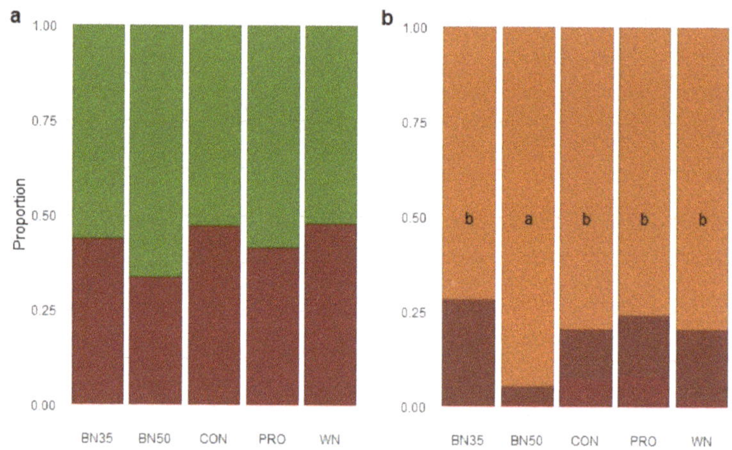

Figure 8. (**a**). Incidence of sunburn at harvest according to treatment (BN35 = Black net 35, BN50 = Black net 50, WN = White net, CON = control and PRO = protectants) SBF = sunburned fruit (brown) HF = healthy fruit (green); (**b**). severity of sunburn at harvest according to treatment (BN35 = Black net 35, BN50 = Black net 50, WN = White net, CON = control and PRO = protectants), the intensity of the color brown represents the intensity of the sunburn: Mild = slightly sunburned fruit (orange), Mod = moderately sunburned fruit (light brown), Sev = heavily sunburned fruit (brown). Bars with different letters indicate significant differences (α = 0.05), absence of letters indicate non-significant differences.

3.3. Fruit Growth and Sprouting

The effect of the treatments on final fruit size depended on the seasons. The largest difference found between the largest and smallest fruits was very small, being 2.5 mm in Control and WN fruits in season 4 (Table 1). The mean final fruit diameter was 61, 71, 68 and 66 mm for seasons 1 to 4, respectively (data not shown).

Table 1. Fruit size difference between smallest and largest size at harvest by season and treatment (BN35 = Black net 35, BN50 = Black net 50, WN = White net, CON = control and PRO = protectants).

	Size of Last Date Evaluated before Harvest			
	Largest Size Treatment	**Smallest Size Treatment**	**Maximum Difference**	
Season 1	BN35	CON	1.4 mm	*
Season 2	PRO	MN50	1.0 mm	Ns
Season 3	WN	BN50	1.5 mm	Ns
Season 4	CON	WN	2.5 mm	*

* Indicates significant difference (α = 0.05) between indicated treatments in the row, ns = not significant.

The seasons showed differences for all sprouting variables (length of shoots, percentage of reproductive shoots and percentage of spurs). The treatment effect was only significant for the length of shoots variable, whereas the interaction was not significant (Table 2). The range of variation for the percentage of reproductive shoots was between 24 and 77% in seasons 2 and 3, respectively. The growth of shoots in seasons 2 and 3 (15 and 17 cm, respectively) was 30% lower than in season 1. Average sprouting percentages of 48 and 57% were recorded for spurs and reproductive shoots, respectively, for all treatments.

Table 2. Effect of season (1, 2 and 3) and treatments (BN35 = Black net 35, BN50 = Black net 50, WN = White net and CON = control) on shoot length, percentage of reproductive structures (%reproductive), and percentage of short internode structures (%spurs) at next sprouting.

	Shoot Length		**%Reproductive ***		**%Spurs ***	
Season						
1	22.30	a	61.99	b	35.02	b
2	15.21	b	24.36	c	55.31	a
3	16.89	b	77.62	a	49.79	a
Treatment						
WN	17.65	b	55.75		50.35	
BN35	15.14	c	59.05		44.64	
BN50	22.01	a	65.56		49.29	
CON	19.11	a	48.76		45.98	

* Evaluated the following spring. Different letters next to values in each group indicate significant differences ($\alpha = 0.05$).

3.4. *Postharvest Evaluation*

3.4.1. Scald Damage

Since the antiscaldant treatment varied for seasons 2 and 3 (DPA and 1-MCP, respectively), the results of the postharvest evaluation were analyzed independently (Figures 9 and 10). In both seasons, the model was significant for field treatments (Control, PRO, BN35, BN50, and WN), fruit condition when entering cold storage (HF and Mild), storage time and postharvest treatment, as well as their interactions. Successive evaluations of all conditions recorded an increase in damage, both in incidence and severity. Fruits without evidence of sunburn (HF) before cold storage had higher proportions of fruits without scald (HF) during storage. Both antiscaldants had similar effects by increasing the proportion of HF and reducing moderate and heavily scalded fruits (MSF and HSF) (Figures 9 and 10).

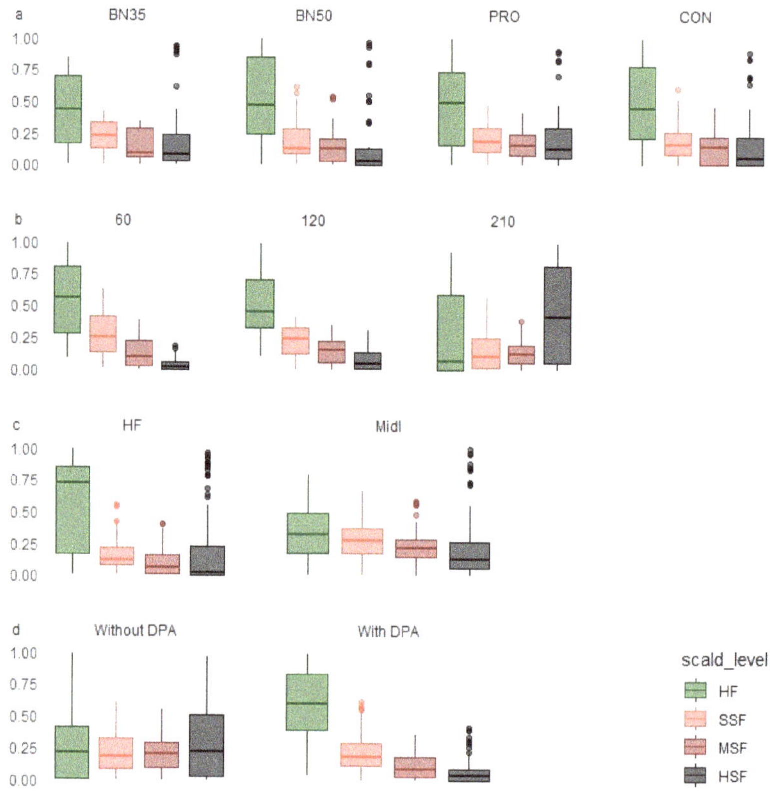

Figure 9. Proportion of fruits at each level of scald (HF healthy fruit, SSF = slightly scalded fruit, MSF = moderately scalded fruit, HSF = heavily scalded fruit) for season 2, according to: (**a**). field treatment (BN35 = Black net 35, BN50 = Black net 50, CON = control and PRO = protectants), (**b**). days of storage, (**c**). sun damage at harvest (HF, healthy fruit; Mild, light sunburn), and (**d**). post-harvest treatment (without DPA or with DPA). The box represents interquartile range (IQR); the upper (Q3) and lower (Q1) quartiles are the ends. The line in the box is the median and the whiskers are the range of the data, excluding outliers. Outliers are values higher than Q3 + 1.5 IQR or smaller than Q1 − 1.5 IQR.

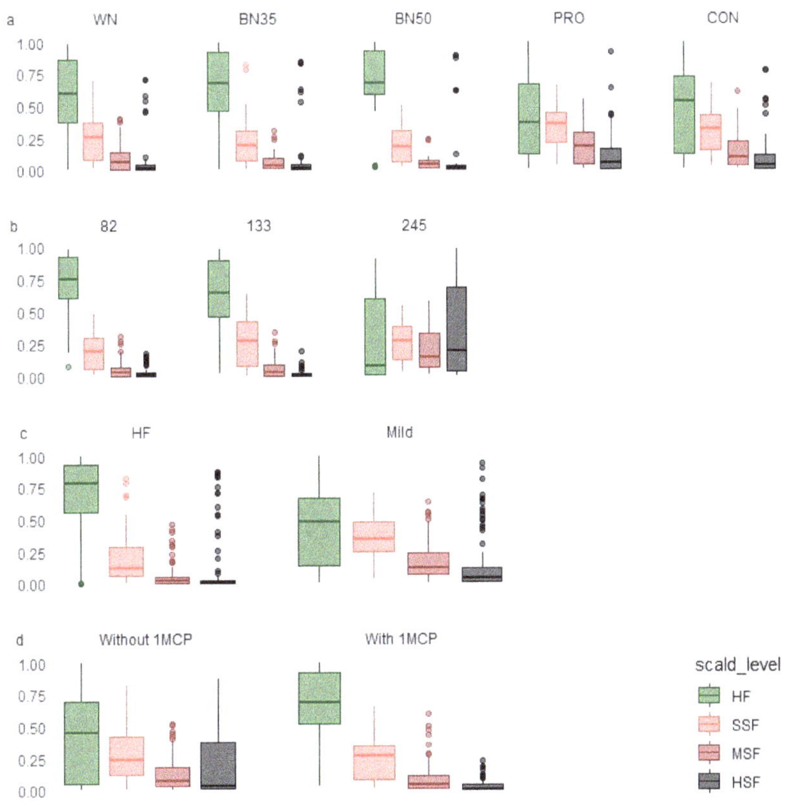

Figure 10. Proportion of fruits at each level of scald (HF healthy fruit, SSF = slightly scalded fruit, MSF = moderately scalded fruit, HSF = heavily scalded fruit) for season 3, according to: (**a**). field treatment (BN35 = Black net 35, BN50 = Black net 50, WN = White net, CON = control and PRO = protectants), (**b**). days of storage, (**c**). sun damage at harvest (HF healthy fruit, SIS light sunburn), and (**d**). post-harvest treatment (without 1-MCP or with 1-MCP). The box represents interquartile range (IQR); the upper (Q3) and lower (Q1) quartiles are the ends. The line in the box is the median and the whiskers are the rang of the data, excluding outliers. Outliers are values higher than Q3 + 1.5 IQR or smaller than Q1 − 1.5 IQR.

3.4.2. Sunscald Damage

An analysis of the interaction between field treatment and sunburn level on the evolution of scald was carried out on fruits treated with antiscaldants. In both seasons, HF had lower scald values than Mild, considering both the proportion of damaged fruits and the index of scalding. BN50 had the lowest scald values in 55% of the generated conditions, defined by evaluation method, season, fruit condition and field treatment. PRO had the lowest values in 66% of the conditions in both seasons (Table 3).

Table 3. Sunscald index (1 to 4) and proportion of sunscald according to season, sun damage condition at harvest (HF = healthy fruit, Mild = slightly sunburned fruit), field treatments (BN35 = Black net 35, BN50 = Black net 50, WN = White net, CON = control and PRO = protectants), and days of storage.

Season	**Harvest Sun Damage**	**Field Treatment**	**Sunscald Index**						**Proportion of Sunscald**					
			Days of Storage											
			60		**120**		**210**		**60**		**120**		**210**	
2	Mild	BN35	1.62	b	1.80	b	2.16	a	0.37	ns	0.47	ns	0.63	ns
		BN50	1.64	b	2.00	b	1.67	b	0.41		0.54		0.46	
		PRO	1.90	a	2.20	a	2.22	a	0.49		0.58		0.57	
		CON	1.81	ab	1.86	b	1.92	ab	0.49		0.50		0.55	
	HF	BN35	1.37	ns	1.39	a	1.40	ns	0.28	a	0.29	a	0.26	ns
		BN50	1.00		1.17	bc	1.20		0.00	d	0.14	b	0.14	
		PRO	1.30		1.30	ab	1.30		0.17	b	0.15	b	0.19	
		CON	1.14		1.10	c	1.30		0.10	c	0.06	c	0.20	
			82		133		245		82		133		245	
3	Mild	WN	1.47	b	1.74	b	1.79	c	0.36	b	0.54	c	0.55	c
		BN35	1.22	c	1.16	d	1.48	d	0.17	c	0.16	e	0.40	d
		BN50	1.48	b	1.48	c	1.63	cd	0.39	b	0.41	d	0.46	d
		PRO	1.93	a	2.26	a	2.39	a	0.61	a	0.87	a	0.87	a
		CON	1.58	b	1.80	b	2.07	b	0.40	b	0.67	b	0.71	b
	HF	WN	1.04	b	1.07	b	1.47	ab	0.04	b	0.07	b	0.28	b
		BN35	1.06	b	1.08	b	1.63	a	0.04	b	0.08	b	0.44	a
		BN50	1.02	b	1.11	b	1.36	b	0.02	b	0.10	b	0.32	b
		PRO	1.13	a	1.32	a	1.52	ab	0.14	a	0.29	a	0.42	a
		CON	1.06	b	1.24	a	1.40	ab	0.06	b	0.23	a	0.37	a

Different letters next to values in each group indicate significant differences ($\alpha = 0.05$). ns = not significant.

4. Discussion

4.1. Microclimate and Sun Damage

In the neotropical climatic conditions of our study, the increase and decrease in air temperature (AT) under WN and BN treatments, respectively, are in agreement with the values reported by other authors. The use of netting and its study differs as the climate of the region changes. In arid, semi-arid or Mediterranean climates (e.g., Washington State, main producing areas of Chile, Israel, South Africa), sunburn is a main objective, while in higher-latitude European conditions netting is mainly used as a protection against hail [11,21,37–39]. Climates with hot and humid summers, on the other hand, are not widely studied. The maximum AT exceeded 35 °C in all treatments, with WN and Control reaching 43 °C. These temperatures required homeostatic control mechanisms of cell metabolism, while also affecting fruit growth and pre- and postharvest quality [26,40,41]. Studies have shown that, under heat stress conditions, trees under photoselective nets had greater leaf-level photosynthetic light-use efficiency compared to the Control [19]. Photoinhibition at the peel chloroplasts and the consequent generation of oxygen free radicals seems to explain sunburn to a large extent [26]. As it has been reported, fruit temperature can be 17 °C higher than air temperature [18]; therefore, damage could even occur at an AT above 29 °C. In the recorded period, the WN treatment presented 2% more days with AT > 29 °C than the Control treatment, while the BN35 and BN50 treatments had 25 and 15% less days with that condition, respectively.

The effect of treatments on fruit surface temperature (FST) varied in relation to the observed effect on AT (Figures 3 and 4). All treatments under netting (WN, BN35 and BN50) reduced fruit temperature compared to treatments without netting (Control and PRO). The BN50 treatment had the highest homogeneity in FST, whereas the Control treatment had the lowest. The AT values recorded for BN35 and BN50 and their relationship with FST are in line with Gindaba and Wand [17], who report that the FST of apples under 20% black shade netting was between 5.4 and 9.7 °C lower on days with AT between 34 and 37 °C, while the average decrease in FST with AT between 30 and 32 °C was 5.6 °C.

Regarding the decrease in radiation in the netting treatments (Figure 5), the results are in accordance with those reported by Bastías and Corelli-Grappadelli [27], who state that black or white nets are generally neutral and reduce a similar radiation profile in the different wavelengths of the PAR range. The transmittance of the BN50 treatment in the 350 to 800 nm wavelength range showed a decrease in radiation values similar to those obtained by Dussi et al. [42] with a 55% black net for PAR. The WN treatment presented transmittance values 20% lower than those presented by Blanke [23] and similar to those presented by Bastías and Corelli-Grappadelli [27] for the PAR range. The behavior of WN in the 350 to 425 nm range differs from that recorded for BN50. The increases in transmittance observed in the values between 700 and 800 nm also differ from those recorded for BN50. The increase in infrared radiation (>740 nm) has a significant caloric contribution [43] and could be related to the increases in air temperature recorded in this treatment. Another element to consider in regard to the changes in radiation is the increase in diffuse radiation transmission, which can increase up to 170%, thus improving light penetration both vertically and horizontally in the canopy [38].

The onset of symptoms in the trial (Figure 1) started at 6 WAFB, when fruits had an average size of 40.5 mm, two weeks earlier than reported by Racsko and Schrader [18]. Between 10 and 15 WAFB, when the maximum rate of damage occurs in treatments without nets, fruit size was 42 to 62 mm (data not shown) (Figure 6). These data are in agreement with those reported by Racsko and Schrader [18], which shows that the occurrence of damage depends on the coupling of predisposing environmental conditions and a certain level of susceptibility with diameters of 45 mm at 7 to 8 WAFB. Although the most widely reported predisposing conditions are high temperatures and radiation [18], Severino et al. [10], for the same seasons and site, associate damage more with water availability than high temperatures.

Season has a significant effect on the incidence of sunburn, both in the assessments of marked fruits and in the harvest assessment (Figures 6–8). Our results show that two of the four studied seasons presented high percentages of sunburn (seasons 2 and 3), while seasons 1 and 4 presented less favorable climatic conditions for sun damage. These data confirm the high interannual variability in the region where the study was conducted [32]. The severity of damage also evidenced the variation among treatments, with the BN50 treatment showing the highest percentage of fruits without damage (without reaching statistical significance) and, among the damaged fruits, the highest percentage of Mild (Figure 8).

Sunburn protectant applications did not reduce neither the incidence (Figure 7) nor the severity of sunburn (Figure 8). The difficulties in achieving a permanent coverage with the product result in periods of time in which unacclimatized tissues, exposed to high FST and radiation conditions, suffer photooxidative stress that can lead to sun damage [44].

4.2. Fruit Growth and Sprouting

The effect of the treatments on fruit growth was of low magnitude and did not modify the commercial category (1.4 to 2.5 mm depending on the season). Similar effects are reported in apple crops under netting in a diverse range of climates [45] and in heat stress studies with temperatures of 29 °C and higher when nets are applied since 4 WAFB (the period of treatment application in this trial) [41]. In turn, all sprouting variables evaluated in the netting and Control treatments (shoot length, percentage of reproductive shoots and percentage of spurs) showed differences related to season. The effect of treatments was only significant for the shoot length variable (Table 2). Apparently, the reductions recorded in the total incident radiation of the treatments under nets did not constitute a limiting factor for net assimilation. Studies conducted in Brazil, with irradiance values similar to those recorded in Uruguay [46], conclude that nets seem to promote the optimization of light interception and carbohydrate partitioning [31]. The average irradiance for the 2006–2015 period in the three months of highest incidence of sun damage [46] is 55% higher in the location of our study and of the works of Bosco et al. [31], compared to the areas where adverse effects of the use of nets are recorded [23]. Changes in the incident light spectrum under netting were not significant enough to modify floral differentiation processes during the summer, nor did they affect bud break in the following season, in accordance with reports by Bastías and Corelli-Grappadelli [27].

4.3. Post-Harvest

Postharvest physiological disorders are determined by many factors. including maturity, nutrition, location of the fruit on the tree, and temperature to which it has been exposed [47]. The factors considered in our study (fruit condition when entering cold storage, field treatments, antiscaldant treatment and storage time) were significant in the model, both individually and for their interactions.

Both antiscaldant products significantly reduced scald damage, as reported in other studies [48]. Fruits entering cold storage without visible damage (HF) showed a lower range of damage than fruit with mild sun damage (Mild). The field treatments (Control, PRO, WN, BN35 and BN50) showed differences in scald and sunscald development for both classifications (HF and Mild) when entering cold storage (Figures 9 and 10).

In fruits with antiscaldant application (mainly expressing sunscald), the performance of treatments varied between seasons, with MN50 showing the best performance and PRO showing the worst (Table 3). Based on the assumption that the existing sun damage at harvest is neither reversible nor controllable, with the postharvest application of antiscaldant products [7,18,49] the results would indicate a different level of damage at harvest and a different sensitivity to sunscald expression depending on the field treatments. This difference in sensitivity could be explained by the variation produced by the nets in the composition and the antioxidant capacity of the fruit's skin [50]. The biochemical modifications triggered in the exposed and acclimatized fruits to withstand the high solar irradiance and high temperatures in the field [7], as well as the photoinhibition in the chloroplasts of the peels and the subsequent generation of oxygen free radicals [26], could induce the disorder later on.

5. Conclusions

The incidence of sunburn was influenced by the season under neotropical conditions. The shade netting reduces the sunburn and sunscald without affecting the growth processes dependent on leaf net assimilation. Protectants, on the other hand, are not effective and can increase sunscald incidence on the fruit during storage. The fruit position within the tree and the initial condition were the main factors affecting sun damage. However, black netting or favorable season conditions can reduce (i.e., reverse) the earlier sunburn symptoms. This shows the need for more studies on the relation between microclimate and physiological conditions at different hierarchical levels (fruit tissue, organ, branch and tree).

Author Contributions: Conceptualization, V.S., M.A.-S., J.A.Y. and J.G.-T.; methodology, V.S., M.A.-S., E.F., J.A.Y. and J.G.-T.; formal analysis, V.S. and E.F.; writing—original draft preparation, V.S.; writing—review and editing, V.S., M.A.-S., S.D., E.F., J.A.Y. and J.G.-T.; visualization, V.S.; supervision, S.D., J.A.Y. and J.G.-T.; project administration, V.S. and M.A.-S.; funding acquisition, V.S., M.A.-S., J.A.Y. and J.G.-T. All authors have read and agreed to the published version of the manuscript.

Funding: Research funded by Agencia Nacional de Investigación e Innovación (RTS_X_2011_22868).

Institutional Review Board Statement: Not applicable.

Informed Consent Statement: Not applicable.

Acknowledgments: We thank Frutisur for their contribution to the funding of this work and for allowing us to use their crops, INIA (Instituto Nacional de Investigación Agropecuaria) and its members Alicia Feippe, Roberto Zoppolo and Danilo Cabrera for their participation in the research project, and CasaAmérica for their contribution to the materials used and their facilities. We also thank Marcia García, Sebastián Galiger and José Luis Álvarez for their contribution to the field and laboratory work.

Conflicts of Interest: The authors declare no conflict of interest.

References

1. Schrader, L.E.; Zhang, J.; Duplaga, W.K. Two Types of Sunburn in Apple Caused by High Fruit Surface (Peel) Temperature. *Plant Health Prog.* **2001**, *2*, 3. [CrossRef]
2. Yuri, J.A.; Torres, C.; Bastías, R.; Neira, A. Golpe de Sol En Manzanas. II. Factores Inductores y Respuestas Bioquímicas. *Agro-Ciencia* **2000**, *16*, 23–32.
3. Piskolczi, M.; Varga, C.; Racskó, J. The Meteorological Causes of Sunburn Injury on the Surface of Apple Fruit (*Malus domestica* Borkh.). *J. Fruit Ornam. Plant Res.* **2004**, *12*, 245–252.
4. Wünsche, J.N.; Greer, D.H.; Laing, W.A.; Palmer, J.W. Physiological and Biochemical Leaf and Tree Responses to Crop Load in Apple. *Tree Physiol.* **2005**, *25*, 1253–1263. [CrossRef] [PubMed]
5. Felicetti, D.A.; Schrader, L.E. Photooxidative Sunburn of Apples; Characterization of a Third Type of Apple Sunburn. *Int. J. Fruit Sci.* **2008**, *8*, 160–172. [CrossRef]
6. Yuri, J.A.; Neira, A.; Quilodran, A.; Razmilic, I.; Motomura, Y.; Torres, C.; Palomo, I. Sunburn on Apples Is Associated with Increases in Phenolic Compounds and Antioxidant Activity as a Function of the Cultivar and Areas of the Fruit. *J. Food Agric. Environ.* **2010**, *8*, 920–925.
7. Hernandez, O.; Torres, C.A.; Moya-León, M.A.; Opazo, M.C.; Razmilic, I. Roles of the Ascorbate—Glutathione Cycle, Pigments and Phenolics in Postharvest 'Sunscald' Development on 'Granny Smith' Apples (*Malus domestica* Borkh.). *Postharvest Biol. Technol.* **2014**, *87*, 79–87. [CrossRef]
8. Schrader, L.; Zhang, J.; Sun, J. Environmental Stresses That Cause Sunburn of Apple. *Acta Hortic.* **2003**, *618*, 397–405. [CrossRef]
9. Meheriuk, M.; Prange, R.K.; Lidster, P.D.; Porritt, S.W. *Postharvest Disorders of Apples and Pears*; Agriculture and Agri-Food: Ottawa, ON, Canada, 1994; ISBN 0662212371.
10. Severino, V.; Arias-Sibillotte, M.; Dogliotti, S.; Frins, E.; Gonzalez-Talice, J.; Yuri, J.A. Climatic and Physiological Parameters Related to the Progress and Prediction of Apple Sunburn Damage in a Neotropical Climate. *Adv. Hortic. Sci.* **2020**, *34*, 431–440. [CrossRef]
11. Szabó, A.; Tamás, J.; Nagy, A. The Influence of Hail Net on the Water Balance and Leaf Pigment Content of Apple Orchards. *Sci. Hortic.* **2021**, *283*, 110112. [CrossRef]
12. Torres, C.A.; León, L.; Sánchez-Contreras, J. Spectral Fingerprints during Sun Injury Development on the Tree in Granny Smith Apples: A Potential Non-Destructive Prediction Tool during the Growing Season. *Sci. Hortic.* **2016**, *209*, 165–172. [CrossRef]
13. Torres, C.A.; Sepúlveda, A.; Leon, L.; Yuri, J.A. Early Detection of Sun Injury on Apples (*Malus domestica* Borkh.) through the Use of Crop Water Stress Index and Chlorophyll Fluorescence. *Sci. Hortic.* **2016**, *211*, 336–342. [CrossRef]
14. Musacchi, S.; Serra, S. Apple Fruit Quality: Overview on Pre-Harvest Factors. *Sci. Hortic.* **2018**, *234*, 409–430. [CrossRef]
15. Parchomchuk, P.; Meheriuk, M. Orchard Cooling with Pulsed Overtree Irrigation to Prevent Solar Injury and Improve Fruit Quality of "Jonagold" Apples. *HortScience* **1996**, *31*, 802–804. [CrossRef]
16. Manja, K.; Aoun, M. The Use of Nets for Tree Fruit Crops and Their Impact on the Production: A Review. *Sci. Hortic.* **2019**, *246*, 110–122. [CrossRef]
17. Gindaba, J.; Wand, S.J.E. Comparative Effects of Evaporative Cooling, Kaolin Particle Film, and Shade Net on Sunburn and Fruit Quality in Apples. *HortScience* **2005**, *40*, 592–596. [CrossRef]
18. Racsko, J.; Schrader, L.E. Sunburn of Apple Fruit: Historical Background, Recent Advances and Future Perspectives. *Crit. Rev. Plant Sci.* **2012**, *31*, 455–504. [CrossRef]
19. Mupambi, G.; Musacchi, S.; Serra, S.; Kalcsits, L.A.; Layne, D.R.; Schmidt, T. Protective Netting Improves Leaf-Level Photosynthetic Light Use Efficiency in 'Honeycrisp' Apple under Heat Stress. *HortScience* **2018**, *53*, 1416–1422. [CrossRef]
20. Kalcsits, L.; Musacchi, S.; Layne, D.R.; Schmidt, T.; Mupambi, G.; Serra, S.; Mendoza, M.; Asteggiano, L. Above and Below-Ground Environmental Changes Associated with the Use of Photoselective Protective Netting to Reduce Sunburn in Apple. *Agric. For. Meteorol.* **2017**, *237–238*, 9–17. [CrossRef]
21. Tanny, J.; Cohen, S.; Grava, A.; Naor, A.; Lukyanov, V. The Effect of Shading Screens on Microclimate of Apple Orchards. *Acta Hortic.* **2009**, *807*, 103–108. [CrossRef]
22. McCaskill, M.R.; McClymont, L.; Goodwin, I.; Green, S.; Partington, D.L. How Hail Netting Reduces Apple Fruit Surface Temperature: A Microclimate and Modelling Study. *Agric. For. Meteorol.* **2016**, *226–227*, 148–160. [CrossRef]
23. Blanke, M.M. The Structure of Coloured Hail Nets Affects Light Transmission, Light Spectrum, Phytochrome and Apple Fruit Colouration. *Acta Hortic.* **2009**, *817*, 177–184. [CrossRef]
24. Solomakhin, A.; Blanke, M. The Microclimate under Coloured Hailnets Affects Leaf and Fruit Temperature, Leaf Anatomy, Vegetative and Reproductive Growth as Well as Fruit Colouration in Apple. *Ann. Appl. Biol.* **2010**, *156*, 121–136. [CrossRef]
25. Bastías, R.M.; Manfrini, L.; Corelli Grappadelli, L. Exploring the Potential Use of Photo-Selective Nets for Fruit Growth Regulation in Apple. *Chil. J. Agric. Res.* **2012**, *72*, 224–231. [CrossRef]
26. Naschitz, S.; Naor, A.; Sax, Y.; Shahak, Y.; Rabinowitch, H.D. Photo-Oxidative Sunscald of Apple: Effects of Temperature and Light on Fruit Peel Photoinhibition, Bleaching and Short-Term Tolerance Acquisition. *Sci. Hortic.* **2015**, *197*, 5–16. [CrossRef]

27. Bastías, R.M.; Corelli-Grappadelli, L. Light Quality Management in Fruit Orchards: Physiological and Technological Aspects. *Chil. J. Agric. Res.* **2012**, *72*, 574–582. [CrossRef]
28. Bosco, L.C.; Bergamaschi, H.; Marodin, G.A.B. Solar Radiation Effects on Growth, Anatomy, and Physiology of Apple Trees in a Temperate Climate of Brazil. *Int. J. Biometeorol.* **2020**, *64*, 1969–1980. [CrossRef]
29. Schrader, L.E. Scientific Basis of a Unique Formulation for Reducing Sunburn of Fruits. *HortScience* **2011**, *46*, 6–11. [CrossRef]
30. Reig, G.; Donahue, D.J.; Jentsch, P. The Efficacy of Four Sunburn Mitigation Strategies and Their Effects on Yield, Fruit Quality, and Economic Performance of Honeycrisp Cv. Apples under Eastern New York (USA) Climatic Conditions. *Int. J. Fruit Sci.* **2020**, *3*, 541–561. [CrossRef]
31. Bosco, L.C.; Bergamaschi, H.; Cardoso, L.S.; Paula, V.A. de Microclimate Alterations Caused by Agricultural Hail Net Coverage and Effects on Apple Tree Yield in Subtropical Climate of Southern Brazil. *Bragantia Camp.* **2018**, *77*, 181–192. [CrossRef]
32. Bernardi, R.E.; Holmgren, M.; Arim, M.; Scheffer, M. Why Are Forests so Scarce in Subtropical South America? The Shaping Roles of Climate, Fire and Livestock. *For. Ecol. Manag.* **2016**, *363*, 212–217. [CrossRef]
33. Zibordi, M.; Domingos, S.; Grappadelli, L.C. Thinning Apples via Shading: An Appraisal under Field Conditions. *J. Hortic. Sci. Biotechnol.* **2009**, *84*, 138–144. [CrossRef]
34. Archbold, D.D.; Nosarzewski, M.; Wu, B.; Vuppalapati, P. Does Availability of Soluble Carbohydrate Reserves Determine Apple Fruit Set? Available online: https://www.actahort.org/books/903/903_110.htm (accessed on 2 August 2021).
35. Corelli-Grappadelli, L.; Lakso, A.N. Fruit Development in Deciduous Tree Crops as Affected by Physiological Factors and Environmental Conditions. Available online: https://www.actahort.org/books/636/636_52.htm (accessed on 2 August 2021).
36. Torres, C.A.; Sepulveda, A.; Gonzalez-Talice, J.; Yuri, J.A.; Razmilic, I. Fruit Water Relations and Osmoregulation on Apples (*Malus domestica* Borkh.) with Different Sun Exposures and Sun-Injury Levels on the Tree. *Sci. Hortic.* **2013**, *161*, 143–152. [CrossRef]
37. Hunsche, M.; Blanke, M.M.; Noga, G. Does the Microclimate under Hail Nets Influence Micromorphological Characteristics of Apple Leaves and Cuticles? *J. Plant Physiol.* **2010**, *167*, 974–980. [CrossRef]
38. Mupambi, G.; Anthony, B.M.; Layne, D.R.; Musacchi, S.; Serra, S.; Schmidt, T.; Kalcsits, L.A. The Influence of Protective Netting on Tree Physiology and Fruit Quality of Apple: A Review. *Sci. Hortic.* **2018**, *236*, 60–72. [CrossRef]
39. Mupambi, G.; Schmeisser, M.; Dzikiti, S.; Reynolds, S.; Steyn, W.J. Ineffectiveness of Foliar S-ABA Application as an Apple Sunburn Suppressant Explained through Effects on Peel Biochemistry and Leaf Ecophysiology. *Sci. Hortic.* **2018**, *232*, 256–263. [CrossRef]
40. Woolf, A.B.; Ferguson, I.B. Postharvest Responses to High Fruit Temperatures in the Field. *Postharvest Biol. Technol.* **2000**, *21*, 7–20. [CrossRef]
41. Flaishman, M.A.; Peles, Y.; Dahan, Y.; Milo-Cochavi, S.; Frieman, A.; Naor, A. Differential Response of Cell-Cycle and Cell-Expansion Regulators to Heat Stress in Apple (*Malus domestica*) Fruitlets. *Plant Sci.* **2015**, *233*, 82–94. [CrossRef] [PubMed]
42. Dussi, M.C.; Giardina, G.; Sosa, D.; González Junyent, R.; Zecca, A.; Reeb, P.R. Shade Nets Effect on Canopy Light Distribution and Quality of Fruit and Spur Leaf on Apple Cv. Fuji. *Span. J. Agric. Res.* **2005**, *3*, 253–260. [CrossRef]
43. Yuri, J.A. Daño Por Sol En Manzanas. *Rev. Frutic.* **2010**, *8*, 2–9.
44. Glenn, D.M.; Yuri, J.A. Photosynthetically Active Radiation (PAR) × ultraviolet Radiation (UV) Interact to Initiate Solar Injury in Apple. *Sci. Hortic.* **2013**, *162*, 117–124. [CrossRef]
45. Bosančić, B.; Mićić, N.; Blanke, M.; Pecina, M. A Main Effects Meta Principal Components Analysis of Netting Effects on Fruit: Using Apple as a Model Crop. *Plant Growth Regul.* **2018**, *86*, 455–464. [CrossRef]
46. PVGIS-SARAH © European Communities, 2001–2017. Available online: https://ec.europa.eu/jrc/en/PVGIS/downloads/SARAH (accessed on 2 August 2021).
47. Ferguson, I.; Volz, R.; Woolf, A. Preharvest Factors Affecting Physiological Disorders of Fruit. *Postharvest Biol. Technol.* **1999**, *15*, 255–262. [CrossRef]
48. Niu, J.P.; Hou, Z.; Ou, Z.; Hui, W. Comparative Study of Effects of Resveratrol, 1-MCP and DPA Treatments on Postharvest Quality and Superficial Scald of 'starkrimson' Apples. *Sci. Hortic.* **2018**, *240*, 516–521. [CrossRef]
49. Contreras, C.; Zoffoli, J.P.; Alcalde, J.A.; Ayala, M. Evolución Del Daño Por Insolación de Manzanas "Granny Smith" Durante El Almacenaje Refrigerado. *Cienc. Investig. Agrar.* **2008**, *35*, 147–157. [CrossRef]
50. Olivares-Soto, H.; Bastías, R.M.; Calderón-Orellana, A.; López, M.D. Sunburn Control by Nets Differentially Affects the Antioxidant Properties of Fruit Peel in 'Gala' and 'Fuji' Apples. *Hortic. Environ. Biotechnol.* **2020**, *61*, 241–254. [CrossRef]

agronomy

MDPI

Article

Growth and Antioxidant Responses of Lettuce (*Lactuca sativa* L.) to Arbuscular Mycorrhiza Inoculation and Seaweed Extract Foliar Application

Farzad Rasouli [1,*], Trifa Amini [1], Mohammad Asadi [2], Mohammad Bagher Hassanpouraghdam [1], Mohammad Ali Aazami [1], Sezai Ercisli [3], Sona Skrovankova [4,*] and Jiri Mlcek [4]

[1] Department of Horticulture, Faculty of Agriculture, University of Maragheh, Maragheh 5518183111, Iran; afirt7903@gmail.com (T.A.); hassanpouraghdam@gmail.com (M.B.H.); aazami58@gmail.com (M.A.A.)
[2] Department of Plant Production and Genetics, Faculty of Agriculture, University of Maragheh, Maragheh 5518183111, Iran; mohammadasadi704@gmail.com
[3] Department of Horticulture, Faculty of Agriculture, Ataturk University, Erzurum 25240, Turkey; sercisli@gmail.com
[4] Department of Food Analysis and Chemistry, Faculty of Technology, Tomas Bata University in Zlín, Vavreckova 275, 76001 Zlin, Czech Republic; mlcek@utb.cz
* Correspondence: farzad.rasouli@maragheh.ac.ir (F.R.); skrovankova@utb.cz (S.S.); Tel.: +98-9141844267 (F.R.); +420-576031524 (S.S.)

Abstract: Biofertilizers, such as arbuscular mycorrhiza fungi (AMF) and seaweed extract (SWE), have been effective in environmental and agricultural ecosystems. In this study, the effects of AMF, SWE, and their co-application were assayed on the growth and antioxidant potential of lettuce plants. The experiment was conducted as a factorial based on a completely randomized design with two factors and four replications under greenhouse conditions. The first factor was AMF (*Glomus mosseae*) at two levels consisting of AMF application (20 g pot^{-1}), and without using AMF; and the second factor was SWE foliar spraying (*Ascophyllum nodosum*) at 0.5, 1.5 and 3 g L^{-1} concentration. The results revealed that the highest root colonization (85%) belonged to AMF and SWE (3 g L^{-1}) × AMF; the lowest colonization rate (65%) was observed for AMF × SWE (0.5 g L^{-1}) treatment. The highest growth parameters (leaf number, shoot and root fresh weight, head diameter), biochemical traits (total soluble proteins, carbohydrates content) and TAA, total antioxidant activity by FRAP method and ascorbic acid, total phenolics, and flavonoids content were obtained with the co-applications. Therefore, the best results of the evaluated traits were achieved with AMF × SWE (3 g L^{-1}). The TAA value was increased three-fold compared to the control. Total phenolics and flavonoids content were 2.24 and 6.59 times higher than the control, respectively. On the other hand, leaf dry weight was decreased with the further growth of the plants. Overall, the co-application of AMF with SWE can be recommended to producers as an alternative and environment-friendly strategy to improve the qualitative and quantitative traits of the lettuce crop.

Keywords: biofertilizer; *Glomus mosseae*; colonization; biostimulant; FRAP

Citation: Rasouli, F.; Amini, T.; Asadi, M.; Hassanpouraghdam, M.B.; Aazami, M.A.; Ercisli, S.; Skrovankova, S.; Mlcek, J. Growth and Antioxidant Responses of Lettuce (*Lactuca sativa* L.) to Arbuscular Mycorrhiza Inoculation and Seaweed Extract Foliar Application. *Agronomy* **2022**, *12*, 401. https://doi.org/10.3390/agronomy12020401

Academic Editor: Pilar Soengas

Received: 13 December 2021
Accepted: 24 January 2022
Published: 5 February 2022

1. Introduction

Lettuce (*Lactuca sativa* L.) belongs to the Asteraceae family and is rich in fiber, vitamins, minerals, and phenolic compounds [1]. In 2019, lettuce production reached 29,134,653 tons in the world from which about 547,590 tons were produced in Iran [2]. Considering the important and valuable nutritional role of lettuce due to its daily consumption and, on the other hand, the excessive application of chemical fertilizers with negative impacts on human health and the environment; there is an effort to find alternative methods for reducing the chemical fertilizers input in lettuce production areas.

The use of chemical fertilizers is widespread throughout the world, leading to soil degradation and environmental pollution. Therefore, the global approach to establishing a sustainable agricultural system has changed with the employment of new management methods. Considering this, it is important to pay attention to the biological and integrated systems, especially biofertilizers, to meet the plant nutritional requirements and reduce the consumption of chemical fertilizers [3]. Bio-fertilizers (bio-stimulants) release their content to make it slowly available to the plants and simultaneously improve soil quality [4]. Therefore, the use of bio-fertilizers has advantages such as removing toxic substances and improving the physicochemical properties of soil [5].

Arbuscular mycorrhiza fungi (AMF) and seaweed extract (SWE) can be useful because they are organic, environmentally friend, cost-effective, and also a rich source of macro-and microelements, vitamins, pro-enzymes, and growth regulators and so play an important role in the sustainable soil fertility [6].

SWEs are a complex mixture of hormones, amino acids, proteins, sugars, lipids, vitamins, humic substances, and phenolic compounds. The organic features of SWEs and their physiological effects have led to their widespread use in the food and pharmaceutical industries [7]. Moreover, SWE contains carbohydrates, organic compounds, and high amounts of nitrogen, phosphorus, potassium, and other minerals that improve soil properties and are easily absorbed by the plants. Thereby, those extracts improve plant growth and the antioxidants pool by activating the respiratory cycle, photosynthesis, and delaying plant aging [8]. Di Mola et al. [9] reported that SWE increased the growth and yield in lettuce and significantly improved fresh and dry weight, stomatal conductance, potassium content, and total antioxidant activity [10]. Jung and Kim [11] reported increases in plant height, chlorophylls and carotenoids content, and total antioxidant activity in lettuce using SWE.

AMF is another natural fertilizer source used in organic farming systems [12]. AMF is one of the biological materials of arable soils that are related to the roots of 90% of plants and improve the effective root area and the ability of P uptake from immobile sources due to phosphatase activity and the insoluble phosphate solubilizing organic compounds release [13]. These microorganisms play an important role in plant nutrition, especially in the soils without humus and poor in P, N and other nutrients so that they can make unabsorbed and inaccessible P available to the plants in the growing medium [14]. In general, AMF symbiosis can play a key role in maintaining soil fertility and stabilizing soil structure when increasing plant water uptake, yield, and quality [15]. There are many reports on various effects of AMF and SWE on plants. Saia et al. [16] showed that AMF increased the content of various phenolic compounds and P, Mg, Fe, Mn, and Zn in lettuce. Tarraf et al. [17] stated that the symbiosis of AMF promoted growth, yield, and dry matter of maize and increased the concentration of P in the shoot. A wide range of studies has shown that AMF is effective in the biosynthesis and accumulation of plant secondary metabolites by modifying the polymorphisms and by stimulating the biosynthesis of polyphenols and can increase the activity of plant enzymes [18]. Moreover, AMF colony formation increases photosynthetic efficiency in plants [15]. Thus, bio-fertilizers can improve the quantity and quality of crops, especially vegetables.

Several studies reported the influence of AMF on lettuce and other vegetables' growth and antioxidant systems [19–21], and others focused on SWE to improve lettuce yield and quality [9]. However, there are no results in the literature on the application of AMF in combination with SWE to improve the growth, biochemical and antioxidant attributes of lettuce. In this context, the present research aimed to evaluate the effects of a SWE (*Aschophyllum nodosum*) and an AMF strain (*Glomus moseae*) and their interaction on the growth, yield, and antioxidant activity of lettuce under greenhouse conditions. The main question was: does the AMF inoculation combined with SWE improve the growth, yield, and antioxidant activity of lettuce?

2. Materials and Methods

This study was conducted in the research greenhouse of the Department of Horticultural Sciences, the University of Maragheh in East Azarbaijan province, Iran, with geographical coordinates of 37° and 23′ north latitude and 46° and 16′ east longitudes and

1485 m above sea level. The temperature regime for the night:day was 18:24 °C, and the relative humidity was around 65–75%. The experiment was arranged as factorial based on a completely randomized design with four replications.

Experimental treatments were: foliar application of SWE at four levels (0.5, 1.5 and 3 g L^{-1}, presented as SWE1, SWE2 and SWE3), AMF inoculation (20 g pot^{-1}), AMF inoculation plus the foliar application of SWE at three levels and control that consisted without AMF × 0 g L^{-1} SWE (control), AMF × 0.5 g L^{-1} SWE, AMF × 1.5 g L^{-1} SWE and AMF × 3 g L^{-1} SWE. Before sowing the seeds, the soil was sterilized to remove the soilborne fungi by autoclaving for 60 min at 121 °C under 1.2 atmospheric pressure. Lettuce seeds were planted in a culture tray containing coco-peat and perlite. The seedlings were transferred at two fully developed leaves stage to 5-L pots. In AMF (inoculation with *Glomus mossae*) treatments 20 g of the autoclaved soil containing mycorrhiza fungal hyphae were added to each pot containing 5 kg soil at transplanting time. The soil was sandy clay loam with pH of 8.16, 1.23% organic carbon, 0.09% total N, 11.05, 570.85, 1.16 and 1.02 mg. kg^{-1} of available P, K, Zn and Fe, respectively. The plants were watered every 3 days and at the beginning of growth, 100 mL, and then 2 weeks later, 500 mL of tap water was given to each pot. Then, foliar application of different concentrations of SWE was started at the six-leaf stage and continued four times at weekly intervals, such that at the first stage 50 mL per plant of SWE solution was sprayed and at the other three stages 100 mL per plant was applied. Control plants were planted in the soil without AMF inoculation, sprayed with distilled water, and irrigated with tap water in the same manner until the harvest.

2.1. Morphological Traits

Plant morphological traits including head diameter, fresh weight of the head, dry weight of leaf, fresh and dry weight of root, and the number of leaves per head were recorded at harvest time. Head diameter was measured using a digital caliper. The weight of the head and root was measured separately by an analytical scale (A&D weighing Japan) (with an accuracy of 0.01 mg). Plant samples were dried in the oven at 75 °C for 48 h and the dry weight of root and leaves were recorded correspondingly.

2.2. Chlorophyll Index Determination

Chlorophyll content expressed as chlorophyll index (SPAD index) was determined in the fully expanded youngest leaves of lettuce using a portable chlorophyll meter (Instruments SPAD-502, Osaka, Japan).

2.3. Root Colonization

After head harvest, the fresh roots of lettuce were taken from the soil and rinsed with tap water to remove the residual soil particles. The root samples were divided into small segments (1 cm) and cleared in hot KOH solution (10%, v/v) for 10 min. The segments were washed with distilled water and then acidified with HCl (2%, v/v) at 25 °C for 20 min, and stained with trypan blue (0.05%) in lactic acid (80%, v/v) for 12 h [22,23]. Finally, the samples were washed with distilled water and stored in a solution containing water, glycerol, and lactic acid (1:1:1, $v/v/v$) [24]. The stained segments were identified and evaluated by an Olympus microscope (BH-2). The organs and hyphae of the fungus, which appeared blue, were recorded as high-quality photos (Figure 1). The percentage of colonization was calculated by the gridline intersection method based on Giovannetti and Mosse [25] so that, for each experimental treatment, the stained roots were cut into 1 cm pieces and randomly placed in a glass plate.

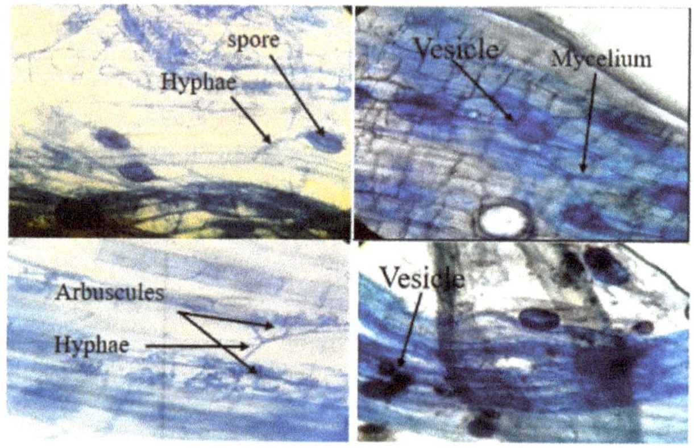

Figure 1. Microscopic images of the stained fragments of lettuce roots to detect the colonization of arbuscular mycorrhiza (*Glomus mosseae*).

2.4. Total Protein Content

Total protein content was recorded using the Bradford method [26]. A 5-fold Coomassie Brilliant Blue (CBB) stock solution was made as a Bradford reagent. We blended 50 mg of CBB with 25 mL of methanol and 50 mL of orthophosphoric acid in a dark bottle and stored this at −24 °C. Different concentrations (0.002−0.01 mg mL^{-1}) of bovine serum albumin were used as standard solutions. Fresh lettuce leaves (1 g) were powdered in 4 mL of 50 mM phosphate buffer solution and then centrifuged at 10,000 rpm at 4 °C for 10 min. The supernatant was used for the total protein content assay. Protein extract (50 µL) was added to 1000 µL of Bradford reagent. The formation of blue color was assessed at the wavelength of 595 nm using the UV−VIS spectrophotometer (Spekol 1500, Analytik Jena, Jena, Germany). Total protein content was calculated as mg g^{-1} FW (fresh weight).

2.5. Total Carbohydrate Content

To measure total carbohydrate content [27], a sample of fresh lettuce leaves (0.2 g) was extracted with 10 mL of 95% ethanol for 1 h in a water bath at 80 °C, and then centrifuged at 12,000 rpm for 10 min. The supernatant was taken and then 1 mL of 0.5% phenol and 5 mL of 98% sulfuric acid were added. The absorption was detected at 483 nm with a spectrophotometer (Spekol 1500, Analytik Jena, Jena, Germany). Total carbohydrates content was recorded as mg. g^{-1} FW.

2.6. Total Antioxidant Activity (TAA)

Total antioxidant activity (TAA) was measured by ferric reducing power (FRAP) assay [28]. The reagents included acetate buffer (300 mM, pH 3.6), TPTZ (2,4,6-tripyridyl-s-triazine, 10 mM) solution in HCl (40 mM) and $FeC1_3.6H_2O$ (20 mM) solutions. The fresh working solution was prepared by mixing these reagents in the volume ratio of 10:1:1. methanolic extract of the sample (100 mL) was mixed with 3 mL of working FRAP reagent. Samples were then incubated at 37 °C in a water bath, and absorbance was recorded after 15 min at 593 nm with a UV−VIS spectrophotometer (Spekol 1500, Analytik Jena, Jena, Germany). L-ascorbic acid was used as a standard solution (100 mM–1000 mM). TAA was calculated as the inactivation of FRAP (%).

2.7. Ascorbic Acid Content

One gram of the leaf lettuce sample was completely digested with a 5% solution of metaphosphoric acid. The extracted sample was centrifuged at 8000 rpm for 20 min at 4 °C. The supernatant was used to measure total ascorbic acid content [29]. To 4 mL of the centrifuged DCIP solution (2,6-dichlorophenolindophenol, 3 mM), 0.5 mL was added to oxidize ascorbic acid to dehydroascorbic acid. The absorbance was measured using a UV-VIS spectrophotometer (Spekol 1500, Analytik Jena, Jena, Germany) at 520 nm. Ascorbic acid content was presented as mg. 100 g^{-1} FW.

2.8. Total Phenolics Content

The Folin–Ciocalteu method by Singleton et al. [30], with gallic acid as a standard was applied to measure total phenolics content. Plant extract (20 µL of 1% acidic methanol solution) and 100 µL of Folin–Ciocalteu reagent (10%) were mixed with 1.59 mL of distilled water and kept for 10 min in the dark. Sodium carbonate (7.5%, 2 mL) was then added into the mixture and placed for 2 h in the dark condition. The absorbance was measured at 765 nm with a UV−VIS spectrophotometer (Spekol 1500, Analytik Jena, Jena, Germany). The content was expressed, using the gallic acid calibration curve, as mg of gallic acid equivalents per 100 g of fresh weight (mg GAE 100 g^{-1} FW).

2.9. Total Flavonoids Content

The aluminum chloride colorimetric method by Chang et al. [31] was used for the measurement of total flavonoids content. A sample of the fresh leaf (1 g), crushed in liquid nitrogen was extracted with 4 mL 96% ethanol. To this extract (500 µL) was added potassium acetate (1 M, 100 µL) and aluminum chloride (10%, 0.1 mL). Then, 1.5 mL of methanol and 2.8 mL of distilled water were added and it was kept for 30 min at 25 °C. Finally, the absorbance was measured using a spectrophotometer (Spekol 1500, Analytik Jena, Jena, Germany) at 415 nm. Total flavonoid content was expressed using the quercetin calibration curve (0, 20, 40, 60, 80 and 100 µg ml^{-1}) as mg quercetin equivalent g^{-1} fresh weight (mg QE g^{-1} FW).

2.10. Statistical Analysis

ANOVA was performed using MSTAT-C ver. 2.1 software. The mean comparisons of the data were analyzed using the least significant difference (LSD) test at 5% probability level. Excel software was used to draw graphs. Pearson's correlation coefficients and a heat map were drawn using Rstudio ver. 14.2.1 software.

3. Results

3.1. Root Colonization Percentage

The results showed that root colonization percentage was significantly affected by the treatments (Figure 2). The highest colonization (85%) belonged to AMF × SWE (3 g L^{-1}) and the lowest percentage (65%) was observed in AMF × SWE 0.5 g L^{-1} (Figure 1).

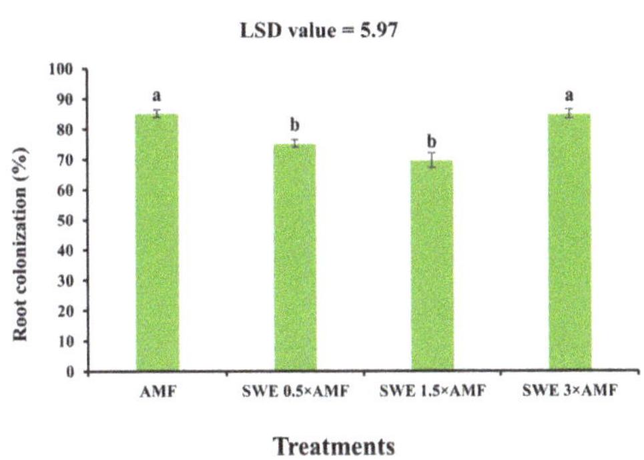

Figure 2. Mean root colonization percentage of lettuce plants affected by arbuscular mycorrhiza (AMF) inoculation × seaweed extract (SWE) foliar treatment. Different letters indicate significant differences at 5% level according to the least significant difference (LSD) test.

3.2. Morphological Traits

The number of leaves and head diameter were significantly affected by AMF × SWE interactions. The highest number of leaves and head diameter were obtained by applying AMF × SWE at concentrations of 1.5 and 3 g L^{-1}. The top number of leaves and head diameter were 44% and 3.6 times more than the control, respectively. The least number of leaves and head diameter were observed in the control (Table 1).

Table 1. The effects of arbuscular mycorrhiza (AMF) inoculation and seaweed (SWE) extract foliar spray on the growth-related traits of lettuce plants.

Fertilization	Number of Leaves	Head Diameter (cm)	Head Fresh Weight (g)	Root Fresh Weight (g)	Head Dry Weight (g)	Root Dry Weight (g)
control	25.5 ± 3.01d	20.75 ± 3.17g	72.5 ± 20.6f	22.38 ± 2.74g	1.42 ± 0.54e	13.88 ± 1.45g
SWE1	28.5 ± 2.96c	29.25 ± 2.39f	134.2 ± 18.90e	33.03 ± 1.26df	2.02 ± 0.035bc	20.9 ± 0.57f
SWE2	30.47 ± 2.75c	34.87 ± 2.57e	173 ± 27.1d	32.25 ± 1.572f	2.2 ± 0.035d	25.88 ± 0.32d
SWE3	34.75 ± 1.62bc	57.53 ± 3.54c	220.5 ± 14.4c	46.49 ± 2.275c	3.02 ± 0.022bc	30.01 ± 0.42c
AMF	31.5 ± 1.83bc	32.1 ± 2.39ef	180.2 ± 22.5d	35.6 ± 6.95ef	2.57 ± 0.041cd	23.4 ± 0.17e
AMF + SWE1	36 ± 2.61ab	43.72 ± 1.95d	199.1 ± 24.2de	42.22 ± 2.351cd	2.37 ± 0.056c	28.38 ± 0.43c
AMF + SWE2	37 ± 2.21a	66.48 ± 3.53b	245.7 ± 18.6b	52.11 ± 1.16b	3.4 ± 0.127b	32.86 ± 0.34b
AMF + SWE3	36.72 ± 3.83a	76.02 ± 3.13a	284.3 ± 17.01a	63.83 ± 3.31a	5.2 ± 0.437a	39.7 ± 2.08a
LSD at 0.05%	8.17	4.49	24.32	4.82	0.587	2.16
Significance						
AMF	ns	**	**	**	**	**
SWE	ns	**	**	**	**	**
AMF × SWE	**	**	*	*	**	**

Control (without arbuscular mycorrhiza fungi and seaweed extract); SWE 1, SWE 2 and SWE 3 (seaweed extract 0.5 g L^{-1}, 1.5 g L^{-1}, and 3 g L^{-1}, respectively), AMF (arbuscular mycorrhiza fungus), AMF × SWE 1, AMF × SWE 2 and AMF × SWE 3 (arbuscular mycorrhiza fungi × seaweed extract 0.5 g L^{-1}, 1.5 g L^{-1}, and 3 g L^{-1}, respectively). Different letters indicate significant differences according to LSD test at $p < 0.05$. ns, * and ** indicate no significant difference, significant at 5% probability level and significant at 1% probability level, respectively.

As shown in Table 1, the application of AMF × SWE treatment had a significant effect on the fresh weight of the lettuce head and roots. The highest dry weight of roots and head was recorded for AMF × SWE, 3 g L^{-1}, and the lowest values were observed in the control.

Moreover, the results showed that the dry weight of lettuce leaf and root was significantly affected by different concentrations of SWE. The highest dry weight of roots and leaves was obtained by applying AMF × SWE at 3 g L^{-1}, which showed an increase of 3.6 and 2.8 times compared to the control, respectively. The lowest dry weight of leaf and roots belonged to the control (Table 1).

3.3. Chlorophyll Index (SPAD)

The SPAD index was significantly increased by the co-application of SWE and AMF. The highest SPAD index (57.15) was obtained by using AMF × SWE, 3 g L^{-1}, while the lowest SPAD index (33.4) was recorded for the control. The top recorded data were 71.10% more than control (Figure 3a).

3.4. Total Protein Content

The total protein content of lettuce leaves was significantly affected by the co-application of AMF and SWE (Table 2). The highest leaf protein content (1.13 mg g^{-1} FW) was obtained for the treatment of AMF × SWE, 3 g L^{-1} which showed an increase of 4.21%, compared to the control. The lowest total protein content (0.268 mg g^{-1} FW) was recorded in the control samples (Figure 3b).

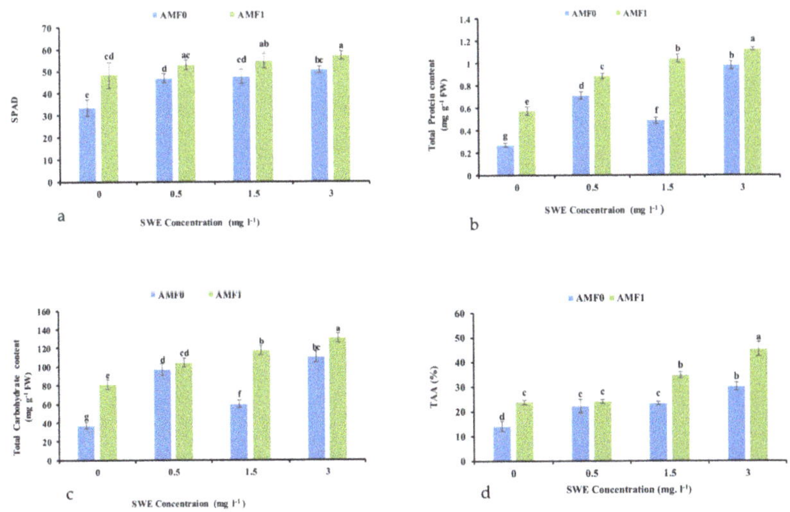

Figure 3. Effect of mycorrhiza (AMF) × seaweed extract (SWE) co-treatments on chlorophyll index (SPAD) (**a**), total proteins content (**b**), total carbohydrates content (**c**), and total antioxidant activity (TAA) (**d**) of lettuce plants. Different letters indicate significant differences according to LSD test at $p < 0.05$. AMF0 and AMF1 refer to without mycorrhiza and with mycorrhiza.

Table 2. Analysis of variance (ANOVA) for the effects of arbuscular mycorrhiza fungi (AMF) inoculation and seaweed extract (SWE) foliar spray on physiological traits of lettuce plant.

S.O.V.	df	Total Protein Content (mg g^{-1} FW)	Carbohydrate Content (mg g^{-1} FW)	TAA (%)	Ascorbic acid Content (mg 100g^{-1} FW)	Total Phenolics Content (mg 100g^{-1} FW)	Total Flavonoids Content (mg g^{-1} FW)
AMF	1	0.146 **	1708.3 **	392.91 **	262.43 **	7699.90 **	14,238.2 **
SWE	3	0.792 **	8346.9 **	663.45 **	961.32 **	19,875.95 **	48,877.1 **
AMF × SWE	3	0.007 **	103.2 *	37.60 *	125.95 **	3326.05 **	1733.1 **
Error	24	0.002	32.06	12.60	10.08	593.75	264.94
C.V.		5.51	6.15	13.05	5.92	12.55	8.82

AMF and SWE refer to arbuscular mycorrhiza fungi and seaweed extract, respectively. S.O.V. and df refer to the source of variation and degree of freedom. *, ** significant at the 5% and 1% probability levels, respectively.

3.5. Total Carbohydrate Content

The co-treatment of AMF and SWE significantly affected the total carbohydrate content of lettuce leaves (Table 2). The highest content of leaf carbohydrates (130.6 mg g^{-1} FW) was observed for AMF × SWE (3 g L^{-1}), which was 3.5 times more than the control (37.02 mg g^{-1} FW) (Figure 3c).

3.6. Total Antioxidant Activity (TAA)

According to Table 2, TAA values were significantly affected by the interaction of SWE and AMF. The highest antioxidant activity (42.24%) measured by the FRAP method was obtained by using AMF × SWE, 3 g L^{-1}. The activity was increased three times compared to the control while the lowest TAA value (14.06%) was recorded for the control sample (Figure 3d).

3.7. Ascorbic Acid Content

The interactions of AMF and SWE significantly affected the content of ascorbic acid as well (Table 2). The highest ascorbic acid content (67.66 mg 100g $^{-1}$ FW) was obtained for AMF × SWE, 3 g L^{-1}, and the least data (29.75 mg 100g^{-1} FW) belonged to the control (Figure 4a).

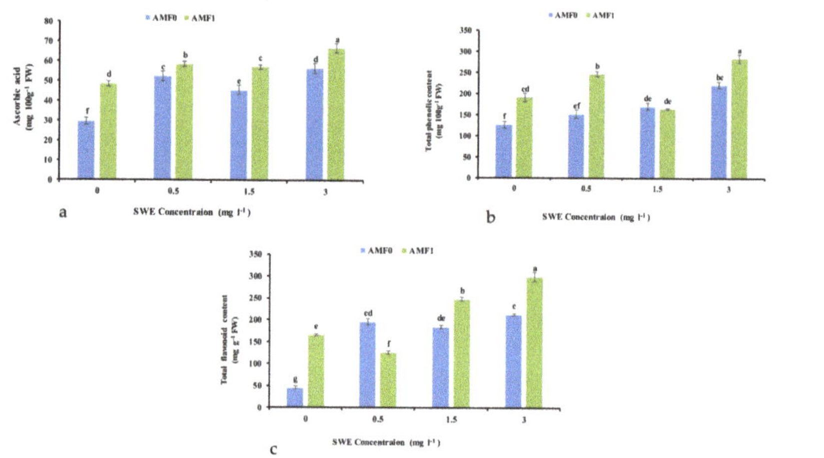

Figure 4. Effect of arbuscular mycorrhiza fungi (AMF) × seaweed extract (SWE) co-treatments on the ascorbic acid content (**a**) total phenolics content (**b**) and total flavonoids content (**c**) of lettuce plants. Different letters indicate significant differences according to the LSD test at $p < 0.05$. AMF0 and AMF1 refer to without and with arbuscular mycorrhiza fungi, respectively.

3.8. Total Phenolic and Flavonoid Content

Also, the content of phenolics and flavonoids in lettuce leaves was affected by the combined application of AMF and SWE (Table 2). The highest values of total phenolics (283.9 mg GAE 100 g^{-1} FW) and flavonoids (300 mg QE g^{-1} FW) were obtained by the co-application of AMF × SWE (3 g L^{-1}). It was 2.24 and 6.59 times higher than the control, respectively. The lowest content of total phenolics and flavonoid was 126.2 mg GAE 100 g^{-1} FW and 45.5 mg QE g^{-1} FW, respectively, which were observed for the control (Figure 4b,c).

3.9. Correlation Matrix and Relative Expressions

The Pearson's correlation of the morphological and biochemical traits is presented in Figure 4. The results revealed a positive significant correlation among phenolics, TAA, flavonoids, root FW, and ascorbic acid and protein content. Also, head diameter significantly correlated to SPAD and shoot FW. On the other hand, shoot DW significantly correlated with leaf number and shoot FW.

The heat map (Figure 5) showed that phenolics, TAA, flavonoids, root FW and DW, ascorbic acid, proteins content, leaf number, carbohydrates content, shoot FW, SPAD, and head diameter had positive compliance with the co-application of SWE and AMF.

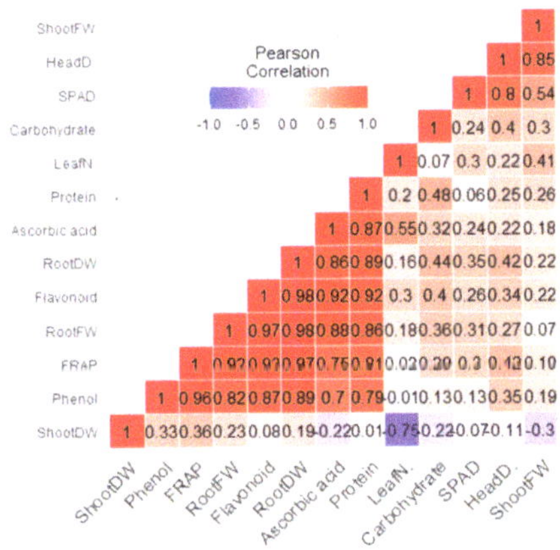

Figure 5. Heat map of Pearson's correlation analysis for the effect of arbuscular mycorrhiza (AMF) inoculation and seaweed extract (SWE) foliar spray on lettuce. Heat map representing head DW (dry weight), root FW (fresh weight), flavonoid (total flavonoids content), root DW, protein (total proteins content), TAA (total antioxidants activity by FRAP method), phenol (total phenolics content), carbohydrate content, head D (head diameter), head FW, SPAD (chlorophyll index), and leaf N (leaf number).

Cluster analysis and dendrograms in a heat map matrix (Figure 6) showed two main groups for the evaluated traits. Group 1 contained phenolics, TAA, flavonoids, root FW and DW, ascorbic acid content, protein content, leaf number, carbohydrates content, shoot FW, SPAD, and head diameter; and group 2 represented shoot DW that had a negative correlation with the higher growth of the plants. In general, cluster analysis of heat maps for treatments showed three main groups. Group 1 contained using AMF with SWE (3 g L^{-1}), group 2 contained, SWE (0.5 and 1.5 g L^{-1}) with and without AMF, and group 3 included the control.

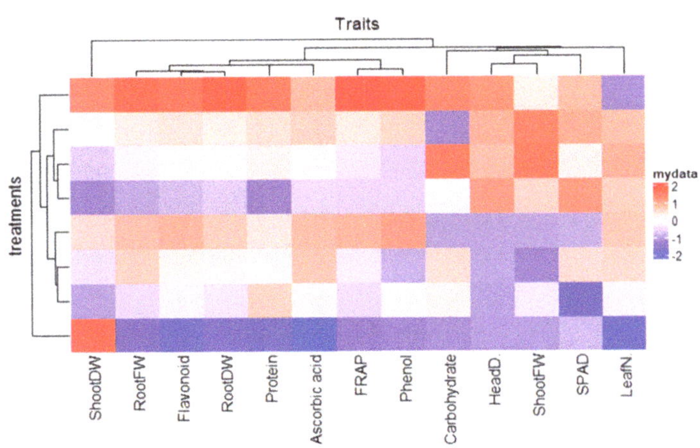

Figure 6. The morphological and biochemical changes in lettuce plants treated with arbuscular mycorrhiza (AMF) and seaweed extract (SWE). Heat map representing head DW (dry weight), root FW (fresh weight), flavonoid (total flavonoids content), root DW, protein (total proteins content), TAA (total antioxidants activity by FRAP method), phenol (total phenolics content), carbohydrates content, head D (head diameter), head FW, SPAD (chlorophyll index) and leaf N (leaf number).

4. Discussion

One of the crucial principles in planning the production of vegetables is to evaluate the effectiveness of diverse plant nutrition strategies. The organic production of vegetables along with a proper nutrition regime, while maintaining the environmental and health standards, will increase the yield and quality of those crops.

The results of this study showed that the highest average number of leaves, head diameter, head, and root fresh weight, and leaf and root dry weight was obtained by the combined application of biological stimulants (AMF and SWE). The increased plant height is a response to AMF, which helps in the appropriate absorption of essential nutrients. Furthermore, the inoculation of this fungus to plant roots causes the production of various hormones, such as auxin and gibberellins, and is effective in increasing the plant height and, consequently, the number of leaves [32]. It also improves water and nutrient uptake, photosynthesis potential, and carbohydrate production by altering the root morphology and thereby increasing shoot and root growth. The symbiosis of AMF plays a vital role in the carbon cycle and can increase the production of growth-promoting molecules [33]. In lettuce, the use of SWE increased plant growth probably due to the availability of significant amounts of growth hormones, amino acids, and macro-and microelements [34]. Following the present results, Dudaš et al. [35] in lettuce and Rouphael et al. [36] in zucchini reported an increase in plant height and number of leaves after the usage of SWE.

The results showed that the co-use of SWE and AMF positively affected the fresh and dry weight of the head and roots. In plants inoculated with AMF, some of the fungal hyphae enter the root system of the plant and reduce abscisic acid levels and increase cytokinins, which expand the root system and enhance water uptake, as well as the secretion of organic acids such as malic acid by extra-root hyphae increasing P uptake and thus improving root development and eventually growth of the aerial parts of the plant [37]. Biological properties of the soil [38] and this symbiosis in the root zone increase the uptake of minerals (P, K, Fe, Zn and Mn) and lead to an increase in dry weight biomass. SWE also improves photosynthetic efficiency by stimulating nutrients translocation and metabolism, phytohormones biosynthesis, proteins accumulation, and delaying the aging process, and can ultimately help to increase the fresh and dry weight of plants [39]. In this regard, Colla et al. [40] reported that foliar application of SWE increased the fresh and dry weight of lettuce.

In the present study, the application of AMF + SWE significantly increased the total proteins content in lettuce. In addition, the higher content of protein in AMF treatment can be attributed to the more nitrogen uptake in the symbiotic behavior of plants and fungi. Nitrogen in organic matter is usually present in the composition of peptides, proteins, and free amino acids. The AMF secretes peptidase and protease into the soil to absorb nitrogen-containing monomers, thus improving leaf proteins content [10]. The increase in proteins content due to the use of SWE can also be a result of the plant's ability to absorb a higher amount of other elements [41].

Consistent with the results of the current study, Sosnowski [42] reported that lettuce leaf carbohydrates content was increased by the combined application of AMF and SWE. Due to an increase in the stomatal conductance and P uptake, the symbiosis of mycorrhizae leads to the accumulation of secondary metabolites, vitamins, minerals, and photosynthetic pigments, and even raises the leaves' carbohydrates content [43]. In addition, AMF fortifies the plant sink for carbohydrates. An increase in carbohydrates content in tomatoes has been reported with the use of AMF [44]. Furthermore, the total carbohydrates content with the use of SWE can be attributed to the increase in chlorophyll index. The use of bio-stimulants raised the amount of soluble carbohydrates in *Vigna radiata* compared to the lack of foliar applications [45].

The ascorbic acid content in lettuce leaves was affected by the application of AMF × SWE. Lettuce is considered a good source of nutrients such as ascorbic acid and carotenoids. Photosynthesis and its products are directly related to the production of ascorbic acid in plants. Bio-fertilizer application improves the photosynthetic potential, N and P absorption, and ultimately leads to an improvement in the ascorbic acid content [46]. Moreover, high amounts of vitamin C can be attributed to the improvements in chlorophyll content in response to the stimuli effects of SWE and AMF. Likewise, Subramanian et al. [47] reported that the highest ascorbic acid content was associated with AMF use in tomatoes. In another study on the spinach plant, Tiruvaimozhi et al. [48] reported an increase in ascorbic acid content using seaweed foliar treatments.

Total antioxidants activity and, phenolics and flavonoids contents in lettuce leaves were also affected by AMF and SWE in this study. The symbiosis of mycorrhiza fungi affects plant metabolism by stimulating the biosynthesis of secondary metabolites and potentially increasing the accumulation of antioxidant compounds in plants. This fungus causes changes in the concentration of phytohormones such as jasmonic acid, gibberellic acid, and cytokinins, which improve the absorption of elements and lead to the production of more antioxidant compounds [49]. Avio et al. [50] reported that mycorrhiza fungus application increased the antioxidant activity of lettuce. Gholinezhad et al. [51] also reported that fungi symbiosis improved antioxidant activity in soybeans. Fan et al. [52] observed a direct relationship between antioxidant capacity and the amount of phenolics in spinach.

Phenolics are a major component for plant cell protection against stressors. In artichoke plants inoculated with mycorrhiza fungus, Avio et al. [50] reported that the highest total phenolics content was obtained by using this fungus. Phenolic compounds appear to increase the symbiosis rate between plant and fungus and lead to the accumulation of secondary metabolites such as carotenoids and polyphenols in host plants by making significant changes in enzymatic activities and the physiological mechanisms involved [53]. Similarly, the increase in phenolics content in SWE treatment can be attributed to the stimulated biosynthesis of growth hormones and nutrient uptake in the roots and the improved polyphenol oxidase activity which increases the accumulation of phenolics and hence the antioxidant activity in the plant [54]. The results of Ashour et al. [55] on red hot pepper are consistent with these findings. Bio-stimulants improve the phenolics accumulation in plants by improving the growth-related traits and enhancing the nutrient uptake and even by enhancing the phenylalanine accumulation in roots.

Flavonoids are a major class of polyphenolic secondary metabolites in plants. De Assis et al. [56] concluded that the levels of phenolics and flavonoids in lemongrass increased with the use of mycorrhiza. In another study, an increase in phenolics, flavonoids, and antioxidant activity of the *Eruca vesicaria* L. plant was obtained by SWE treatment [57]. The results of Mahmoud et al. [58] on the red radish plant are consistent with our results as well.

Fungal hyphae can penetrate the very small pores that even the root hairs are not able to penetrate, increasing the absorption of water and nutrients and improve plant growth, and ultimately increasing crop yield [59]. The fungus provides the required carbon from the host roots and, in turn, increases the uptake of nutrients, especially phosphorus, by the host plant [60]. AMF coexistence can play an important role in maintaining soil fertility and stabilizing soil structure while increasing plant water uptake and product quality. In particular, enhancing the symbiotic activity with AMF is a way to improve food production at the lowest economic and environmental costs [13]. Mycorrhiza improves soil structure by coating a viscous glycoprotein called glomalin, which plays a key role in the formation of soil aggregates and large pores for better hyphae growth. These pores ease air and water penetration and also help to prevent progressive soil erosion [61]. Roots inoculated with AMF have richer secretions, and extra-root hyphae create an appropriate condition for the growth of some beneficial bacteria. Also, the extra-root hyphae of the AMF cause soil particles to stick together and the formation of soil aggregates that improve the airflow in the soil, which is essential for the growth and multiplication of soil bacteria [62]. Thus, in the rhizosphere of these plants, the population of plant growth-promoting rhizobacteria (PGPR), nitrogen-fixing bacteria, and several Gram-positive bacteria is higher which can inhibit the growth of pathogens. The enhancement in the population of the beneficial bacteria has been considered as a factor in helping to reduce the population of *Fusarium* or

Phytophthora pathogens [63]. Also, Atayese [64] showed that in peanut plants inoculated with *Glomus mosseae* the grain yield was increased up to 22% compared to non-symbiotic plants. Boomsma and Vyn [65] stated that inoculation with AMF caused extensive changes in root morphological characteristics, especially lateral root growth. Mycorrhiza fungus expands the root surface area up to 40 times and increases the population of beneficial soil bacteria [66].

Mycorrhizae regulate plant physiological functions such as leaf water potential, stomatal conductance, photosystem II efficiency, and carbon dioxide stabilization and this even occurs under stressful conditions [67]. It also improves nitrogen uptake by increasing the activity of nitrogen-absorbing enzymes such as glutamine synthase, and ultimately leads to an increase in proteins and amino acids content [10]. These fungi identify their host with the signals released from the plant root and coexist with it and, in the absence of the host root, they cannot form hypha and complete their life cycle [68,69]. Colony formation with AMF increases photosynthetic efficiency in plants as well [61].

5. Conclusions

The results showed that the co-application of SWE and AMF had a positive effect on the morphological and biochemical traits of lettuce. The sole application of AMF and SWE, as well as their combined application, significantly improved the percentage of colonization, number of leaves, head diameter, fresh and dry weight of roots, and dry weight of leaves compared to the control. Moreover, total proteins and carbohydrates content, antioxidant activity, ascorbic acid content, and phenolics and flavonoids content were responsive to the combined application of treatments. The application of AMF and SWE alone improved most of the evaluated traits, however, the co-application of AMF and SWE improved the traits more than their sole application. Therefore, we recommend the co-application of the tested biostimulants to improve the growth parameters and quality attributes of lettuce plants. Organic crop production may be one of the most important environmental challenges of sustainable agricultural systems. These bio-stimulants are eco-friendly alternative strategies to improve the quality and yield of vegetables that may pave the way for meeting the nutritional demands of the growing population of the world.

Author Contributions: Conceptualization, data curation, F.R. and T.A.; formal analysis and methodology, F.R., T.A., M.A. and S.E.; project administration, F.R. and M.A.A.; visualization, F.R., M.A., M.B.H. and J.M.; writing original draft, M.A., T.A. and M.B.H.; writing-review and editing, F.R., M.A., M.B.H., S.S. and J.M. All authors have read and agreed to the published version of the manuscript.

Funding: This study was funded and carried out by the University of Maragheh, Iran, and partly supported by the project. This work was also supported by the internal grant of Tomas Bata University in Zlin (No. IGA/FT/2022/004).

Institutional Review Board Statement: Not applicable.

Informed Consent Statement: Not applicable.

Data Availability Statement: All new research data were presented in this contribution.

Acknowledgments: We thank the help from the Central Laboratory of the University of Maragheh for the technical assistance.

Conflicts of Interest: The authors declare that they have no conflict of interest.

References

1. Kim, M.J.; Moon, Y.; Tou, J.C.; Mou, B.; Waterland, N.L. Nutritional value, bioactive compounds and health benefits of lettuce (*Lactuca sativa* L.). *J. Food Compost. Anal.* **2016**, *49*, 19–34. [CrossRef]
2. FAOSTAT. UN Food and Agriculture Organization, Statistics Division. Production of Lettuce and Chicory by Countries. 2019. Available online: https://www.fao.org/faostat/en/#home (accessed on 4 March 2021).
3. Agarwal, P.; Gupta, R.; Gill, I.K. Importance of biofertilizers in agriculture biotechnology. *Ann. Biolog. Res.* **2018**, *9*, 1–3.
4. Yakhin, O.I.; Lubyanov, A.A.; Yakhin, I.A.; Brown, P.H. Biostimulants in plant science: A global perspective. *Front. Plant Sci.* **2017**, *7*, 2049. [CrossRef] [PubMed]

5. Rouphael, Y.; Colla, G. Biostimulants in agriculture. *Front. Plant Sci.* **2020**, *11*, 40. [CrossRef]
6. Du Jardin, P. Plant biostimulants: Definition, concept, main categories and regulation. *Sci. Hortic.* **2015**, *196*, 3–14. [CrossRef]
7. Battacharyya, D.; Babgohari, M.Z.; Rathor, P.; Prithiviraj, B. Seaweed extracts as biostimulants in horticulture. *Sci. Hortic.* **2015**, *196*, 39–48. [CrossRef]
8. Türkmen, M.; Su, A. The effect of sea lettuce (*Ulva lactuca*) liquid fertilizer and zeolite combinations on the development of cucumber (*Cucumis sativus*). *TURJAF.* **2019**, *7*, 1021–1027. [CrossRef]
9. Di Mola, I.; Cozzolino, E.; Ottaiano, L.; Giordano, M.; Rouphael, Y.; Colla, G.; Mori, M. Effect of vegetal-and seaweed extract-based biostimulants on agronomical and leaf quality traits of plastic tunnel-grown baby *lettuce* under four regimes of nitrogen fertilization. *Agronomy* **2019**, *9*, 571. [CrossRef]
10. Chrysargyris, A.; Xylia, P.; Anastasiou, M.; Pantelides, I.; Tzortzakis, N. Effects of *Ascophyllum nodosum* seaweed extracts on lettuce growth, physiology and fresh-cut salad storage under potassium deficiency. *J. Sci. Food Agric.* **2018**, *98*, 5861–5872. [CrossRef]
11. Jung, H.Y.; Kim, J.K. Complete reutilization of mixed mackerel and brown seaweed wastewater as a high-quality biofertiliser in open-flow *lettuce* hydroponics. *J. Clean. Prod.* **2020**, *247*, 119081. [CrossRef]
12. Varma, A.; Prasad, R.; Tuteja, N. *Mycorrhiza-Nutrient Uptake, Biocontrol, Ecorestoration*; Springer: Berlin/Heidelberg, Germany, 2018; pp. 1–539.
13. Santander, C.; Aroca, R.; Ruiz-Lozano, J.M.; Olave, J.; Cartes, P.; Borie, F.; Cornejo, P. arbuscular mycorrhiza effects on plant performance under osmotic stress. *Mycorrhiza* **2017**, *27*, 639–657. [CrossRef] [PubMed]
14. Cobb, A.B. From Soil Ecology to Human Nutrition: Crop Symbiosis with Arbuscular Mycorrhizal Fungi in Agroecosystems. Ph.D. Thesis, Oklahoma State University, Stillwater, OK, USA, 2016; pp. 10–127.
15. Bona, E.; Cantamessa, S.; Massa, N.; Manassero, P.; Marsano, F.; Copetta, A.; Lingua, G.; D'Agostino, G.; Gamalero, E.; Berta, G. Arbuscular mycorrhizal fungi and plant growth-promoting pseudomonads improve yield, quality and nutritional value of *tomato*: A field study. *Mycorrhiza* **2017**, *27*, 1–11. [CrossRef] [PubMed]
16. Saia, S.; Colla, G.; Raimondi, G.; Di Stasio, E.; Cardarelli, M.; Bonini, P.; Rouphael, Y. An endophytic fungi-based biostimulant modulated lettuce yield, physiological and functional quality responses to both moderate and severe water limitation. *Sci. Hortic.* **2019**, *256*, 108595. [CrossRef]
17. Tarraf, W.; Ruta, C.; Tagarelli, A.; De Cillis, F.; De Mastro, G. Influence of arbuscular mycorrhizae on plant growth, essential oil production and phosphorus uptake of *Salvia officinalis* L. *Ind. Crops Prod.* **2017**, *102*, 144–153. [CrossRef]
18. Kaur, S.; Suseela, V. Unraveling arbuscular mycorrhiza-induced changes in plant primary and secondary metabolome. *Metabolites* **2020**, *10*, 335. [CrossRef] [PubMed]
19. Baslam, M.; Garmendia, I.; Goicoechea, N. Arbuscular mycorrhizal fungi (AMF) improved growth and nutritional quality of greenhouse-grown lettuce. *J. Agric. Food Chem.* **2011**, *59*, 5504–5515. [CrossRef] [PubMed]
20. Baslam, M.; Goicoechea, N. Water deficit improved the capacity of arbuscular mycorrhizal fungi (AMF) for inducing the accumulation of antioxidant compounds in lettuce leaves. *Mycorrhiza* **2012**, *22*, 347–359. [CrossRef]
21. Avio, L.; Sbrana, C.; Giovannetti, M.; Frassinetti, S. Arbuscular mycorrhizal fungi affect total phenolics content and antioxidant activity in leaves of oak leaf lettuce varieties. *Sci. Hortic.* **2017**, *224*, 265–271. [CrossRef]
22. Phillips, J.M.; Hayman, D. Improved procedures for clearing roots and staining parasitic and vesicular-arbuscular mycorrhizal fungi for rapid assessment of infection. *Trans. Br. Mycol. Soc.* **1970**, *55*, 158–161. [CrossRef]
23. Koske, R.E.; Gemma, J.N. A modified procedure for staining roots to detect VA mycorrhizas. *Mycol. Res.* **1989**, *92*, 486–505. [CrossRef]
24. McGonigle, T.; Miller, M.; Evans, D.; Fairchild, G.; Swan, J. A new method which gives an objective measure of colonization of roots by vesicular arbuscular mycorrhizal fungi. *New Phytol.* **1990**, *115*, 495–501. [CrossRef] [PubMed]
25. Giovannetti, M.; Mosse, B. An evaluation of techniques for measuring vesicular arbuscular mycorrhizal infection in roots. *New Phytol.* **1980**, *84*, 489–500. [CrossRef]
26. Bradford, M.M. A rapid and sensitive method for the quantitation of microgram quantities of protein utilizing the principle of protein-dye binding. *Anal. Biochem.* **1976**, *72*, 248–254. [CrossRef]
27. Chen, Q.; Wang, J.; Rayson, G.; Tian, B.; Lin, Y. Sensor array for carbohydrates and amino acids based on electrocatalytic modified electrodes. *Anal. Chem.* **1993**, *65*, 251–254. [CrossRef]
28. Benzie, I.E.F.; Strain, J.J. The ferric reducing ability of plasma (FRAP) as a measure of antioxidant power: The FRAP assay. *Anal. Biochem.* **1996**, *239*, 70–76. [CrossRef] [PubMed]
29. Smirnoff, N. Ascorbic acid: Metabolism and functions of a multi-facetted molecule. *Curr. Opin. Plant Biol.* **2000**, *3*, 229–235. [CrossRef]
30. Singleton, V.L.; Orthofer, R.; Lamuela-Raventós, R.M. [14] Analysis of total phenols and other oxidation substrates and antioxidants by means of folin-ciocalteu reagent. *Methods Enzymol.* **1999**, *299*, 152–178. [CrossRef]
31. Chang, C.C.; Yang, M.H.; Wen, H.M.; Chern, J.C. Estimation of total flavonoid content in propolis by two complementary colorimetric methods. *J. Food Drug. Anal.* **2002**, *10*, 178–182.
32. Moradi, S.; Pasari, B.; Talebi, R. Study of the effects of mycorrhiza, fulvic acid, seaweed extract and urea on physiological traits and leaf yield of *tobacco* (*Burley* 21). *Eur. J. Environ. Sci.* **2019**, *9*, 33–40. [CrossRef]
33. Kowalska, I.; Konieczny, A.; Gastol, M. Effect of mycorrhiza and the phosphorus content in a nutrient solution on the yield and nutritional status of lettuce grown on various substrates. *J. Elem.* **2015**, *20*, 631–642. [CrossRef]

34. Choi, J.; Summers, W.; Paszkowski, U. Mechanisms underlying establishment of arbuscular mycorrhizal symbioses. *Annu. Rev. Phytopathol.* **2018**, *56*, 135–160. [CrossRef]
35. Dudaš, S.; Šola, I.; Sladonja, B.; Erhatić, R.; Ban, D.; Poljuha, D. The effect of biostimulant and fertilizer on "low input" lettuce production. *Acta Bot. Croat.* **2016**, *75*, 253–259. [CrossRef]
36. Rouphael, Y.; De Micco, V.; Arena, C.; Raimondi, G.; Colla, G.; De Pascale, S. Effect of Ecklonia maxima seaweed extract on yield, mineral composition, gas exchange, and leaf anatomy of zucchini squash grown under saline conditions. *J. Appl. Phycol.* **2017**, *29*, 459–470. [CrossRef]
37. Kinany, S.E.; Achbani, E.; Faggroud, M.; Ouahmane, L.; Hilali, R.E.; Haggoud, A.; Bouamri, R. Effect of organic fertilizer and commercial arbuscular mycorrhizal fungi on the growth of micropropagated date palm cv. *Feggouss. J. Saudi Soc. Agric. Sci.* **2019**, *18*, 411–417. [CrossRef]
38. Zhang, F.; Liu, M.; Li, Y.; Che, Y.; Xiao, Y. Effects of arbuscular mycorrhizal fungi, biochar and cadmium on the yield and element uptake of *Medicago sativa*. *Sci. Total Environ.* **2019**, *655*, 1150–1158. [CrossRef] [PubMed]
39. Shukla, P.S.; Mantin, E.G.; Adil, M.; Bajpai, S.; Critchley, A.T.; Prithiviraj, B. *Ascophyllum nodosum*-based biostimulants: Sustainable applications in agriculture for the stimulation of plant growth, stress tolerance, and disease management. *Front. Plant Sci.* **2019**, *10*, 655. [CrossRef] [PubMed]
40. Colla, G.; Cardarelli, M.; Bonini, P.; Rouphael, Y. Foliar applications of protein hydrolysate, plant and seaweed extracts increase yield but differentially modulate fruit quality of greenhouse tomato. *HortScience* **2017**, *52*, 1214–1220. [CrossRef]
41. Calvo, P.; Nelson, L.; Kloepper, J.W. Agricultural uses of plant biostimulants. *Plant Soil* **2014**, *383*, 3–41. [CrossRef]
42. Sosnowski, J.; Jankowski, K.; Wiśniewska-Kadżajan, B.; Malinowska, E.; Kolczarek, R.; Czeluściński, W.; Radzka, E. Shaping of chemical composition and digestibility of Medicago× varia T. Martyn under the influence of Ecklonia maxima seaweed extract. *Fresenius. Environ. Bull.* **2015**, *24*, 881–887.
43. Bao, X.; Wang, Y.; Olsson, P.A. Arbuscular mycorrhiza under water—Carbon–phosphorus exchange between rice and arbuscular mycorrhizal fungi under different flooding regimes. *Soil Biol. Biochem.* **2019**, *129*, 169–177. [CrossRef]
44. González-González, M.F.; Ocampo-Alvarez, H.; Santacruz-Ruvalcaba, F.; Sánchez-Hernández, C.V.; Casarrubias-Castillo, K.; Becerril-Espinosa, A.; Hernández-Herrera, R.M. Physiological, ecological, and biochemical implications in tomato plants of two plant biostimulants: Arbuscular mycorrhizal fungi and seaweed extract. *Front. Plant Sci.* **2020**, *11*, 999. [CrossRef] [PubMed]
45. Paul, J.; Yuvaraj, P. Effect of seaweed liquid fertilizer of *Colpomenia sinuosa* (Mert. ex Roth) Derbes & Solier (Brown Seaweed) on *Vigna radiata* (L.) R. Wilczek, *Koothankuzhi*, Tirunelveli district, Tamil Nadu, India. *Int. J. Pure App. Biosci.* **2014**, *2*, 177–184.
46. Patel, Z.; Bhalerao, P.P.; Gaikwad, S.S.; Sachin, A.J. Effect of foliar application of chemicals on growth and yield of garlic (*Allium sativum* L.) var. GG-4. *Int. J. Chem. Stud.* **2017**, *5*, 1035–1037.
47. Subramanian, K.S.; Santhanakrishnan, P.; Balasubramanian, P. Responses of field grown tomato plants to arbuscular mycorrhizal fungal colonization under varying intensities of drought stress. *Sci. Hortic.* **2006**, *107*, 245–253. [CrossRef]
48. Tiruvaimozhi, Y.V.; Varma, V.; Sankaran, M. Nitrogen fixation ability explains leaf chemistry and arbuscular mycorrhizal responses to fertilization. *Plant Ecol.* **2018**, *219*, 391–401. [CrossRef]
49. Jiang, Q.Y.; Tan, S.Y.; Zhuo, F.; Yang, D.J.; Ye, Z.H.; Jing, Y.X. Effect of *Funneliformis mosseae* on the growth, cadmium accumulation and antioxidant activities of Solanum nigrum. *Appl. Soil Ecol.* **2016**, *98*, 112–120. [CrossRef]
50. Avio, L.; Maggini, R.; Ujvári, G.; Incrocci, L.; Giovannetti, M.; Turrini, A. Phenolics content and antioxidant activity in the leaves of two artichoke cultivars are differentially affected by six mycorrhizal symbionts. *Sci. Hortic.* **2020**, *264*, 109153. [CrossRef]
51. Gholinezhad, E.; Darvishzadeh, R.; Moghaddam, S.S.; Popović-Djordjević, J. Effect of mycorrhizal inoculation in reducing water stress in sesame (*Sesamum indicum* L.): The assessment of agrobiochemical traits and enzymatic antioxidant activity. *Agric. Water Manag.* **2020**, *238*, 106234. [CrossRef]
52. Fan, D.; Hodges, D.M.; Zhang, J.; Kirby, C.W.; Ji, X.; Locke, S.J.; Prithiviraj, B. Commercial extract of the brown seaweed Ascophyllum nodosum enhances phenolic antioxidant content of spinach (*Spinacia oleracea* L.) which protects *Caenorhabditis elegans* against oxidative and thermal stress. *Food Chem.* **2012**, *124*, 195–202. [CrossRef]
53. De Sousa Oliveira, M.; da Silva Campos, M.A.; de Albuquerque, U.P.; da Silva, F.S.B. Arbuscular mycorrhizal fungi (AMF) affect biomolecules content in Myracrodruon urundeuva seedlings. *Ind. Crops Prod.* **2013**, *50*, 244–247. [CrossRef]
54. Basavaraja, P.K.; Yogendra, N.D.; Zodape, S.T.; Prakash, R.; Ghosh, A. Effect of seaweed sap as foliar spray on growth and yield of hybrid maize. *J. Plant Nutr.* **2018**, *41*, 1851–1861. [CrossRef]
55. Ashour, M.; Hassan, S.M.; Elshobary, M.E.; Ammar, G.A.; Gaber, A.; Alsanie, W.F.; El-Shenody, R. Impact of commercial seaweed liquid extract (TAM®) biostimulant and its bioactive molecules on Growth and Antioxidant Activities of Hot Pepper (*Capsicum annuum*). *Plants* **2021**, *10*, 1045. [CrossRef] [PubMed]
56. de Assis, R.M.A.; Carneiro, J.J.; Medeiros, A.P.R.; de Carvalho, A.A.; da Cunha Honorato, A.; Carneiro, M.A.C.; Bertolucci, S.K.V.; Pinto, J.E.B.P. Arbuscular mycorrhizal fungi and organic manure enhance growth and accumulation of citral, total phenols, and flavonoids in *Melissa officinalis* L. *Ind. Crops Prod.* **2020**, *158*, 112981. [CrossRef]
57. Hassan, S.M.; Ashour, M.; Soliman, A.A.; Hassanien, H.A.; Alsanie, W.F.; Gaber, A.; Elshobary, M.E. The potential of a new commercial seaweed extract in stimulating morpho-agronomic and bioactive properties of *Eruca vesicaria* (L.) Cav. *Sustainability* **2021**, *13*, 4485. [CrossRef]
58. Mahmoud, S.H.; Salama, D.M.; El-Tanahy, A.M.; Abd El-Samad, E.H. Utilization of seaweed (*Sargassum vulgare*) extract to enhance growth, yield and nutritional quality of red radish plants. *Ann. Agric. Sci.* **2019**, *64*, 167–175. [CrossRef]

59. Hao, Z.; Xie, W.; Chen, B. arbuscular mycorrhizal symbiosis affects plant immunity to viral infection and accumulation. *Viruses* **2019**, *11*, 534. [CrossRef]
60. Berger, F.; Gutjahr, C. Factors affecting plant responsiveness to arbuscular mycorrhiza. *Curr. Opin. Plant Biol.* **2021**, *59*, 101994. [CrossRef]
61. Talaat, N.B.; Shawky, B.T. Protective effects of arbuscular mycorrhizal fungi on wheat (*Triticum aestivum* L.) plants exposed to salinity. *Environ. Exp. Bot.* **2014**, *98*, 20–31. [CrossRef]
62. Jannoura, R.; Joergensen, R.G.; Bruns, C. Organic fertilizer effects on growth, crop yield, and soil microbial biomass indices in sole and intercropped peas and oats under organic farming conditions. *Eur. J. Agron.* **2014**, *52*, 259–270. [CrossRef]
63. Tarkka, M.T.; Frey-Klett, P. Mycorrhiza Helper Bacteria. In *Mycorrhiza*; Springer: Berlin/Heidelberg, Germany, 2008; pp. 113–132.
64. Atayese, M.O. Field response of groundnut (*Arachis hypogea* L.) cultivars to mycorrhizal inoculation and phosphorus fertilizer in Abeokuta, South West Nigeria. *Am.-Eurasian J. Agric. Tur. Environ. Sci.* **2007**, *2*, 16–23.
65. Boomsma, C.R.; Vyn, T.J. Maize drought tolerance: Potential improvements through arbuscular mycorrhizal symbiosis. *Field Crops Res.* **2008**, *108*, 14–31. [CrossRef]
66. Urcoviche, R.C.; Gazim, Z.C.; Dragunski, D.C.; Barcellos, F.G.; Alberton, O. Plant growth and essential oil content of *Mentha crispa* inoculated with arbuscular mycorrhizal fungi under different levels of phosphorus. *Ind. Crops Prod.* **2015**, *67*, 103–107. [CrossRef]
67. Kiran, S.; Kuşvuran, Ş.; Özkay, F.; Ellialtioğlu, Ş.Ş. Change in physiological and biochemical parameters under drought stress in salt-tolerant and salt-susceptible eggplant genotypes. *Turk. J. Agric. For.* **2019**, *43*, 593–602. [CrossRef]
68. Lehmann, A.; Veresoglou, S.D.; Leifheit, E.F.; Rillig, M.C. Arbuscular mycorrhizal influence on zinc nutrition in crop plants—A meta-analysis. *Soil Biol. Biochem.* **2014**, *69*, 123–131. [CrossRef]
69. Subaşi, İ. Seed fatty acid compositions and chemotaxonomy of wild Crambe (Brassicaceae) taxa in Turkey. *Turk. J. Agric. For.* **2020**, *44*, 662–670. [CrossRef]

agronomy

Article

Using HPLC–MS/MS to Assess the Quality of Beet, Mizuna, Lettuce and Corn Salad after Juglone and Walnut Leaf Extract Treatments

Aljaz Medic *, Tilen Zamljen, Mariana Cecilia Grohar, Ana Slatnar, Metka Hudina and Robert Veberic

Department of Agronomy, Biotechnical Faculty, University of Ljubljana, Jamnikarjeva 101, 1000 Ljubljana, Slovenia; tilen.zamljen@bf.uni-lj.si (T.Z.); marianacecilia.grohar@bf.uni-lj.si (M.C.G.); ana.slatnar@bf.uni-lj.si (A.S.); metka.hudina@bf.uni-lj.si (M.H.); robert.veberic@bf.uni-lj.si (R.V.)

* Correspondence: aljaz.medic@bf.uni-lj.si; Tel.: +386-40240201

Abstract: The present study was carried out to investigate the yield, quality, and metabolomic responses of four different vegetable crops to treatments with pure juglone standard and walnut (*Juglans regia* L.) leaf extract at soil concentrations found in walnut orchards. A total of 60 phenolic compounds were identified and quantified, some for the first time in these crop vegetables. *Beta vulgaris* L. and *Lactuca sativa* L. were less susceptible to juglone. For crop quality, *B. vulgaris* showed the least effects of the different treatments. Both *Brassica rapa* L. var. *japonica* and *Valerianella locusta* Laterr. showed lower yields, even at the lower juglone concentration, and reduced quality, so their cultivation in juglone-containing soils should be avoided. This study also investigated leaf quality at different ages and the quality and yield of these crop vegetables grown under the influence of allelochemicals, to determine the influence of allelochemicals on metabolomics and, thus, on the uptake of phenolic compounds considered to be beneficial to human health.

Keywords: allopathy; *Beta vulgaris* L.; *Brassica rapa* L. var. *japonica*; *Lactuca sativa* L.; phenolic compounds; *Valerianella locusta* Laterr.

Citation: Medic, A.; Zamljen, T.; Grohar, M.C.; Slatnar, A.; Hudina, M.; Veberic, R. Using HPLC–MS/MS to Assess the Quality of Beet, Mizuna, Lettuce and Corn Salad after Juglone and Walnut Leaf Extract Treatments. *Agronomy* **2022**, *12*, 347. https://doi.org/10.3390/agronomy12020347

Academic Editor: Gerardo Fernández Barbero

Received: 7 January 2022
Accepted: 28 January 2022
Published: 29 January 2022

1. Introduction

Consumption of fresh or processed vegetables is an essential part of the human diet and has been associated with many health benefits (e.g., reduced diet-related diseases and risk of obesity) [1]. The most important factors that contribute to the nutritional and health benefits of vegetables are vitamins and phytochemicals. To date, over 5000 phytochemicals have been identified. Phytochemicals are usually classified into three major classes: (i) phenols, (ii) terpenes, and (iii) nitrogen-containing compounds [2]. Of these phytochemicals, phenols, or phenolic compounds, are the best-studied and most abundant group of phytochemicals and are associated with various health benefits [3,4]. Phenolic compounds also have important roles in the quality of vegetables, as they affect their taste, appearance, and stability [5].

The concentration and types of phenolic compounds vary within different vegetables and plant tissues [6]. Phenolic compounds can be unique and found only in one crop or cultivar, or they can be present across several varieties. They also have important roles in plant defence against pathogens, predators, and biotic and abiotic stresses [7,8]. If a plant is under stress, its content of phenolic compounds increases as a response to the stress [8], and therefore, the plant uses energy and nutrients meant for growth and other primary functions to produce these defensive compounds [9–11]. As higher levels of phenolic compounds are associated with higher vegetable quality, plant stress is considered beneficial to some degree [9]. The problem is that the higher the levels of phenolic compounds, the lower the growth of the plants and, therefore, the lower the yield, so there is the need for a balance between quality and yield. While some stress factors can be controlled through agronomic practices (e.g., irrigation, pest and disease control, fertilisation), others are more difficult or are not possible to control (e.g., plant residues and fungi in the soil) [9].

Plant residues can release allelochemicals into the soil, some of which have positive effects on plant growth, while others have negative effects. In some plants, allelochemicals can cause deformity, chlorosis, and wilting, thus reducing vitality, slowing down or preventing germination, hindering growth and development, and increasing susceptibility to disease, which can lead to their collapse [12]. In agriculture, this is especially problematic when one crop follows another. The greatest problems occur when walnut (*Juglans regia* L.) orchards are replaced by other crops because walnuts contain one of the first and most studied allelochemicals—juglone [13]. Juglone is released into the soil and affects the growth of surrounding crops and of crops planted after the walnut trees are cut down, with effects lasting for years [12].

Although clearing an old orchard to make way for new varieties or crop fields is common practice, little is known about the short- and long-term effects of allelochemicals that might still be present in the soil, and how they might affect the quality and yield of future crops [12]. There have been some studies on the effects of juglone, but most have focused on seed germination rather than mature crop yield or quality [11,13,14].

Therefore, the aim of the present study was to investigate the yield, quality, and molecular responses to known concentrations of juglone alone and in walnut leaf extract with concentrations of juglone normally found in the soil of walnut orchards [15], based on four different vegetable crops: beet (*Beta vulgaris* L.), mizuna (*Brassica rapa* L. var. *japonica*), lettuce (*Lactuca sativa* L.), and corn salad (*Valerianella locusta* Laterr.). The goal was also to determine whether juglone is really the crucial and only allelochemical in walnut, or whether there are other allelochemicals that have remained hidden in the shadow of juglone, as suggested more recently [13]. The data obtained represent an important basis for determining which crop varieties are susceptible to juglone and/or walnut leaf allelochemicals, and how these allelochemicals affect the yield and metabolomics of selected crops that influence human health and nutrition.

2. Materials and Methods

2.1. Plant Material

To determine whether pure juglone and a juglone-containing leaf extract have the same effects on different crop vegetables, or whether the effects are specific, four commonly cultivated crop vegetables were used: beet (*Beta vulgaris* L.), mizuna (*Brassica rapa* L. var. *japonica*), lettuce (*Lactuca sativa* L.), and corn salad (*Valerianella locusta* Laterr.). The juglone concentrations used were based on our previous germination study [13]. The plants were treated using (i) two control treatments, with K1 as the juglone extraction medium and vehicle control (0.17% dimethylsulphoxide (DMSO), 0.17% ethanol in H_2O) and K2 as the water control; (ii) positive control pure juglone treatments, prepared for final juglone concentrations of 1 mM and 10 µM in extraction medium; (iii) leaf juglone extract, prepared for the final juglone concentration of 10 µM in extraction medium. The juglone was dissolved in the extraction medium, as it is only partially soluble in water (52 mg/L), and thus, the required concentration of 1 mM control juglone cannot be achieved in water alone. As previously noted [13], studies that have used >100 µM juglone dissolved in water are questionable at best.

2.2. Growing Conditions

The experiment was conducted using nutrient film technique (NFT) hydroponics systems in a greenhouse, to better control the environment (especially to control the soil as a medium) and the juglone concentrations in the water. Five NFT systems were used, one for each treatment. Each NFT system had 4 rows with 10 plants of the same crop vegetable grown in each row, for a total of 5 biological repetitions per measure (5 for metabolomics studies and 5 for yield determination).

The plants were grown from seed in a greenhouse. After the appearance of the third leaf, the roots of the seedlings were washed to remove the growth substrate, and the seedlings were placed in plastic pots filled with rockwool. Overall, 50 evenly grown plants per crop vegetable were transferred to an NFT system, where they were grown for 3 weeks with added nutrients, as reported previously [16]. After this acclimatisation in the NFT system for 3 weeks, the treatments were added. When the crops reached technological maturity, they were collected and further analysed.

2.3. Chemicals and Plant Material

The plants were grown from seeds obtained as follows: *Beta vulgaris* L. 'Delta'; *Brassica rapa* L. var. *japonica* 'Mizuna grun'; *Lactuca sativa* L. 'Grazer Krauthäuptel 2 Treibstamm'; *Valerianella locusta* Laterr. 'Verte de Cambrai' (Austrosaat AG, Wien, Austria).

The leaf extract was prepared in the extraction medium, with HPLC–mass spectrometry (MS) used to determine the juglone content as accurately as possible. The control juglone and leaf extract dilutions were prepared as previously described [13].

The following standards were used: *p*-coumaric acid, ferulic acid, quercetin-3-*O*-galactoside, quercetin-3-*O*-glucoside, kaempferol-3-*O*-glucoside, apigenin-7-glucoside (Fluka Chemie GmbH, Buchs, Switzerland); neochlorogenic acid (3-caffeoylquinic acid), chlorogenic acid (*trans*-5-caffeoylquinic acid), cryptochlorogenic acid (4-caffeoylquinic acid), sinapic acid, caffeic acid, gallic acid, luteolin-7-glucoside, juglone (5-hydroxy-1,4-naphthoquinone) (Sigma-Aldrich Chemie GmbH, Steinheim, Germany); gluconapin and isorhamnetin-3-*O*-glucoside (Extrasynthese, Genay, France).

A Milli-Q water purification system (Millipore, Bedford, MA, USA) was used to bidistil and purify the water used in the preparation of the samples. The acetonitrile and formic acid used for the mobile phases for MS analysis were HPLC–MS grade (Fluka Chemie GmbH, Buchs, Switzerland). The methanol used for the phenolic compound extraction was HPLC–MS grade (Sigma-Aldrich, Steinheim, Germany).

2.4. Sampling of the Plants

When the plants had reached technological maturity, they were collected and analysed. First, all of the roots were cut from the plants. Once removed from the plants, the roots and leaves were weighed to determine the yield per plant for the different treatments. To determine the dry weights, five sets of roots and leaves per crop vegetable and treatment were placed in an oven at 105 °C, to constant mass. The remaining five leaves per crop vegetable and treatment were divided into three categories: (i) young leaves (undeveloped leaves); (ii) semi-old leaves (the remaining fully developed leaves); (iii) old leaves (four outer fully developed leaves). Thereafter, the fresh leaves were immediately frozen with liquid nitrogen and stored at −20 °C until further analysis.

2.5. Extraction of the Phenolic Compounds

The protocol for extraction of the individual phenolic compounds was as previously described [13]. Briefly, 200 mg of previously lyophilised samples of *B. vulgaris*, *B. rapa* var. *japonica*, *L. sativa*, and *V. locusta* were extracted at a tissue:solution ratio of 1:100 (w/v). The phenolics extraction medium used was 80% methanol and 3% formic acid in water.

2.6. Preparation of *J. regia* Leaf Extract

Leaves for the leaf extract were obtained on 10 September 2020, from a 24-year-old *J. regia* tree grown at a planting density of 10 m × 10 m. It belonged to the French cultivar 'Franquette', which has been most frequently studied and used in research. Leaves were collected from the south side of the tree, from the middle part of the canopy, placed in a paper bag, frozen with liquid nitrogen, and then lyophilised. After lyophilisation, the leaves were ground with an automatic grounder (IKA A11 Basic, IKA-Werke GmbH & Co., KG, Köln, Germany) and added to the extraction medium containing 0.17% DMSO and 0.17% ethanol in H_2O. The extraction was performed in an ultrasonic bath filled with ice

(Sonis 4; Iskra Pio, Sentjernej, Slovenia) for 60 min. The extraction was then centrifuged at 10,000× *g* for 10 min at 4 °C (5810 R; Eppendorf, Hamburg, Germany). A sample was taken from the extract (i.e., supernatant) for quantification of juglone by HPLC–MS. The samples were then diluted to prepare the required leaf extract containing 10 μM juglone (referred to here as 'leaf juglone'). Parallel juglone solutions were prepared from the juglone dissolved in the extraction medium (0.17% DMSO, 0.17% ethanol in H_2O) at the required final juglone concentrations (1 mM and 10 μM), referred to here as 'control juglone'.

2.7. HPLC–Mass Spectrometry Analysis of Individual Phenolic Compounds

The individual phenolic compounds were analysed on a UHPLC system (Vanquish; Thermo Scientific, Waltham, MA, USA). The diode detector used was at 350 nm for flavonols and 280 nm for the other phenolic compounds. The spectra were recorded between 200 nm and 600 nm. A C18 column (Gemini 150 × 4.60 mm; 3 μm; Phenomenex, Torrance, CA, USA) operated at 25 °C was used to separate the phenolic compounds. Solvent A was 0.1% formic acid with 3% acetonitrile in bi-distilled water (*v*/*v*/*v*), and solvent B was 0.1% formic acid with 3% bi-distilled water in acetonitrile (*v*/*v*/*v*). The flow rate of elution was 0.6 mL/min. The gradient, washing, and reconditioning of the column between samples were similar to those described previously [17], with minor modifications. The gradient used was as follows: 0–15 min, 5–20% B; 15–20 min, 20–30% B; 20–25 min, 30–50% B; 25–30 min, 50–90% B; 30–35 min, 90–100% B; 35–45 min, 100–5% B; 46–50 min, 5% B.

Identification of the phenolic compounds was achieved by tandem MS (MS/MS; LCQ Deca XP MAX; Thermo Scientific, Waltham, MA, USA), with heated electrospray ionisation operated in negative ion mode. The parameters were as follows: sheath temperature, 320 °C; sheath gas, 50 arb; auxiliary gas, 20 arb; ion spray voltage, 3.5 kV; capillary temperature, 320 °C; capillary voltage, 10.0 V; tube lens, −68 V. Scans were performed from *m*/*z* 50 to 2000. The collision energy was 35 eV, with helium used as the collision gas to achieve collision-induced dissociation. The Xcalibur 2.2 software (Thermo Fischer Scientific Institute, Waltham, MA, USA) was used for data acquisition.

Known compounds were identified and quantified using external standards, with the literature data and MS fragmentation used for identification of the unknown compounds. The quantification of unknown compounds was based on a similar standard. Total flavonols, total flavones, total hydroxycinnamic acids, and total analysed phenolics content (TAPC), which represents the sum of all of the identified phenolics, are expressed as g/kg fresh weight, while individual phenolic compounds are expressed as mg/kg fresh weight.

2.8. Statistical Analysis

The data were collated using Microsoft Excel 2016 and R commander (Package Rcmdr) version 2.7.1. (Team, R.D.C., 2008, Stanford, CA, USA). For each methodology, five biological repetitions were performed. Data are expressed as means ± standard error (SE). One-way analysis of variance (ANOVA) with Tukey tests was used to determine significant differences between treatments, and statistical means were calculated at the 95% confidence level to determine the significance of the differences.

3. Results and Discussion

3.1. Identification of Individual Phenolic Compounds in the Crop Vegetables

A total of 60 phenolic compounds were identified based on the previous literature data: 15 for *B. vulgaris*; 15 for *B. rapa* var. *japonica*; 17 for *L. sativa*; 13 for *V. locusta*. Some of these were identified for the first time in these crop vegetables. Of these 60 compounds, 9 were identified, and their fragmentation was confirmed using standards. The remaining compounds were tentatively identified according to their pseudo-molecular ions (i.e., $[M-H]^-$) and specific fragmentation patterns (i.e., MS^2, MS^3). The phenolic compounds identified, their fragmentation, and the standards used to express them, are given in Table 1. Representative chromatograms of the phenolic compounds identified can also be seen in the Supplementary Materials.

Table 1. Tentative identification of the 60 phenolic compounds from *B. vulgaris*, *B. rapa* var. *japonica*, *L. sativa*, and *V. locusta*, and the standards used.

Source	Compound	Rt (min)	[M-H]$^-$ (*m/z*)	MS^2 (*m/z*)	MS^3 (*m/z*)	Expressed as
Beta vulgaris L.	*p*-Coumaroylcaffeic acid	11.09	337	119 (100), 179 (42), 163 (12)		*p*-Coumaric acid
	p-Coumaric acid hexoside	12.17	325	163 (100), 145 (83), 119 (8), 235 (8)		*p*-Coumaric acid
	Ferulic acid hexoside	13.12	355	193 (100), 217 (54), 175 (29)		Ferulic acid
	Vitexin hexoside	14.85	593	311 (100), 341 (25)	283 (100), 191 (3)	Apigenin-7-glucoside
	Ferulic acid	15.13	193	149 (100), 178 (72), 134 (48)		Ferulic acid
	Ferulic acid derivative 1	17.48	443	267 (100), 193 (9)	249 (100), 113 (72), 175 (14)	Ferulic acid
	Ferulic acid derivative 2	17.79	639	443 (100)	267 (100), 193 (10)	Ferulic acid
	Vitexin hexoside	18.23	577	413 (100)	293 (100)	Apigenin-7-glucoside
	Luteolin dihexoside	18.89	609	285 (100), 257 (4)	257 (100), 241 (45), 229 (37), 151 (32) 213 (26)	Luteolin-7-glucoside
	Vitexin pentoside	18.89	563	413 (100)	293 (100)	Apigenin-7-glucoside
	Vitexin (apigenin-*C*-hexoside isomer)	19.47	431	311 (100)	283 (100), 191 (3)	Apigenin-7-glucoside
	Isorhamnetin dihexoside	19.93	639	315 (100), 300 (17)	300 (100), 287 (4)	Isorhamnetin-3-glucoside
	Vitexin hexoside derivative	20.19	651	607 (100)	457 (100), 293 (3)	Apigenin-7-glucoside
	Isorhamnetin rutinoside	20.68	609	315 (100), 300 (13)	300 (100), 287 (5)	Isorhamnetin-3-glucoside
	Malonyl pentosylvitexin	21.59	649	605 (100)	455 (100), 293 (5)	Apigenin-7-glucoside
Brassica rapa L. var. *japonica*	Neochlorogenic acid (3-caffeoylquinic acid)	9.27	353	191 (100), 179 (46), 135 (7)		Neochlorogenic acid
	Kaempferol-3-*O*-diglucoside-7-*O*-glucoside	10.34	771	609 (100)	285 (100), 284 (100), 429 (98), 257 (10), 179 (3)	Kaempferol-3-glucoside
	Gluconapin	11.57	372	259 (100), 275 (29), 194 (20), 130 (7)	139 (100) 97 (39), 199 (13), 241 (7)	Gluconapin
	Kaempferol-3-*O*-caffeoyldiglucoside-7-*O*-glucoside	12.19	933	771 (100)	609 (100)	Kaempferol-3-glucoside
	Kaempferol-3-*O*-sinapoyldiglucoside-7-*O*-glucoside	13.42	977	815 (100)	609 (100)	Kaempferol-3-glucoside
	Sinapoyl glycoside	13.42	385	223 (100), 247 (48), 205 (40)	164 (100), 208 (44), 179 (35)	Sinapic acid
	Kaempferol diglucoside	13.57	609	447 (100), 285 (13), 284 (2)	284 (100), 285 (46), 151 (5)	Kaempferol-3-glucoside
	Kaempferol-3-*O*-feruoylglucoside-7-*O*-glucoside	14.07	947	785 (100)	623 (100), 309 (92), 591 (43)	Kaempferol-3-glucoside
	Isorhamnetin-3-*O*-glucoside-7-*O*-glucoside	14.07	639	477 (100), 315 (9), 300 (1)		Isorhamnetin-3-glucoside

Table 1. *Cont.*

Source	Compound	Rt (min)	[M-H]$^-$ (m/z)	MS^2 (m/z)	MS^3 (m/z)	Expressed as
	Caffeoyl malate	17.50	295	179 (100), 133 (1)		Caffeic acid
	Hydroxyferuoyl malate	17.89	325	209 (100), 133 (23), 165 (5)		Ferulic acid
	Kaempferol hexoside derivative	19.92	567	447 (100)	285 (100), 284 (31)	Kaempferol-3-glucoside
	Coumaroyl malate	21.41	279	163 (100), 133 (19)		*p*-Coumaric acid
	Sinapoyl malate	22.06	339	223 (100)	164 (100), 208 (79), 179 (53)	Sinapic acid
	Feruloyl malate	22.24	309	193 (100), 133 (7)		Ferulic acid
Lactuca sativa L.	Dihydroxybenzoic acid hexoside	8.69	315	153 (100), 108 (19)		Gallic acid
	Esculetin glucoside	9.87	339	177 (100)		Gallic acid
	Chlorogenic acid (5-caffeoylquinic acid)	12.47	353	191 (100), 179 (4), 135 (1)		Chlorogenic acid
	Galloyl hexoside	13.13	331	313 (100), 168 (61), 125 (19), 169 (24)		Gallic acid
	Cryptochlorogenic acid (4-caffeoylquinic acid)	14.21	353	191 (100), 179 (3), 135 (1)		Cryptochlorogenic acid
	Sinapoyl hexoside derivative	15.38	431	385 (100)	223 (100), 179 (41), 208 (27)	Sinapic acid
	cis 5-*O*-*p*-Coumaroylquinic acid	15.91	337	191 (100), 163 (8)		*p*-Coumaric acid
	trans 5-*O*-*p*-Coumaroylquinic acid	17.24	337	191 (100), 163 (7)		*p*-Coumaric acid
	Caffeoyl malate	17.60	295	179 (100), 133 (53)		Caffeic acid
	Quercetin-3-*O*-galactoside	19.34	463	301 (100), 300 (3)		Quercetin-3-*O*-galactoside
	Quercetin-3-*O*-glucoside	20.68	463	301 (100), 300 (19)		Quercetin-3-*O*-glucoside
	Kaempferol-3-*O*-glucuronide	21.42	463	285 (100), 284 (41)		Kaempferol-3-glucoside
	Quercetin-3-*O*-glucuronide	21.96	477	301 (100), 300 (4)		Quercetin-3-*O*-glucoside
	Quercetin 3-(6″-malonylglucoside)	22.55	549	505 (100)	301 (100), 300 (57)	Quercetin-3-*O*-glucoside
	Quercetin-3-(6″-acetylglucoside)	23.70	505	301 (100), 300 (64)		Quercetin-3-*O*-glucoside
	Caffeoyltartaric acid hexoside 1	26.63	473	293 (100), 311 (99)		Caffeic acid
	Caffeoyltartaric acid hexoside 2	29.21	473	311 (100), 293 (87)		Caffeic acid
Valerianella locusta Laterr.	4-Hydroxyphenylaoyl glucoside derivative	7.16	359	313 (100)	151 (100), 269 (2), 185 (1)	Gallic acid
	Chlorogenic acid (5-caffeoylquinic acid)	12.37	353	191 (100), 179 (4), 135 (1)		Chlorogenic acid
	Cryptochlorogenic acid (4-caffeoylquinic acid)	14.10	353	191 (100), 179 (3), 135 (1)		Cryptochlorogenic acid
	cis 5-*O*-*p*-Coumaroylquinic acid	15.89	337	191 (100), 163 (8)		*p*-Coumaric acid
	cis 5-*O*-Feruoylquinic acid	16.88	367	191 (100), 173 (3)		Ferulic acid
	trans 5-*O*-*p*-Coumaroylquinic acid	17.22	337	191 (100), 163 (7)		*p*-Coumaric acid
	trans 5-*O*-Feruoylquinic acid	18.02	367	191 (100), 173 (2)		Ferulic acid

Table 1. *Cont.*

Source	Compound	Rt (min)	$[M-H]^-$ (*m/z*)	MS^2 (*m/z*)	MS^3 (*m/z*)	Expressed as
	Luteolin-7-rutinoside	19.62	593	285 (100)	285 (100), 241 (24), 217 (14), 199 (16), 175 (17), 151 (6)	Luteolin-7-glucoside
	Diosmetin apiosylglucoside	21.74	593	299 (100), 284 (17)	284 (100)	Luteolin-7-glucoside
	Diosmin (diosmetin-7-*O*-rutinoside)	22.07	607	299 (100), 284 (24)		Luteolin-7-glucoside
	Dicaffeoylquinic acid	22.58	515	353 (100), 179 (2)		Caffeic acid
	Apigenin-rutinoside	24.75	577	531 (100), 269 (98)		Apigenin-7-glucoside
	Caffeic acid hexoside derivative	25.48	637	535 (100), 341 (23)	161 (100), 179 (57), 341 (57)	Caffeic acid

Rt, retention time; $[M-H]^-$, pseudo-molecular ion identified in negative ion mode; (), relative abundance of fragment ions.

For the 25 hydroxycinnamic acids, 4 were identified through the use of standards and their fragmentation: neochlorogenic acid (3-caffeoylquinic acid); cryptochlorogenic acid (4-caffeoylquinic acid); chlorogenic acid (*trans*-5-caffeoylquinic acid); ferulic acid. The remaining 21 hydroxycinnamic acids were identified through their typical fragmentation patterns. *p*-coumaric acid and caffeic acid derivatives were identified through their fragmentation patterns of MS^n ion m/z 163 and 119 and m/z 179, as reported by Liu et al. [18]. *cis* 5-*O*-*p*-coumaroylquinic acid, *trans* 5-*O*-*p*-coumaroylquinic acid, caffeoyltartaric acid hexoside, and dicaffeoylquinic acid were identified through their fragmentation patterns as previously reported for *V. locusta* and *L. sativa* by Hernández et al. [19] and Abu-Reidah et al. [20]. Coumaroyl malate, sinapoyl malate, sinapoyl glycoside, feruloyl malate, and caffeoyl malate were identified by their specific fragmentation patterns as reported for pak choi (*Brassica campestris* L. ssp. *chinensis* var. *comunis*) by Harbaum et al. [21], which are described here for the first time for *B. rapa* var. *japonica*. Ferulic acid derivatives were identified through their fragmentation patterns of MS^n ion m/z 193 and 175, as reported by Vieira et al. [22], and *cis* 5-*O*-feruoylquinic acid and *trans* 5-*O*-feruoylquinic acid were identified through their fragmentation patterns as reported by Hernández et al. [19]. There were only two hydroxybenzoic acids identified, dyhydroxybenzoic acid hexoside and galloyl hexoside, both of which were previously identified by Abu-Reidah et al. [20].

All of the 11 flavones identified were in *B. vulgaris* and *V. locusta*. Four flavones (luteolin-7-rutinoside, diosmetin apiosylglucoside, diosmin, and apigenin-rutinoside) were identified in *V. locusta*, all of which were previously reported by Hernández et al. [19], and seven (vitexin hexoside 1 and 2, vitexin pentoside, luteolin dihexoside, vitexin, vitexin hexoside derivate, and malonyl pentosylvitexin) in *B. vulgaris*. Vitexin, vitexin hexoside, vitexin pentoside, and malonyl pentosylvitexin were previously reported in *B. vulgaris* by Vissers et al. [23] and AbdEl-Ghffar et al. [24], while, luteolin dihexoside and vitexin hexoside derivatives are reported here for the first time in *B. vulgaris*.

There were 13 flavonols identified, 2 in *B. vulgaris*, 7 in *B. rapa* var. *japonica*, and 6 in *L. sativa*. In *B. vulgaris*, both of the flavonols identified (isorhamnetin dihexoside, rutinoside) are reported here for the first time in *B. vulgaris*, and these followed the typical fragmentation pattern MS^n m/z 315 and 300, MS^{n+1} m/z 300, 287 for isorhamnetin. In *B. rapa* var. *japonica*, of the seven flavonols identified, only kaempferol hexoside derivative is reported for the first time, which followed the kaempferol typical fragmentation pattern MS^n m/z 285 and 284. The other flavanols identified in *B. rapa* var. *japonica* were previously identified in pak choi (*Brassica campestris* L. ssp. *chinensis* var. *comunis*) and curly kale (*Brassica oleracea* L. Convar. *acephala* var. *sabellica*) by Harbaum et al. [21] and Olsen et al. [25], and here for the first time in *B. rapa* var. *japonica*. In *L. sativa*, all six flavonols identified were previously reported by Medic et al. [13].

3.2. Effects of the Juglone Treatments on the Crop Vegetable Yields

The control effects of juglone (1 mM, 10 μM) and the leaf extract with 10 μM juglone were not the same across these different crop vegetables, as was also seen previously in germination studies by Medic et al. [13]. The effects of the different treatments on the crop yields and root weights are shown in Figure 1 and in Supplementary Materials.

Interestingly, the highest crop yields and root weights were obtained for the 10 μM leaf juglone treatments for *B. rapa* var. *japonica* and *L. sativa*, and for the K2 control as the second-highest for *B. vulgaris* and *V. locusta*. This might be because the leaf extracts contain other nutrients and beneficial allelochemicals besides juglone that can stimulate plant growth and increase crop yields. In contrast to our previous seed germination study [13], where the 10 μM juglone leaf extract was not different from the control treatments, here, the 10 μM leaf juglone even showed higher yields, compared with the other treatments. Indeed, this would appear logical considering that most biostimulants are produced from plant waste [26]. However, higher concentrations of juglone in the leaf extract might have inhibitory effects also on crop yields, as observed in the germination trial by Medic et al. [13], where 1 mM juglone leaf extracts significantly inhibited seed germination and seedling growth. Therefore, further studies are needed before a conclusion can be drawn.

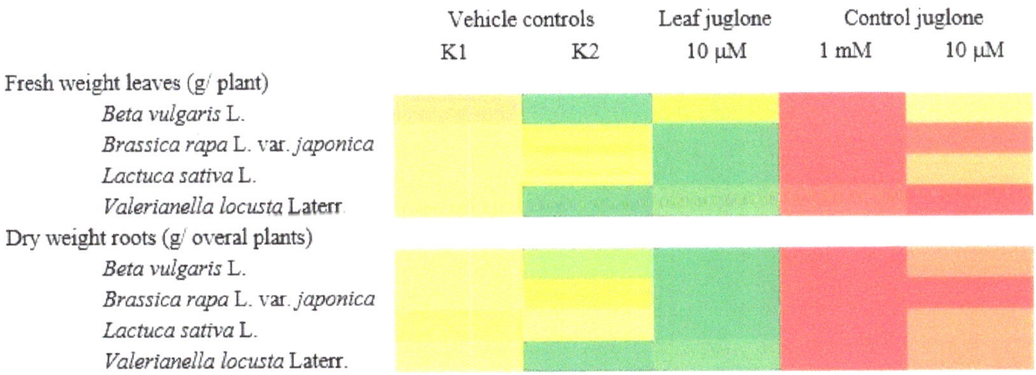

Figure 1. Heat map showing leaf fresh weights and root dry weights, from highest (green) to lowest (red) values between treatments (columns). K1: extraction medium control (0.17% DMSO, 0.17% ethanol in H_2O); K2: water control.

It is known that juglone is absorbed from the soil through the roots and, therefore, acts first on the roots of the plants. It penetrates the plasma membrane of the root cells and induces depolarisation by blocking the K^+ channels, which inhibits root and, consequently, shoot, nutrient uptake, and growth [27]. This explains the lowest yields for all of these crop vegetables, as well as the lowest root weights, which were seen for 1 mM, followed by 10 μM, control juglone treatments. Overall, juglone showed allelopathic effects on the yields of all of these crop vegetables. However, the yields of *B. vulgaris* and *L. sativa* appeared to be less affected by the lower concentration of the control juglone (10 μM) than that of *B. rapa* var. *japonica* and *V. locusta* and would, therefore, be a better choice for cultivation in soils where juglone is still present.

3.3. *Effects of the Juglone Treatments on the Crop Vegetable Quality*

Looking at the effects of the juglone treatments with the same 10 μM concentration of control juglone and leaf juglone on the crop quality in terms of TAPC, it can be seen that these were not the same across the crop vegetables (Figures 2–5), as also previously reported by Medic et al. [13] and Ercisli and Turkkal [28]. While *L. sativa* and *V. locusta* showed similar responses, the responses of *B. vulgaris* and *B. rapa* var. *japonica* were almost contrary. All of the data that showed significant differences are further detailed in the Supplementary Materials.

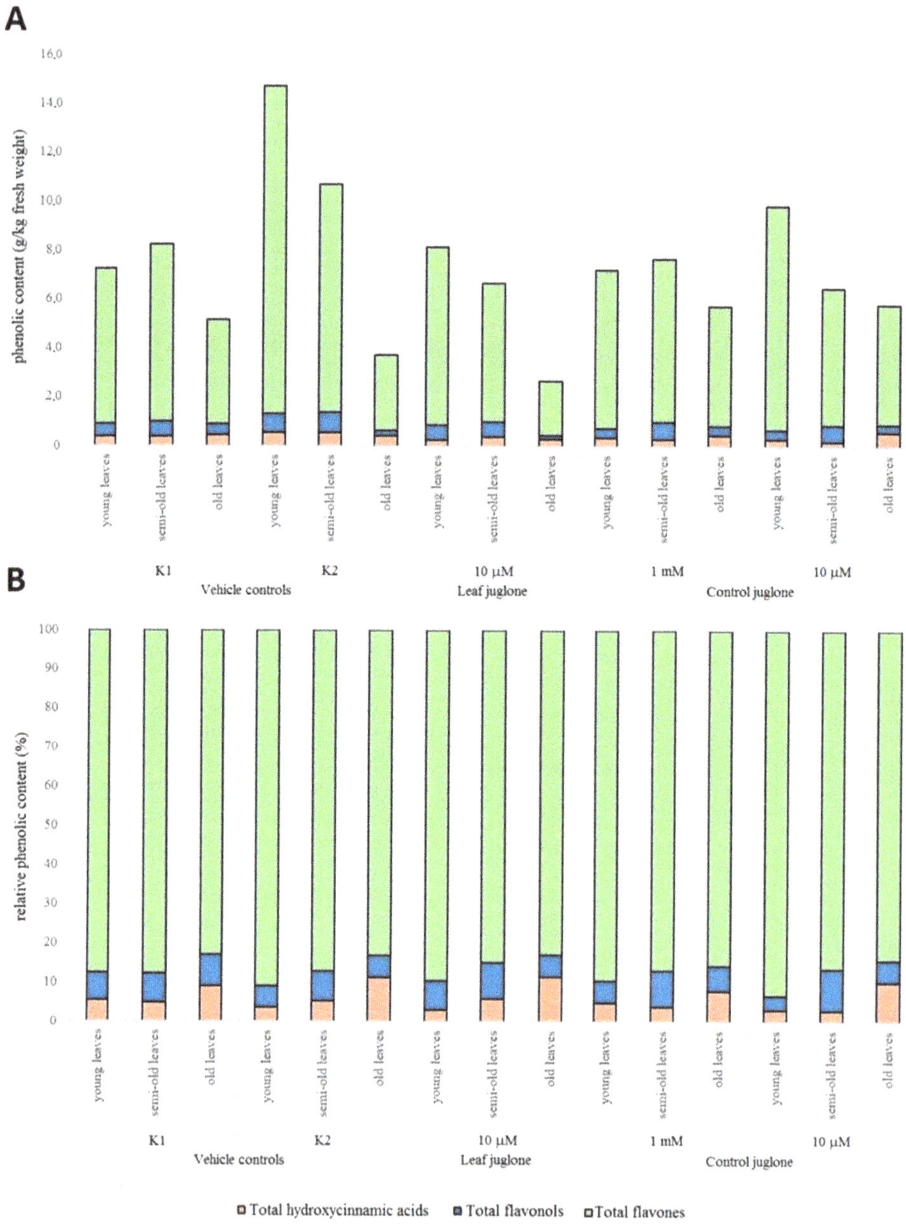

Figure 2. Contents of the total phenolic groups identified in *B. vulgaris* expressed relative to fresh weight (**A**) and as proportions of the total phenolic groups identified (**B**). K1: extraction medium control (0.17% DMSO, 0.17% ethanol in H_2O); K2: water control. Young leaves, undeveloped leaves; semi-old leaves, remaining fully developed leaves; old leaves, four outer fully developed leaves.

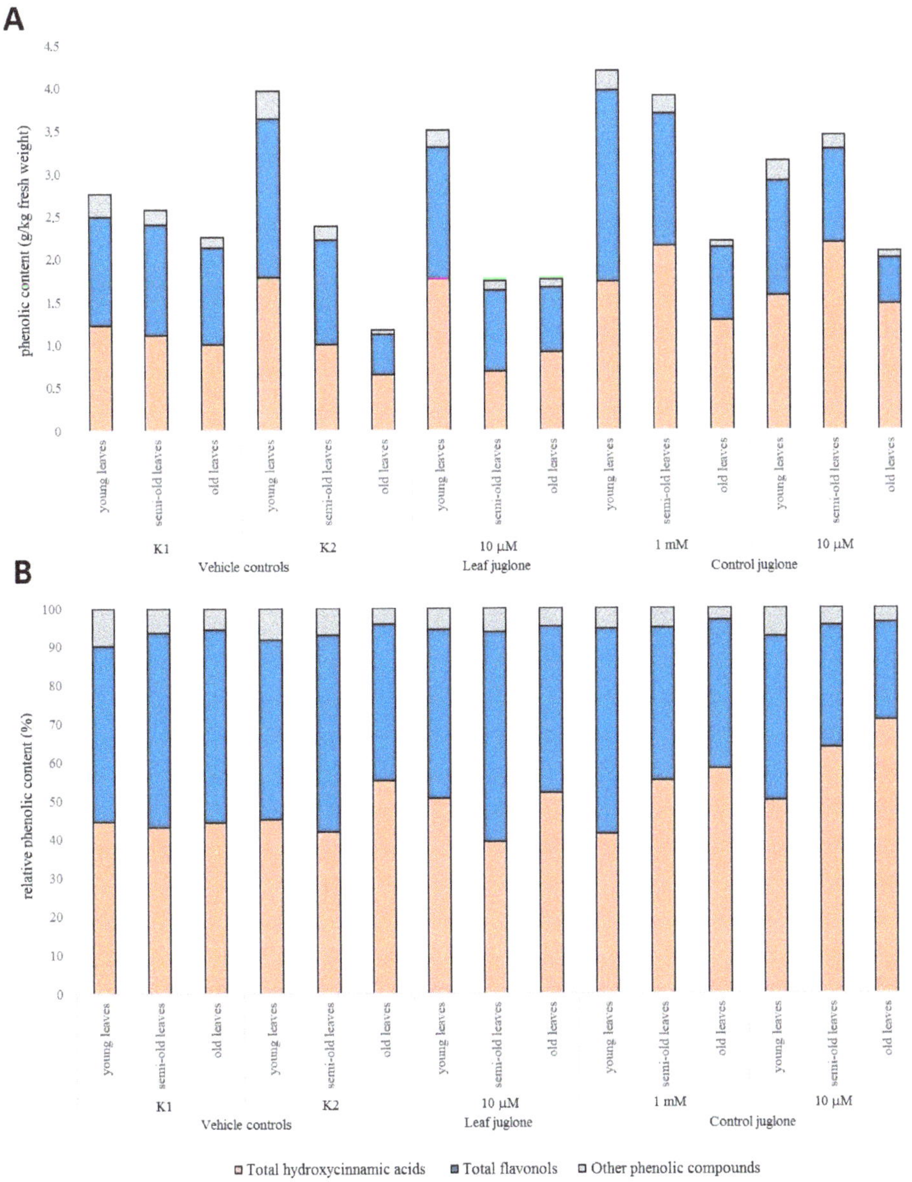

Figure 3. Contents of the total phenolic groups identified in *B. rapa* var. *japonica* expressed relative to fresh weight (**A**) and as proportions of the total phenolic groups identified (**B**). K1: extraction medium control (0.17% DMSO, 0.17% ethanol in H_2O); K2: water control. Young leaves, undeveloped leaves; semi-old leaves, remaining fully developed leaves; old leaves, four outer fully developed leaves.

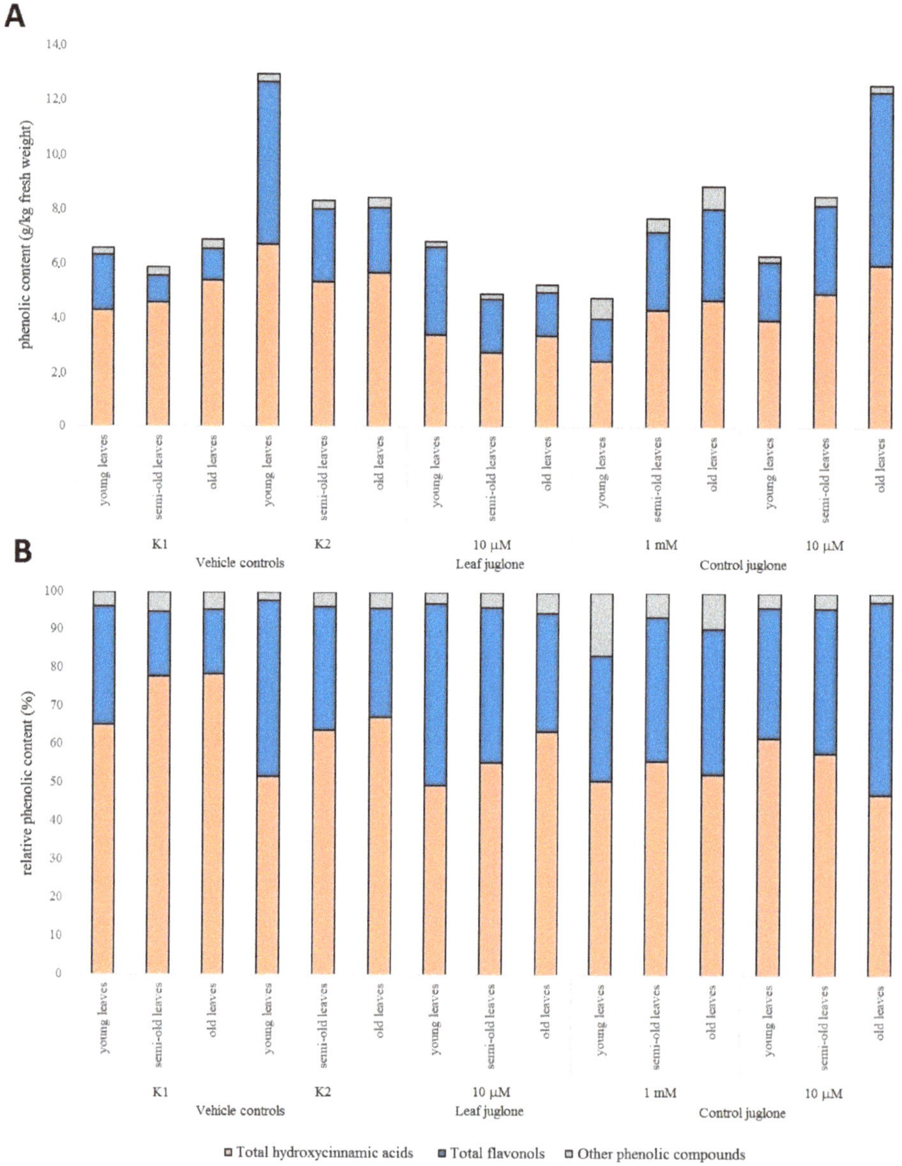

Figure 4. Contents of the total phenolic groups identified in *L. sativa* expressed relative to fresh weight (**A**) and as proportions of the total phenolic groups identified (**B**). K1: extraction medium control (0.17% DMSO, 0.17% ethanol in H_2O); K2: water control. Young leaves, undeveloped leaves; semi-old leaves, remaining fully developed leaves; old leaves, four outer fully developed leaves.

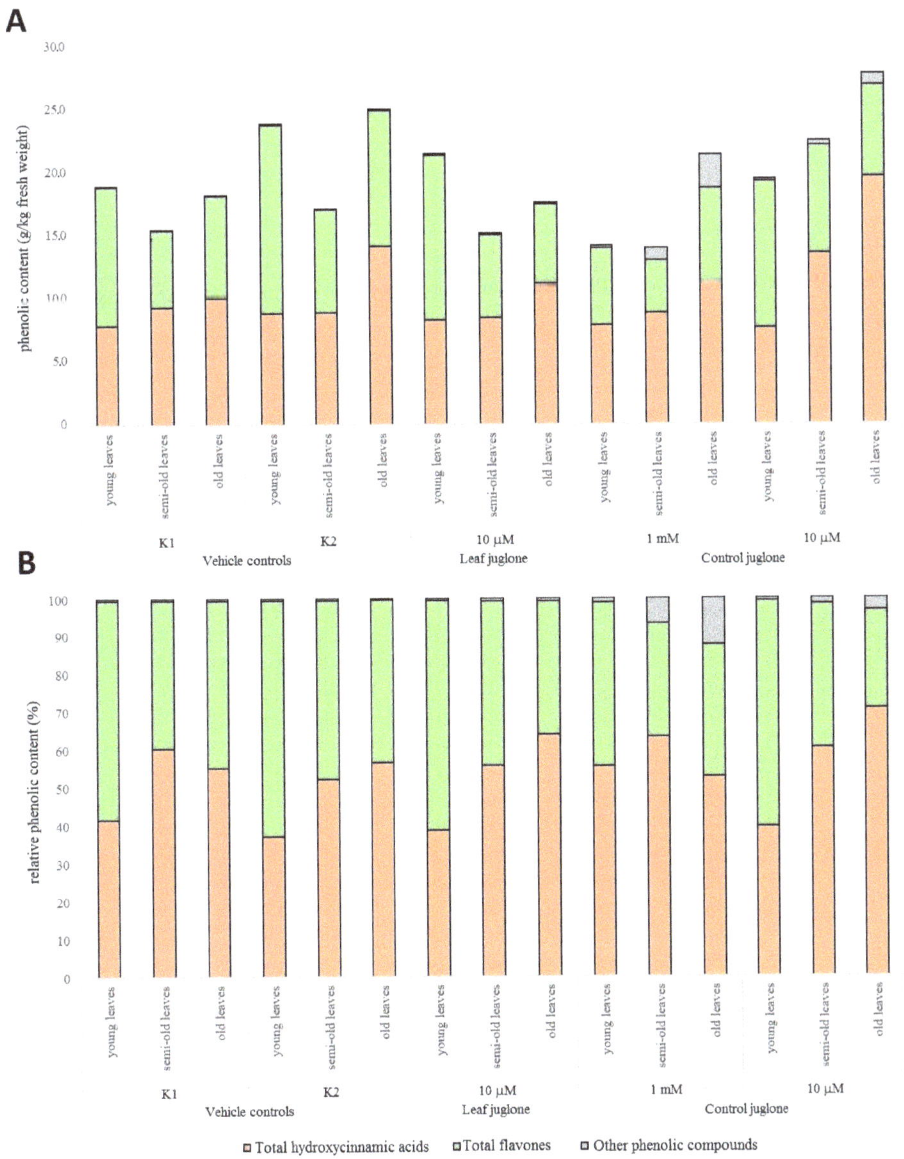

Figure 5. Contents of the total phenolic groups identified in *V. locusta* expressed relative to fresh weight (**A**) and as proportions of the total phenolic groups identified (**B**). K1: extraction medium control (0.17% DMSO, 0.17% ethanol in H_2O); K2: water control. Young leaves, undeveloped leaves; semi-old leaves, remaining fully developed leaves; old leaves, four outer fully developed leaves.

As can be seen in Figure 2A, the extraction medium control (K1; with vehicles used for the control juglone and leaf juglone treatments) affected the quality of *B. vulgaris*, compared with K2 (water control), with lower TAPC in the younger and semi-old leaves. There were no differences between the K1, 1 mM and 10 µM control juglone, and 10 µM leaf juglone treatments, except for TAPC in older leaves treated with 10 µM leaf juglone, which had lower TAPC than for the other treatments. Figure 2B shows that for *B. vulgaris*, there were no differences in the proportions of total hydroxycinnamic acid, total flavonols, and flavones, indicating that the quality of *B. vulgaris* was not greatly affected by the 1 mM and 10 μM control juglone and 10 μM leaf juglone treatments. In terms of quality, the younger leaves tended to have the highest TAPC and, therefore, the highest quality, while the older leaves had lower TAPC, which is usually the case for all plants [29]. The higher TAPC in the younger leaves is the result of the plant defence mechanisms, as leaves with higher TAPC are better protected against bacterial infections than older leaves [30]. The TAPC of *B. vulgaris* was consistent with previous measures of Vissers et al. [23]; however, most of the phenolic compounds identified, as well as the highest contents seen in the present study, were flavones, compared with the phenolic acids reported by Vissers et al. [23].

The vehicle effect (i.e., K1 vs. K2) on quality seen for *B. vulgaris* was not seen for *B. rapa* var. *japonica*, *L. sativa*, or *V. locusta*. As shown in Figure 3A, in *B. rapa* var. *japonica*, there were no clear trends seen for TAPC, and thus, 1 mM and 10 μM control juglone and 10 μM leaf juglone did not have any effects on the crop quality. However, if the contents of hydroxycinnamic acids and flavonols are considered, it can be seen that the 1 mM and 10 μM control juglone treatments resulted in higher hydroxycinnamic acids contents in the semi-old leaves, while the semi-old leaves treated with 10 μM leaf juglone had lower hydroxycinnamic acids content, compared with both controls (K1, K2), as can be seen in Figure 3B. In addition, the treatments with 1 mM and 10 μM control juglone also affected the flavonols content, with increased total flavonols in the older leaves, compared with the young leaves, which was not seen for the control or 10 μM leaf juglone treatments. Overall, in the semi-old leaves, 10 μM leaf juglone decreased the hydroxycinnamic acids content, while the flavonols content increased, in contrast to the 1 mM and 10 μM control juglone treatments, for which in the semi-old leaves, the hydroxycinnamic acids content increased and the flavonols content decreased. This suggests that other allelochemicals are present in *J. regia* that can have actions similar to those of juglone, as also previously indicated by Medic et al. [13]. The TAPC of *B. vulgaris* was consistent with that of Harbaum et al. [21] in *Brassica campestris* L. *chinensis* var. *communis* and higher than previously reported in *B. rapa* var. *japonica* by Khanam et al. [31]. Most of the phenolic compounds identified in the present study, as well as the highest contents, were for the hydroxycinnamic acids, similar to a previous report by Khanam et al. [31].

Similar to *B. rapa* var. *japonica*, in *L. sativa* the majority of the identified and quantified phenolic compounds represented the hydroxycinnamic acids, followed by the flavonols, which is in agreement with Abu-Reidah et al. [20] and Ribas-Agustí et al. [32]. TAPC was slightly lower than previously reported by Santos et al. [33]. Figure 4 shows that the 1 mM and 10 μM control juglone treatments affected the metabolic response of *L. sativa* in the same way as for *B. rapa* var. *japonica*, while the 10 μM leaf juglone treatment had no effects on the secondary metabolites. The highest TAPCs were seen for the older leaves in the 1 mM and 10 μM control juglone treatments, compared with the younger leaves, mainly due to the higher flavonols and hydroxycinnamic acids contents. Considering the relative values of the phenolic compounds (Figure 4B), it can be seen that the treatments with 1 mM and 10 μM control juglone resulted in higher proportions of hydroxycinnamic acids and lower proportions of flavonols in the younger leaves, compared with the older leaves, contrary to other treatments (K1, K2, 10 μM leaf juglone), in which the proportions of hydroxycinnamic acids were higher in the older leaves, and proportions of flavonols were higher in the younger leaves. This has been observed previously for *B. vulgaris*, *B. rapa* var. *japonica*, and *V. locusta*, and it suggests a uniform metabolomics response of these crop vegetables to produce hydroxycinnamic acids when affected by the allelochemical juglone.

As can be seen in Figure 5, both of the 1 mM and 10 μM control juglone treatments had effects on the quality of *V. locusta*, while there were no differences between the controls and the 10 μM leaf juglone treatments. The metabolic responses of *V. locusta* were similar to those of *L. sativa*, with the older leaves showing the highest levels of TAPC, mainly due to the increase in the hydroxycinnamic acids content in the older leaves treated with juglone. Apart from this difference in the hydroxycinnamic acids in the 1 mM and 10 μM control juglone treated plants in the old leaves, the major difference was in the contents of the other phenolic compounds, which were higher in older leaves, compared with younger leaves. The TAPC of *V. locusta* is in agreement with that reported by Hernández et al. [19]. Most of the phenolic compounds identified in the present study, as well as the highest contents, were hydroxycinnamic acids, followed by flavones, as previously reported by Hernández et al. [19].

4. Conclusions

As also previously reported by Medic et al. [13], we can confirm that each of these crop vegetables responds differently to the allelochemicals, and although juglone appears to be the most toxic of the allelochemicals, it might not be the only one in these leaf extracts of *J. regia*. Here, the treatment with the leaf extract (i.e., 10 μM leaf juglone) resulted in the highest yields and comparable qualities in terms of TAPC, compared with the other treatments. This either means that this leaf extract of *J. regia* also contains beneficial allelochemicals that can stimulate growth, or that the leaf extract is simply a good source of additional minerals and nutrients that stimulate growth without affecting the metabolic responses of the plants themselves. The possibility that different concentrations of *J. regia* leaf extracts have different biostimulatory effects on such crop yields should be further investigated.

In addition to the identification and quantification of the phenolic compounds in these crop vegetable leaves of different ages, the main objective of this study was to investigate the yields, quality, and metabolic responses of these different crop vegetables, so as to determine which of them cannot thrive in soils containing juglone or other *J. regia* allelochemicals. From the data obtained, it can be concluded that *B. vulgaris* and *L. sativa* are more suitable for planting in soils where walnuts were previously grown, since, although they were affected by the 1 mM control juglone, they showed fewer negative effects for the 10 μM control juglone in terms of crop yields. In terms of crop quality, the smallest effects of the different treatments were seen for *B. vulgaris*. Both *B. rapa* var. *japonica* and *V. locusta* showed lower yields even for the lower of the control juglone treatments (i.e., 10 μM), and the quality was also affected, so cultivation in juglone-containing soils should be avoided.

Apart from identifying and quantifying 60 compounds from *B. vulgaris*, *B. rapa* var. *japonica*, *L. sativa*, and *V. locusta*, with some reported for these crop vegetables for the first time, this study also serves as a basis for the selection of more suitable crops in the early years of planting of crops in the soils where walnuts were previously planted. The quality of the leaves of these crop vegetables at different ages and the quality and yields of these selected crop vegetables grown under the influence of allelochemicals were also investigated, to determine the influence of allelochemicals on their metabolomics and, thus, on the uptake of phenolic compounds considered to be beneficial to human health. The data obtained could be relevant for future studies on crop management in different soils and the use of allelochemicals to modify the phytochemical composition of vegetables.

Supplementary Materials: The following supporting information can be downloaded at: https://www.mdpi.com/article/10.3390/agronomy12020347/s1, Figure S1: Representative full scan for the HPLC–MS analysis, and identification of the phenolic compounds for *Beta vulgaris* L., Figure S2: Representative full scan for the HPLC–MS, and identification of the phenolic compounds for *Brassica rapa* L. var. *japonica*, Figure S3: Representative full scan for the HPLC–MS, and identification of the phenolic compounds for *Lactuca sativa* L., Figure S4: Representative full scan for the HPLC–MS, and identification of the phenolic compounds for *Valerianella locusta* Laterr., Table S1: Yields for the different crop vegetables in terms of leaf fresh weight and dry matter, and root dry weight, Table S2: Individual phenolic compounds quantified in *Beta vulgaris* L., Table S3: Individual compounds quantified in *Brassica rapa* L. var. *japonica*, Table S4: Individual compounds quantified in *Lactuca sativa* L., Table S5: Individual compounds quantified in *Valerianella locusta* Laterr.

Author Contributions: Conceptualisation, A.M. and A.S.; data curation, A.M. and T.Z.; formal analysis, A.M.; funding acquisition, A.M., M.H. and R.V.; investigation, A.M., T.Z. and M.C.G.; methodology, A.M.; project administration, R.V.; resources, A.M., M.H. and R.V.; software, A.M.; supervision, A.S. and R.V.; validation, A.S. and R.V.; visualisation, A.M., T.Z. and A.S.; writing—original draft preparation, A.M.; writing—review and editing, T.Z., A.S., M.H. and R.V. All authors have read and agreed to the published version of the manuscript.

Funding: This study is a part of programme P4-0013-0481, which is funded by the Slovenian Research Agency (ARRS) and the infrastructural centre IC RRC-AG (IO-0022-0481-001).

Institutional Review Board Statement: Not applicable.

Informed Consent Statement: Not applicable.

Data Availability Statement: Part of the data presented in this study are available in Supplementary Materials here. The remaining data presented in this study are available on request from the corresponding author. The remaining data are not publicly available due to privacy.

Conflicts of Interest: The authors declare no conflict of interest.

References

1. Ahmed, S.; Shanks, C.B. Quality of Vegetables Based on Total Phenolic Concentration Is Lower in More Rural Consumer Food Environments in a Rural American State. *Int. J. Environ. Res. Public Health* **2017**, *14*, 924. [CrossRef] [PubMed]
2. Liu, R.H. Health-Promoting Components of Fruits and Vegetables in the Diet. *Adv. Nutr. Int. Rev. J.* **2013**, *4*, 384S–392S. [CrossRef] [PubMed]
3. Medic, A.; Jakopic, J.; Solar, A.; Hudina, M.; Veberic, R. Walnut (*J. regia*) Agro-Residues as a Rich Source of Phenolic Compounds. *Biology* **2021**, *10*, 535. [CrossRef] [PubMed]
4. Wolfe, K.L.; Kang, X.; He, X.; Dong, M.; Zhang, Q.; Liu, R.H. Cellular Antioxidant Activity of Common Fruits. *J. Agric. Food Chem.* **2008**, *56*, 8418–8426. [CrossRef] [PubMed]
5. Di Noia, J. Defining Powerhouse Fruits and Vegetables: A Nutrient Density Approach. *Prev. Chronic Dis.* **2014**, *11*, E95. [CrossRef]
6. Medic, A.; Zamljen, T.; Hudina, M.; Veberic, R. Identification and Quantification of Naphthoquinones and Other Phenolic Compounds in Leaves, Petioles, Bark, Roots, and Buds of *Juglans regia* L., Using HPLC–MS/MS. *Horticulturae* **2021**, *7*, 326. [CrossRef]
7. Ahmed, S.; Stepp, J.R. Beyond yields: Climate change effects on specialty crop quality and agroecological management. *Elem. Sci. Anthr.* **2016**, *4*, 92. [CrossRef]
8. Medic, A.; Solar, A.; Hudina, M.; Veberic, R. Phenolic Response to Walnut Anthracnose (*Ophiognomonia leptostyla*) Infection in Different Parts of *Juglans regia* Husks, Using HPLC–MS/MS. *Agriculture* **2021**, *11*, 659. [CrossRef]
9. Li, Z.-H.; Wang, Q.; Ruan, X.; Pan, C.-D.; Jiang, D.-A. Phenolics and Plant Allelopathy. *Molecules* **2010**, *15*, 8933–8952. [CrossRef]
10. Wink, M. Introduction: Biochemistry, Role and Biotechnology of Secondary Metabolites. *Annu. Plant Rev. Online* **2018**, *3*, 1–17. [CrossRef]
11. Zubay, P.; Kunzelmann, J.; Ittzés, A.; Zámboriné, É.N.; Szabó, K. Allelopathic effects of leachates of *Juglans regia* L., *Populus tremula* L. and juglone on germination of temperate zone cultivated medicinal and aromatic plants. *Agrofor. Syst.* **2021**, *95*, 431–442. [CrossRef]
12. Islam, A.K.M.M.; Widhalm, J.R. Agricultural Uses of Juglone: Opportunities and Challenges. *Agronomy* **2020**, *10*, 1500. [CrossRef]
13. Medic, A.; Zamljen, T.; Slatnar, A.; Hudina, M.; Veberic, R. Is Juglone the Only Naphthoquinone in *Juglans regia* L. with Allelopathic Effects? *Agriculture* **2021**, *11*, 784. [CrossRef]
14. Aliskan, I.K.; Terzi, I. Allelopathic effects of walnut leaf extracts and juglone on seed germination and seedling growth. *J. Hortic. Sci. Biotechnol.* **2001**, *76*, 436–440. [CrossRef]
15. de Scisciolo, B.; Leopold, D.J.; Walton, D.C. Seasonal patterns of juglone in soil beneath *Juglans nigra* (black walnut) and influence of *J. nigra* on understory vegetation. *J. Chem. Ecol.* **1990**, *16*, 1111–1130. [CrossRef] [PubMed]
16. Maršić, N.K.; Mikulič-Petkovšek, M.; Hudina, M.; Veberič, R.; Slatnar, A. Leafy Asian vegetables cultivated on a floating hydroponic system and a substrate culture, during the autumn period in greenhouse. *Acta Hortic.* **2021**, *1320*, 413–420. [CrossRef]
17. Medic, A.; Jakopic, J.; Hudina, M.; Solar, A.; Veberic, R. Identification and quantification of the major phenolic constituents in *Juglans regia* L. peeled kernels and pellicles, using HPLC–MS/MS. *Food Chem.* **2021**, *352*, 129404. [CrossRef]
18. Liu, P.; Li, L.; Song, L.; Sun, X.; Yan, S.; Huang, W. Characterisation of phenolics in fruit septum of *Juglans regia* Linn. by ultra performance liquid chromatography coupled with Orbitrap mass spectrometer. *Food Chem.* **2019**, *286*, 669–677. [CrossRef]
19. Hernández, V.; Botella, M.; Hellín, P.; Cava, J.; Fenoll, J.; Mestre, T.; Martínez, V.; Flores, P. Phenolic and Carotenoid Profile of Lamb's Lettuce and Improvement of the Bioactive Content by Preharvest Conditions. *Foods* **2021**, *10*, 188. [CrossRef]

20. Abu-Reidah, I.M.; Contreras, M.D.M.; Arraez-Roman, D.; Carretero, A.S.; Fernández-Gutiérrez, A. Reversed-phase ultra-high-performance liquid chromatography coupled to electrospray ionization-quadrupole-time-of-flight mass spectrometry as a powerful tool for metabolic profiling of vegetables: *Lactuca sativa* as an example of its application. *J. Chromatogr. A* **2013**, *1313*, 212–227. [CrossRef]
21. Harbaum, B.; Hubbermann, E.M.; Wolff, C.; Herges, R.; Zhu, Z.; Schwarz, K. Identification of Flavonoids and Hydroxycinnamic Acids in Pak Choi Varieties (*Brassica campestris* L. ssp. *chinensis* var. *communis*) by HPLC–ESI-MSnand NMR and Their Quantification by HPLC–DAD. *J. Agric. Food Chem.* **2007**, *55*, 8251–8260. [CrossRef] [PubMed]
22. Vieira, V.; Pereira, C.; Pires, T.C.; Calhelha, R.C.; Alves, M.J.; Ferreira, O.; Barros, L.; Ferreira, I.C. Phenolic profile, antioxidant and antibacterial properties of *Juglans regia* L. (walnut) leaves from the Northeast of Portugal. *Ind. Crop. Prod.* **2019**, *134*, 347–355. [CrossRef]
23. Vissers, A.; Kiskini, A.; Hilgers, R.; Marinea, M.; Wierenga, P.A.; Gruppen, H.; Vincken, J.-P. Enzymatic Browning in Sugar Beet Leaves (*Beta vulgaris* L.): Influence of Caffeic Acid Derivatives, Oxidative Coupling, and Coupled Oxidation. *J. Agric. Food Chem.* **2017**, *65*, 4911–4920. [CrossRef] [PubMed]
24. El-Ghffar, E.A.A.; Hegazi, N.M.; Saad, H.H.; Soliman, M.M.; El-Raey, M.A.; Shehata, S.M.; Barakat, A.; Yasri, A.; Sobeh, M. HPLC–ESI–MS/MS analysis of beet (*Beta vulgaris*) leaves and its beneficial properties in type 1 diabetic rats. *Biomed. Pharmacother.* **2019**, *120*, 109541. [CrossRef] [PubMed]
25. Olsen, H.; Aaby, K.; Borge, G.I.A. Characterization and Quantification of Flavonoids and Hydroxycinnamic Acids in Curly Kale (*Brassica oleracea* L. Convar. *acephala* Var. *sabellica*) by HPLC-DAD-ESI-MSn. *J. Agric. Food Chem.* **2009**, *57*, 2816–2825. [CrossRef] [PubMed]
26. Colla, G.; Rouphael, Y. Biostimulants in horticulture. *Sci. Hortic.* **2015**, *196*, 1–2. [CrossRef]
27. Duroux, L.; Delmotte, F.M.; Lancelin, J.-M.; Kéravis, G.; Jay-Allemand, C. Insight into naphthoquinone metabolism: β-glucosidase-catalysed hydrolysis of hydrojuglone β-d-glucopyranoside. *Biochem. J.* **1998**, *333*, 275–283. [CrossRef]
28. Ercisli, S.; Turkkal, C. Allelopathic effects of juglone and walnut leaf extracts on growth, fruit yield and plant tissue composition in strawberry cvs. 'Camarosa' and 'Sweet Charlie'. *J. Hortic. Sci. Biotechnol.* **2005**, *80*, 39–42. [CrossRef]
29. Liu, Z.; Bruins, M.E.; de Bruijn, W.J.; Vincken, J.-P. A comparison of the phenolic composition of old and young tea leaves reveals a decrease in flavanols and phenolic acids and an increase in flavonols upon tea leaf maturation. *J. Food Compos. Anal.* **2020**, *86*, 103385. [CrossRef]
30. Dadáková, K.; Heinrichová, T.; Lochman, J.; Kašparovský, T. Production of Defense Phenolics in Tomato Leaves of Different Age. *Molecules* **2020**, *25*, 4952. [CrossRef]
31. Khanam, U.K.S.; Oba, S.; Yanase, E.; Murakami, Y. Phenolic acids, flavonoids and total antioxidant capacity of selected leafy vegetables. *J. Funct. Foods* **2012**, *4*, 979–987. [CrossRef]
32. Ribas-Agustí, A.; Gratacós-Cubarsí, M.; Sárraga, C.; García-Regueiro, J.-A.; Castellari, M. Analysis of Eleven Phenolic Compounds Including Novel p-Coumaroyl Derivatives in Lettuce (*Lactuca sativa* L.) by Ultra-high-performance Liquid Chromatography with Photodiode Array and Mass Spectrometry Detection. *Phytochem. Anal.* **2011**, *22*, 555–563. [CrossRef] [PubMed]
33. Santos, J.; Oliveira, M.; Ibáñez, E.; Herrero, M. Phenolic profile evolution of different ready-to-eat baby-leaf vegetables during storage. *J. Chromatogr. A* **2014**, *1327*, 118–131. [CrossRef] [PubMed]

MDPI
St. Alban-Anlage 66
4052 Basel
Switzerland
Tel. +41 61 683 77 34
Fax +41 61 302 89 18
www.mdpi.com

Agronomy Editorial Office
E-mail: agronomy@mdpi.com
www.mdpi.com/journal/agronomy

www.ingramcontent.com/pod-product-compliance
Lightning Source LLC
Chambersburg PA
CBHW040837250926
51850CB00007BA/1341
* 9 7 8 3 0 3 6 5 5 7 8 0 9 *